In Search of Human Nature

University Campus **Oldham**
Higher Education at Oldham College

Books by Carl N. Degler

Out of Our Past:
The Forces That Shaped Modern America

The Age of the Economic Revolution, 1876–1900

Affluence and Anxiety, 1945 to the Present

Neither Black Nor White:
Slavery and Race Relations in Brazil and the United States

The Other South:
Southern Dissenters in the Nineteenth Century

Place Over Time:
The Continuity of Southern Distinctiveness

At Odds:
*Women and the Family in America
from the Revolution
to the Present*

In Search
of Human Nature

The Decline and Revival of Darwinism
in American Social Thought

CARL N. DEGLER

New York   Oxford
OXFORD UNIVERSITY PRESS
1991

Oxford University Press

Oxford   New York   Toronto
Delhi   Bombay   Calcutta   Madras   Karachi
Petaling Jaya   Singapore   Hong Kong   Tokyo
Nairobi   Dar es Salaam   Cape Town
Melbourne   Auckland

and associated companies in
Berlin   Ibadan

Copyright © 1991 by Carl N. Degler

Published by Oxford University Press, Inc.,
200 Madison Avenue, New York 10016

Oxford is a registered trademark of Oxford University Press

Library of Congress Cataloging-in-Publication Data
Degler, Carl N.
In search of human nature:
biology and culture in American social science,
1880 to present / Carl N. Degler.
p. cm.   Includes index.
ISBN 0-19-506380-5
1. Social Darwinism—United States—History.
2. Sociobiology—History.   I. Title.
HM106.D37   1991   304.5—dc20
90-39903

3 5 7 9 8 6 4 2

Printed in the United States of America
on acid-free paper

To Eric Voorhees Degler,
born October 5, 1989,
on whom culture and biology
are already working

Preface

Like most white Americans of my sex and class (the son of a fireman) and my generation (born in 1921) I came into a world that soon made me a racist and a sexist. And then, like most well-educated people of that generation, as I grew up I repudiated both race and sex as explanations for differences in the behavior of human beings. Indeed, I spent a good deal of my youth and adulthood arguing by voice and in print against biology as a source of human behavior, not only in regard to race and sex, but in other respects as well. How and why that sea change occurred in my thinking concerned me only peripherally. I knew there had been a time when biology was thought to be an important way of explaining why social groups differed, why some people were considered better than others. But that was another time. In my new outlook it was a given that the repudiation of biology had resulted from a penetrating, perhaps even lengthy scientific investigation of biology's inadequacy in accounting for the ways in which human groups differed. In ruling out biology as a cause of human differences, I thought of myself as defending a truth as solidly established as the heliocentric universe. Human nature, I believed, was constructed over time, not inherited from time. I had no trouble accepting Karl Marx's famous remark that man made his own history, but not entirely as he pleased, meaning that history may limit us at times, but biology has little to say about our social behavior.

The story I tell in these pages is how Americans like me, that is, students of human nature—social scientists—made the momentous shift from believing that biology explained some human actions to seeing culture or human experience—history, if you will—as the primary if not the sole source of the differential behavior of human beings. What kinds of evidence and arguments were used to bring that shift in outlook, who made them, and why, are among the questions I seek to answer in the first part of the book. More is involved, of course, than identifying and explicating the crucial ideas or even the advocates of those ideas. As in any study of the acceptance of a new paradigm or way of thinking, the crucial historical question is why did others accept and then begin to work within the new dispensation that some innovative leaders were propounding? Why did so many repudiate the traditional in favor of the novel?

That part of the story was not easy to find answers for. As all historians know, "why" questions on the grand scale are the most fundamental, but also the most difficult to document fully. I have made some suggestions and offer some supporting evidence, but much of the story, I fear, remains recalcitrantly undocumented; too much of it remains in the heads of the dead and the living alike. What the available evidence does seem to show is that ideology or a philosophical belief that the world could be a freer and more just place played a large part in the shift from biology to culture. Science, or at least certain scientific principles or innovative scholarship also played a role in the transformation, but only a limited one. The main impetus came from the wish to establish a social order in which innate and immutable forces of biology played no role in accounting for the behavior of social groups. Individuals certainly differed in ability and achievement, but those differences derived from their individual inheritances, not from the biology of the social group to which they may have belonged. To the proponents of culture the goal was the elimination of nativity, race, and sex, and any other biologically based characteristic that might serve as an obstacle to an individual's self-realization.

That an ideological purpose should thus shape the answers to what might otherwise seem to be a scientific question is not a novel idea. Scientists, social and natural, are human beings and for that reason alone, if no other, their investigations have been known to be initiated, even directed by unspoken values or hopes. Those scholars who have exposed and criticized the past misuse of biological ideas in social science have quite properly called attention to the large part ideology played in fostering those uses of biology. But much less widely acknowledged is the view that ideology also underpinned the repudiation of biology in social science and encourages the present widespread acceptance of culture as the alternative explanation of the behavior of human beings.

Today, in the thinking of citizens and social scientists alike the deeply held assumption is that culture has severed for good the linkage between human behavior and biology. The conviction is that human beings in their social behavior, alone among animals, have succeeded in escaping biology. The irony is heavy here. For that belief is accompanied by another deeply held conviction: that human beings, like all other living things, are products of the evolution that Charles Darwin explained with his theory of natural selection. The irony is almost palpable inasmuch as Darwin entertained no doubt that behavior was as integral a part of human evolution as bodily shape. And that is where Book III enters. It seeks to tell the story of how biological explanations have begun to return to social science. This return of biology stems from several things, among which is a renewed recognition of the pertinence of Darwinian evolutionary theory, and the apparent relevance of an explosion of knowledge about animal behavior to the search for human nature.

It is important to recognize that this "return of biology" is not simply a revival of repudiated ideas, like racism, sexism, or eugenics. The new evolutionary approach to social science has no more place for them than has the currently dominant cultural interpretation. Nor is "the return" an attempt to order society according to hierarchies or other normative outlooks. Social Darwinism was definitely killed, not merely scotched; the story told here of the return of biology does not resuscitate Herbert Spencer. Rather, the true aim of those social scientists who advocate a "return" is to place, once again, the study of human nature within evolution, to ask how human beings fit into that framework which Darwin laid down over a century ago and which very few social scientists consciously repudiate—except when the behavior of human beings is included in it.

The story is still unfolding. Many natural scientists as well as social scientists resist the "return," denounce those who point to it, sometimes misunderstand its nature, and sometimes even fear to acknowledge it. Which is only to say that the search for human nature has always been unfinished and thus ongoing, and so it remains in this book.

A substantial part of my scholarly work over the years has been concerned with race and sex in American history, but the history of the transformation in ideas, both scientific and social, about the relation between biology and human behavior was not something I knew much about. At the same time I had long entertained an interest in ethology or the study of animal behavior. Thus my work on this book can be said to have begun just about ten years ago, when I was given the opportunity by a fellowship at the Center for Advanced Study in the Behavioral Sciences at Stanford to learn and think about the relation between animal and human behavior. This preface offers me a chance to thank the Center for its generous hospitality, scholarly support, and its congenial, often critical group of fellows who helped me to take my first steps on what turned out to be a long and arduous road to discovery.

My education in this new, yet related area of research was continued and substantially advanced by fellowships during the academic year 1983–84 from the Stanford Humanities Center and the National Endowment for the Humanities to which institutions I am greatly indebted. The fellowships not only offered support that allowed me to escape from competing professional obligations, but also provided me with congenial critics, whose encouragement I deeply appreciated. In the course of that year, I presented some of my preliminary findings in the Samuel Paley Lectures at Hebrew University in Jerusalem. The supportive reception given those lectures, which were my first public foray into this topic, was a major source of encouragement to me. I am most grateful for that and to a wonderful visit to Israel, in the course of which the hospitality of the faculty at Hebrew University was manifested in a most heartwarming manner to me and my wife, Catherine.

I wish to thank, too, the German Historical Institute, which honored me in 1988 with an invitation to offer the Annual Lecture, another occasion that helped me to clarify before a critical audience my thoughts, in this case, on the ideas of Franz Boas and Alfred Kroeber.

Most recently, I am indebted to St. Catherine's College, Oxford, for awarding me a Christiansen Fellowship for the Michaelmas term, 1989. Aside from permitting me to spend eight weeks in the company of a congenial and stimulating group of scholars, the fellowship provided an ideal place in which to work upon the final revisions of this book.

Over the years, various scholars, the works of some of whom are discussed in the pages which follow, have been kind enough to respond to my queries and quests for information and advice. Thus I am more than pleased to express my thanks to Otto Klineberg, William Durham, Roger Masters, Jerome Kagan, Melvin Konner, Hamilton Cravens, David Barash, Ruth Hubbard, Joseph Lopreato, Deborah Rhode, Mary Ann Kevles, Daniel J. Kevles, Edward O. Wilson, Rosalind Rosenberg, Steven Goldberg, Robert J. Richards, Lee Ellis, Pierre L. Van den Berghe, Martin Tweedale, and Henry L. Minton. None of this host of helpers is to be held responsible for the errors, either in judgment or fact, that surely will pop up in the pages that follow. Not all of them approve of what is included in this book. All of them, however, have contributed in one way or another to help me clarify my ideas.

Charles Kimball, a young scholar just beginning his career in history, has more than earned my thanks by expertly carrying out some research for me at a crucial stage in the writing of this book. (No typist is thanked here since the modern technology of wordprocessing allowed me to escape from having to incur that familiar obligation of authors.)

Finally, but never least, I want to thank Sheldon Meyer, my editor at Oxford University Press, whose steady support and enthusiasm for this book has been of the kind all authors wish for, but do not always receive. So, bless you, Sheldon. My thanks also go to Stephanie Sakson-Ford for her work on the manuscript.

Stanford   C.N.D.
April, 1990

Contents

I

Biology Acknowledged

1

Invoking the Darwinian Imperative

" What must we do to be human?" is a question as old as humanity itself.
Margaret Mead, 1949

Intellectual progress usually occurs through sheer abandonment of questions, an abandonment that results from their decreasing vitality and a change of urgent interest. We do not solve them; we get over them. . . . Doubtless the greatest dissolvent in contemporary thought of old questions, the greatest precipitants of new methods, new intentions, new problems is the one effected by the scientific revolution that found its climax in the "Origin of Species."
John Dewey, 1909

Anthropologists tell us that primitive people commonly call themselves simply "people" or "folk," thereby seeing other human beings who may stumble upon them as something other than people. Such ethocentrism, as we would denominate terminology like that today, is the root of the concept of human nature. It defines the group by separating it from other living things, most especially animals. Historian Keith Thomas, in his fascinating book *Man and the Natural World,* has shown how persistent that delineation of differences has been. At times it has mean an emphasis upon man's superiority as in Aristotle's chain of being with human beings at the top; at other occasions, as in Aesop's fables, it has bestowed human wit and even human intelligence and values on animals. If the Bible validated human beings' sovereignty over animals, it also provided a basis, Thomas observes, for human objections to cruelty toward animals on the ground that they were God's creatures, which had been placed under human care. The Romantics went even further when they referred to animals as friends and brothers for whom pain was as excruciating as for human beings.

Yet underlying all these varieties of human response to animals was a polarity, a difference that was never quite eliminated, a barrier that was never crossed. Even today, to call a person an "animal" is the worst of insults, just as to have sex with an animal was the seventeenth century's most heinous sexual crime. Reclassifying human enemies as "animals" was often necessary before they could be comfortably massacred. In short, we have been fascinated as well as threatened by our fellow animate creatures. The very existence of zoos attests to the fascination, just as the prevalence of pets reflects our need to cherish while dominating them. Animals have always provided a primary means of defining ourselves, of determining what it means to have a human nature.

The very phrase "human nature" is itself filled with implications. From time immemorial human beings have used that phrase in thinking about themselves. Historian Merle Curti has traced the history of the term back to the early seventeenth century, and then only in America, yet the result of his labors fills a four-hundred-page book. Today in conversation the phrase bespeaks relative persistence over time, of a nature not easily altered yet not easily ignored. "That's human nature for you," goes the phrase, in explanation of how some one failed to act as well as she ought. The implication is that there is some power or some innate character that limits even as it defines a human being. The pre-Socratics were convinced that man was inherently selfish, while Aristotle emphasized the social or cooperative nature of human beings. At the same time, human nature has never been seen as totally immutable; religion, schools, and parents depend upon a potentiality for change as they seek to shape and reshape that nature, fully relying on the clear assumption that it can indeed be made better.

The very emphasis upon a resistant, enduring "nature" added a dimension to the relation between animals and man. It implied an embeddness in nature, a characteristic of all living things, of something that was not only beyond the will of animals but in many ways beyond the will and hopes of human beings, too. Human beings, along with animals, were controlled, rather than being in control. The very word "nature" emphasized naturalness, implicitly denying to a large degree humankind's independence of the natural world. The very similarities between animals and human beings proclaimed the power of nature even as human beings simultaneously asserted their superiority over their fellow creatures of nature.

This general acceptance of the power of nature in defining human beings came into question in the course of the Enlightenment of the eighteenth century when circumstances or surroundings seemed increasingly to offer a complementary or even alternative explanation as to why people behaved as they did and why they differed from one another. The sources of the new view are not difficult to uncover. The new outlook owes much to Europe's frenetic exploration of the world outside itself that began with the Great Discoveries. Three centuries of exploration brought home as never before the tremendous diversity of human behavior and life patterns within environments and under circumstances dramatically different from those of Europe. It was not the first time men had discerned a casual connection between circumstances and behavior, but surely it was the first time the evidential basis for such a conclusion was on such a vast scale and scope. Out of that large laboratory of human experience was born the conflict between nature and nurture.

At no time, of course, was the subsequent controversy a clean or sharply defined set of alternatives. Those who emphasized the shaping power of surroundings, or environment, could still find a place for nature, as that quintessential man of the Enlightenment Thomas Jefferson certainly did when he doubted that Afro-Americans and white Americans were equal in mental abilities. Yet on the whole, even in the early nineteenth century, intellectuals and public figures alike placed more emphasis upon nurture than upon nature. And even those who perceived nurture as the weaker of the two still recognized a certain malleability in nature that left environment with substantial influence.

Around mid-century Charles Darwin gave to that recognition of nature's malleability a novel, even a paradoxical, interpretation. It was paradoxical because in showing that nature changed over time, Darwin simultaneously made nature a more powerful influence on human thought. Darwin, to be sure, was not the originator of the idea of evolution. For, like environmentalism itself, evolution was a child of the eighteenth-century Enlightment. Although some ancient Greek thinkers had recognized change as inherent in life, few in antiquity and virtually none in early Christian Europe took seriously the idea that the present had emerged out of the past. Aristotle certainly knew that an adult developed from a tiny embryo but he made no use of that knowledge in thinking about society or the universe. Christianity, in its turn, simply postulated a world of living things as a creation of six days, an accomplishment that remained unchanged thereafter. Like Aristotle, Christians knew that change occurred, as when pagan Rome became Christian, or when Islam overthrew medieval Christianity in Spain. They perceived, however, no process in the change; they recognized causality, to be sure, but not continuity. Their world outlook could accommodate precipitant change, even catastrophe, but not slow, small, almost indistinguishable changes over time. The natural world existed now as it had been since its creation, except as the Creator chose to alter it for His own purposes, as with the Flood. Evolution in short, was beyond their ken.

The concept of evolution, which seems so obvious to us today, emerged only in the eighteenth century. Immanuel Kant's assertion that the universe was the product of slow change over eons of time was among the earliest examples of an evolutionary outlook. Another was the recognition by geologists that the earth, too, had a history, that it had not always been as it appeared. Once there had been mountains where now there were plains, seas where deserts now stood. Others applied the idea of slow change over time to living nature, seeing an evolution of animals from simple to complex forms. Among such proponents of animal evolution was Erasmus Darwin, the grandfather of Charles. In more ways than one, in short, Charles Darwin's work is best seen as the culmination rather than the initiation of a line of thought that saw evolutionary change in man and nature. Yet simply because Darwin was the culmination, he shaped men's thinking about evolution and man's relation to animals. He rephrased, as no one before him, what it meant to be human.

Because Darwin set the framework within which American social scientists of the late nineteenth century pursued their effort to understand human behavior and human nature, we need briefly to examine that Darwinian framework. What was the Darwinian imperative to which American social scientists responded? And why did they find his framework so compelling?

Charles Darwin's great authority as an evolutionist stemmed first of all from his having provided what turned out to be the most convincing explanation as to how evolution worked, how the plants and animals of the natural world had achieved their present forms. The mechanism, the principle of natural selection, to which his book *The Origin of Species* was devoted to proving, was simple enough. Its essential argument was that the enormous variety of living things had resulted from the interaction of three principles. The first was that all organisms reproduce, the second was that even within a given species each organism differed slightly from any other, and the third was that all organisms competed for survival. If the environment changed or if new organisms entered the habitat of established organisms, then those organisms that best adapted to the changed situation would gradually outbreed those less well adapted. In turn, the adjustment or adaptation to the environment might involve sufficient change in the descendants of the organism to bring a new species into being.

The engine of change was called *natural* selection on the ground that nature was accomplishing what human breeders of domestic animals achieved when they bred a new kind of horse or dog by systematically mating those individual animals which displayed the traits the breeders sought to develop in the offspring. Nature, Darwin always stressed, harbored no comparable purpose; in fact, it had no aim at all. The immense diversity of living organisms resulted, according to Darwin, not from a plan of purpose but from the accidents of history, from those changes in climate, weather, geology, and food supply, or the increase or decrease in the presence of enemies to which an animal or plant might be subjected. There was no place in the Darwinian world of natural selection for Creation or any supernatural force, and that, of course, was why many Victorians found Darwin's writings a danger to traditional Christianity, or any religion for that matter. There was, however, a place for environment as a source of change. For in Darwin's concept of natural selection, environment called the tune to which an organism must adapt or adjust; failure to do so meant extinction in the long run. Ironically enough, this way of looking at Darwinian natural selection would prove to be influential in giving an environmental along with a biological cast to the thinking of American social scientists.

Nowhere within the *Origin of Species* is evolution or natural selection applied to human beings. Only on the next to the last page of his book does Darwin become so bold as to mention human beings at all, and then it is to suggest only that "much light will be thrown on the origin of man and his history."[1] Within twelve years, however, his second contribution to evolutionary theory appeared in the form of *The Descent of Man.* There he explicitly included human beings in evolution, depicting them, along with all other animals, as shaped by natural selection.

Today we know that Darwin had included man in evolution long before he wrote the *Origin of Species,* much less the *Descent of Man.* Several times in his notebooks of the late 1830s, when he began to rough out his theory of natural selection, he was already seeing connections between human beings and animals. "I will never allow that because there is a chasm between man . . . and animals that man has different origins," he wrote in 1838. In his notebook on Metaphysics he twice made clear his belief in the continuity between human beings and animals. "Origin of man now proved. Metaphysics must flourish.— He who understands baboon would do more toward metaphysics than Locke." A few pages later a second reference to John Locke and his *tabula rasa* theory of human knowledge appears: "(The monkeys understand the affinities of man better than the boasted philosopher himself.)" And then, at length, at another place, Darwin again interlocked man and beast. "If we choose to let conjecture run wild, then animals, our fellows brethren in pain, disease, suffering, and famine—our slaves in the most laborious work, our companions in our amusements—they may partake of our origin in one common ancestor—we may be all melted together."[2]

As these early jottings imply, from the beginning Darwin discerned in the linkage between animals and human beings something much more profound than a simple morphological continuity. It is true that he spent the first two chapters of the *Descent of Man* showing how the physical shape of human beings was derived from animal ancestors. Thereafter, however, he filled three chapters with reasons for seeing the mental and moral habits of human beings descendent from animal forebears. "There is no fundamental difference between man and the higher mammals in their mental faculties," he boldly announced. Yes, he conceded, "the difference between the mind of the lowest man and that of highest animal is immense." Animals may exhibit certain abilities and feelings that are similar to those found in human beings but a quality like "disinterested love for all the living, the most noble attribute of man," he recognized was quite beyond their capacity. "Nevertheless," he concluded, "the difference in mind between man and the higher animals, great as it is, certainly is one of degree and not of kind." He then proceeded to fill his pages with examples of the wide range of animal actions and feelings that mimicked, if not equalled, those of human beings. Animals reasoned, gave evidence of wonder, curiosity, dread, and joy. In animals he discerned abstractions, self-consciousness, and mental individuality, as he phrased it, along with a sense of visual and aural beauty. The last he defended by calling attention, among other examples, to the bower birds of the tropics, the male of which elaborately decorated a nest to attract a mate, and to the birds whose beautiful "songs" were directly related to the effort of males to gain access to a female.[3]

Underlying Darwin's anthropomorphism was his determination to demonstrate as often and as thoroughly as possible the continuity between the so-called lower animals and human beings. "It may be freely admitted," he wrote in the *Descent,* "that no animal is self-conscious, if by this term it is implied that he reflects on such points, as whence he comes or wither he will go, or what is life and death and so forth." But, he asked, can we be sure that an old dog "with some power of imagination, as shewn by his dreams, never reflects on his past pleasures or pains in the chase?" Would this not be a "form of self-consciousness?" he remarked, answering his own question. The thought recommended itself to him when he compared the life of the dog with that of "the hard-working wife of a degraded Australian savage, who used very few abstract words, and cannot count above four." Can she, he wondered, "exert her self-consciousness, or reflect on the nature of her own existence"?[4]

For those critics who thought it disreputable to make such comparisons or to see human beings as descended from animals, Darwin had a ready answer. No one doubts that we are descended from savages, he began, alluding to contemporary anthropological conceptions of social evolution. Yet those denizens of Tierra del Fuego whom he had met on the voyage of the *Beagle* were without culture, arts, clothes, and government. Anyone who has seen such people, he contended, cannot "feel much shame if forced to acknowledge that the blood of some more humble creature flows in his veins. For my part," he stoutly maintained, "I would as soon be descended from that . . . old baboon, who descending from the mountains, carried away in triumph his young comrade from a crowd of astonished dogs—as from a savage who delights to torture his enemies, offers up bloody sacrifices, practises infanticide without remorse, treats his wives like slaves, knows no decency, and is haunted by the grossest superstitions."⁵

Because of Darwin's determination to demonstrate at all cost the truth of the continuity between animals and human beings, he devoted Chapter 4 of the *Descent* to the evolution of man's moral sense. As he was well aware, the demonstration was essential to his larger argument because upon man's moral sense opponents of human evolution rested their case for the uniqueness of humanity. Darwin conceded that only human beings could act morally, for only they were moral, by which he meant that only a human being was "capable of comparing his past and future actions or motive, and of approving or disapproving of them. We have no reason to suppose," he continued, "that any of the lower animals have this capacity." When a dog rescues a child or takes charge of an orphan monkey, he observed, we do not consider it a moral act. But a similar act by a human being is described as moral simply because human beings are, by definition, moral.⁶ Darwin, in sum, had no intention of making animals into moral creatures, but he deliberately sought to locate the roots or origins of human moral consciousness in man's animal ancestry.

The root of human morality he found in the social instincts that caused animals to cooperate with one another for what he called "the general good," a term he defined as the "rearing of the greatest number of individuals in full vigour and health, with all their faculties perfect, under the conditions to which they are subjected." (In the late twentieth century, ethologists and sociobiologists would label this definition "reproductive success.") Since the social instincts of "both man and the lower animals have no doubt been developed by nearly the same steps," Darwin suggested, "it would be advisable . . . to use the same definition in both cases, and to take as the standard of morality, the general good or welfare of the community, rather than the general happiness," as his contemporary utilitarians argued.*

To utilitarians like John Stuart Mill, for example, human morality consisted at bottom of making conscious choices based on the principle of the greatest happiness for the greatest number of human beings. It was, as Darwin pointedly observed, an individualistic rather than a social basis of moral behavior. And

---

* In the *Descent* (pp. 489–90) Darwin took pains to explain why he departed from the utilitarian explanation for moral behavior. He saw that explanation as arguing that any action "must be associated with some pleasure or displeasure. But man," Darwin objected, "seems often to act impulsively, that is from instinct or long habit, without any consciousness of pleasure, in the same manner as does probably a bee or ant, when it blindly follows its instincts. Under circumstances of extreme peril, as during a fire, when a man endeavors to save a fellow-creature without a moment's hesitation, he can hardly feel pleasure; still less has he time to reflect on the dissatisfaction which he might subsequently experience if he did not make the attempt. Should he afterwards reflect over his own conduct, he would feel that there lies within him an impulsive power widely different from a search after pleasure or happiness; and this seems to be the deeply planted social instinct."

because human beings alone were said to follow it, Darwin took exception to it. For, as he said, even when "a man risks his life to save that of a fellow-creature, it seems almost more correct to say that he acts for the general good, rather than for the general happiness of mankind." Happiness and welfare usually coincide in an individual, he conceded, and happiness certainly enhances sociality. So he did not think individual happiness could be ignored. But the social instinct, he insisted, "together with sympathy (which leads to our regarding the approbation and disapprobation of others) . . . served as the primary impulse and guide" for human beings just as it did for animals. His fundamental objection to the utilitarian principle of individual happiness was precisely its emphasis upon self-interest. By emphasizing the good of the group, he reminded his critics, "the reproach is removed of laying the foundation of the noblest part of our nature in the base principle of selfishness."[17]

Animals may not have reached the high moral level of human beings, Darwin recognized but his understanding of human moral evolution convinced him that "any animal whatever endowed with well-marked social instincts, the parental and filial affections being here included, would inevitably acquire a moral sense or conscience as soon as its intellectual powers had been as well, or nearly as well developed as in man." For he further recognized that human morality arose "either directly or indirectly much more through the effects of habit, the reasoning power, instruction, religion, etc., than through natural selection." But he could not forbear to add immediately that to natural selection "may be safely attributed the social instincts, which afforded the basis for the development of the moral sense," in the first place. Historian of science Robert Richards has recently characterized Darwin's conception of the origin of morality in the following way: "Aristotle believed that men were by nature moral creatures; Darwin demonstrated it."[8]

In writing the *Descent,* Darwin's intention had been to demonstrate the continuity not only of behavior and morality between animals and human beings but of emotions as well. The *Descent* swelled to such a size, however, that he had to postpone that demonstration to yet a third book on evolution, *The Expression of the Emotions in Man and Animals,* which he published in 1873. Although today ethologists and sociobiologists can see in that book Darwin's astute recognition of animal communication, for Darwin the book was simply an additional body of evidence to support his fundamental proposition of the continuity of behavior, morals, and emotions between animals and human beings.

Where, in all of this discussion of the continuity between animals and human beings, one might ask, does social Darwinism figure? Should we think of Darwin as a social Darwinist? It is quite true that scattered throughout the *Descent of*

Man are passages suggesting that Darwin believed that the principle of survival of the fittest *justified* as well as explained the social hierarchy in human affairs. "Man, like every other animal," he wrote in one place, "has no doubt advanced to his present high condition through a struggle for existence" and if that advance is to continue he "must remain subject to a severe struggle. Otherwise he would sink into indolence, and the more gifted men would not be more successful in the battle of life than the less gifted." Or as he remarked in another place: "The inheritance of property by itself is very far from an evil; for without the accumulation of capital the arts could not progress; and it is chiefly through their power that the civilised races have extended, and are now everywhere extending their range, so as to take the place of the lower races." He denied that the "moderate accmulation of wealth" interfered with "the process of selection," for when poor men become moderately rich, their offspring gain advantages. "The presence of a body of well-instructed men, who have not to labour for their daily bread," he believed, "is important to a degree which cannot be overestimated." For they are the people who carry on all the "high intellectual work" of the society, upon which "material progress of all kinds mainly depends, not to mention other and higher advantages."[9] Darwin, in sum, was hardly free from the accoutrements of social Darwinism.

The man who coined the key phrase in social Darwinism—survival of the fittest—was the contemporary English philosopher Herbert Spencer, who was an advocate of evolution before the publication of the *Origin of Species*. Darwin, it is true, later accepted the phrase as a kind of shorthand for natural selection, but even so, historians of the period have generally pointed to Spencer rather than Darwin as the chief proponent of social Darwinism. Spencer expressed himself in ways that well qualified him as a social Darwinist. Again and again he publicly advocated that governments and other institutions keep their regulatory hands off what he liked to think of as the "natural" processes of the social order. Let nature take its course, Spencer counseled; do not interfere with the operation of the principle of the survival of the fittest. "To aid the bad in multiplying is, in effect," he wrote in 1874 in the *Study of Sociology*, "the same as maliciously providing for our descendants a multitude of enemies." He thought that those organizations of society which undertake "in a wholesale way to foster good-for-nothings" commit an "unquestionable injury." For they "put a stop to that natural process of elimination by which society continually purifies itself."[10] Such statements have understandably prompted some historians of ideas to contend that "social Spencerism" is a more accurate label for the concept than social Darwinism.

More recently some historians have begun to doubt that Darwin's version of evolution was ever drawn upon to any significant degree in defense of the status quo. For, as that indefatigable Darwinian among American social scientists, Lester Frank Ward, complained at the time, "I have never seen any distinctively Darwinian principle appealed to in the discussion of 'social Darwinism.' It is therefore wholly inappropriate to characterize as social Darwinism the *laissez faire* doctrine of political economists." Ward was quite correct that those publicists and businessmen whom historians and contemporaries alike denominated social Darwinists rarely referred to natural selection or to Darwin. More likely, as Ward implied, they were drawing upon the ideas of a couple of other Britishers, namely, Thomas Malthus and Adam Smith.[11]

The central point is that Darwin himself, despite his social Darwinist remarks quoted earlier, was only peripherally interested in defending the status quo. Always his aim was to demonstrate the application of natural selection to human evolution, and the continuity between animals and human beings on every level of existence. He never doubted that after the emergence of *Homo sapiens* human experience and reason would improve the behavior and morals of human beings beyond the level reached by any animal.

Nowhere does that conviction come through more forcefully than in his remarks that follow his expression of social Darwinist ideas. "The weak members of civilised societies" do indeed, he remarked, propagate their kind. "No one who has attended to the breeding of domestic animals will doubt that this must be highly injurious to the race of men." Then he quickly added a major qualifier: "the aid which we feel impelled to give to the helpless is mainly an incidental result of the instinct of sympathy, which was originally acquired as part of the social instincts, but subsequently rendered . . . more tender and more widely diffused. Nor could we check our sympathy, even at the urging of hard reason," he cautioned, "without deterioration in the noblest part of our nature." Drawing an analogy with a surgeon who hardens his heart to accomplish his good work, Darwin went on to say that "if we were intentionally to neglect the weak and helpless, it would only be for a contingent benefit, with an overwhelming present evil. We must therefore bear the undoubtedly bad effects of the weak surviving and propagating their kind," he concluded.[12]

He then explained why he subscribed to that view. Although "the self-regarding virtues . . . now appear to us so natural and to be thought innate," they were not valued by man early in his evolution. For contrary to what many people in his own time seemed to think, "man can generally and readily distinguish between the higher and lower [or selfish] moral rules." That is so because "the higher are founded on the social instincts, and relate to the welfare of others. They are supported by the approbation of our fellow-men and by reason. The lower rules," on the other hand, "relate chiefly to self, and arise from public opinion, matured by experience and cultivation; for they are not practised by

rude tribes." Thus, despite the often alleged pessimism of his application of evolution to human beings, in Darwin's estimation the future was bright. "There
is no cause to fear that the social instincts will grow weaker," he reassured his
readers, "and we may expect that virtuous habits will grow stronger becoming
fixed perhaps by inheritance. In this case," he optimistically concluded, "the
struggle between our higher and lower impulses will be less severe, and virtue
will be triumphant."[13] In short, if Darwin was a social Darwinist, he was not a
dark or gloomy one.

In any event, whether called social Darwinism, or social Spencerism, the
defense of the social and economic hierarchy of nineteenth-century America
that the doctrine was intended to accomplish held little appeal for the men and
women who were shaping the emerging fields of sociology, psychology, economics, and anthropology at the end of the century. The aim of social Darwinism was frankly conservative; the rising social scientists were not. Sociologists
like Albion Small of the University of Chicago, Franklin A. Giddings of Columbia, Charles A. Cooley of the University of Michigan, and Edward Ross of the
University of Wisconsin were reformers to a man. For them, the new social
science was intended to shape a fresh and just world for Americans. Evolution
to them, as for Herbert Spencer, and to a less extent for Darwin, too, was
actually another word for progress. The purpose and point of social science
were to hasten and channel the achievement of that goal. Scholarship and
reform would work like hand in glove toward an improved and worthy society.

Anthropologist John Wesley Powell's indignant response to Herbert Spencer's
conception of social evolution was characteristic of the outlook of these rising
social scientists. "Man does not compete with plants and animals for existence,"
Powell insisted in 1888, "for he emancipates himself from that struggle by the
invention of arts; and again, man does not compete with his fellow-man for
existence, for he emancipates himself from the brutal struggle by the invention
of institutions." Animal evolution, Powell admitted, may arise out of the struggle for existence, but "human evolution arises out of the endeavor to secure
happiness; it is a conscious effort for improvement in condition."[14] (This, of
course, was not evolution according to Darwin, either. Darwin's whole purpose
had been to disprove the very discontinuity between animals and human beings
that Powell insisted upon. But then, as the early sociologist Lester Frank Ward
complained, few social scientists of the time rightly understood Darwin however closely they may have wanted to follow him.)

Sociologist Edward A. Ross came closer to the true Darwinian message when
he remarked in 1901 that "under the tutelage of Darwinism the world returns
again to the idea that *might* as evidence of fitness has something to do with
right." But he, too, despite his Darwinian outlook, refused to accept the darker

side of things. "Yet with it all dwell vastly richer human sympathies and a far more haunting consciousness of the corporate self," he concluded. An equally forthright repudiation of social Darwinism was set forth by the prominent sociologist and self-proclaimed Darwinist Charles Ellwood of the University of Missouri in 1910. "The rich and economically successful are . . . by no means to be confused with the biologically fit," he told his students. "On the contrary, many of the economically successful are such simply through artificial advantageous circumstances," Ellwood pointed out. "From the standpoint of biology and sociology they are often among the less fit, rather than the more fit elements of society."[15]

The goal of the new social scientists was to show that evolution and Darwinism encouraged cooperation and cohesion in society rather than conflict between groups, as social Darwinism taught. And since many of the rising social scientists of the time were inclined toward environmental explanations for behavior, they generally opposed social Darwinism just because it extolled conflict and competition. Simply because they were Darwinians they looked to environment in shaping human behavior and the social order. Under Darwinian evolution animals had been shown to adapt to changing environments, so these early social scientists easily moved to the view that if one changed the social environment human behavior would adapt or adjust to it.

For some of them, that outlook affected even their conception of racial influences on behavior. It is true that in late nineteenth-century America racism was endemic. Most white Americans took it for granted that black people, Amerindians, or immigrants from Asia were morally as well as intellectually inferior to whites. The assertion or inference that a person's particular biological makeup shape his or her capability is, of course, the essence of racism. But that has to be differentiated from a related belief—hereditarianism. It contends that biology determines the behavior of human beings as racism does, but hereditarianism does not draw invidious comparisons between human groups. Darwin's conception of evolution itself, as we have seen, rested upon the assertion that human nature was built upon a foundation of animal biology, that the mind as well as the body of man was biologically traceable in large part to animal ancestors. That was hereditarianism, not racism, because it lacked the invidious comparison among races that is the essential and distinguishing element in the concept of racism. Indeed, in the *Descent of Man* Darwin devoted a number of pages to refuting some contemporary American anthropologists who had developed a theory of racist evolution, a concept worth examining briefly.

The theory of racist evolution contended that the differences between the races were so profound as to mark the races as incipiently different species rather than simply varieties of humanity. Darwin, however, was too learned a biologist to accept their arguments even if he was not a modern egalitarian on racial matters. "Although the existing races of man differ in many respects," he pointed out, "yet if their whole structure be taken into consideration they are found to resemble each other closely in a multitude of points." His conclusion that they were closely alike physically, he noted, held "good with equal or greater force with respect to the numerous points of mental similarity between the most distinct races of man." American Indians, Negroes, and Europeans "are as different from each other in mind as any three races that can be named," he observed, "yet I was incessantly struck, whilst living with the Fuegians on board the 'Beagle,' with the many little traits of character, shewing how similar their minds were to ours; and so it was with a full-blooded negro with whom I happened once to be intimate."[16]

Darwin's principal reason for rejecting the idea that the races were different species was that he could not figure out how natural selection could have separated the races. The physical differences between races, Darwin thought, could not be accounted for by natural selection because "none of the differences between the races of man are of any direct or special service to him." He even sought to prove that there was no connection between climate and dark skin color.[17] In short, for Darwin race was outside evolution.*

Despite his doubts about the importance of race, as historian Nancy Stepan has cogently argued, Darwin still left an opening for racist conclusions to be drawn from his work. By acknowledging different levels of human societies—he spoke frequently of savages and lower races who were intermediate between animals and civilized people—he implicitly accepted a hierarchy of human beings. More important than that, he wove that acceptance of hierarchy into his conception of races. When he wrote that he could identify no advantages in the different *physical* aspects of human races, he simultaneously specified that "the intellectual and moral or social factulties must of course be excepted from this remark." As a consequence, Nancy Stepan has reminded us, the nineteenth-century anthropologist who believed in Darwinian evolution with its acceptance of a hierarchy of human races would find it difficult "to be a Darwinist without being a social Darwinist," that is, a racist. For it is true, as historian Robert Bannister has argued, that the most prevalent form of social Darwinism at the turn of the century was actually racism, that is, the idea that one people might be superior to another because of differences in their biological natures. Thanks to Darwin's acceptance of the idea of hierarchy among human societies

---

*Darwin himself explained the physical differences between the races, at least in part, by reference to his theory of sexual selection. See Darwin, *Origin,* p. 908. To my knowledge, however, no American social scientists, despite their acceptance of Darwinian beliefs in regard to other matters, drew upon that explanation for racial differences.

the spread and endurance of a racist form of social Darwinism owes more to Charles Darwin than to Herbert Spencer.[18]

Reflections of Darwinian racism appear in the writings of American social scientists in the waning decades of Darwin's century. Writing against Chinese immigration in 1882, for example, sociologist Gerrit L. Lansing drew upon both Darwin and hereditarianism in calling attention to "some of the laws of development of races and species, for they hold equally well with man as with other members of the animal kingdom." In discussing races, he noted, "physical characters alone have usually been employed, but those mental traits which are made manifest in habits and customs . . . are equally constant." Sociologist William Elwang in 1904 denied outright that better education could lift blacks to the level of whites. "The trouble with the negro is not merely that he is ignorant," Elwang explained. Better schools would remedy that. "The difficulty is more radical and lies embedded in the racial character, in the very conditions of existence. The negro race," Elwang asserted, "lacks those elements of strength that enable the Caucasian to hold its own, and win its way and bring things to pass. Negroes cannot create civilizations," he simply concluded. His explanation was almost straight out of Darwin. "Their's is a child-race, left behind in the struggle for existence because of original unfavorable inheritance of physical and mental conditions that foredoom to failure their competition on equal terms with other races."[19]

Less racist but no less hereditarian an explanation for behavior could be heard from even a black social scientist in the early years of the twentieth century. "The Negro is primarily an artist," W. E. B. DuBois began an article in a national magazine in 1913. "This means that the only race which has held at bay the life destroying forces of the tropics, has gained therefrom in some slight compensation a sense of beauty, particularly for sound and color, which characterizes the race. The Negro blood which flowed in the veins of many of the mightiest of the Pharoahs accounts for much in its origins," DuBois believed, "to the development of the large strain of Negro blood which manifested itself in every grade of Egyptian society."[20]

The point here is not that social scientists at the opening of the twentieth century were as hereditarian or racist as Americans in general. Instead, it is that they more or less viewed race as a contributory but not necessarily as the primary explanation for human behavior. In the professional literature of the time questions of race were not of central importance. When George Stocking, a modern historian of anthropology, surveyed twenty-three social science journals published between 1890 and 1915 he found relatively few articles treating race as an explanation for behavior. The years in which the subject figured most prominently in professional journals, he found, were between 1910 and 1915, but even then the proportion of articles dealing with race was no more than 5 percent of the total.[21]

Yet when professional sociologists, for example, did touch the subject, it was clear that they, like their less highly trained contemporaries, were far from immune to the idea that certain people were biologically inferior to others. There was an ambivalence about the matter, to be sure, an ambivalence that would prove to be of importance when racist ideas came under attack.

The ambivalence was clearly evident in Columbia University sociologist Franklin Giddings's *Principles of Sociology,* first published in 1896. Giddings did not devote much space to race or Afro-Americans, but when he did raise the subject, he was careful to deny that simple lack of opportunity prevented the "lower races"—by which he meant Chinese and Amerindians—from achieving as much as the dominant white race. After all, he pointedly wrote, "they have been in existence ... much longer than the European race, and have accomplished immeasurably less. We are, therefore warranted," he concluded, "in saying that they have not the same inherent abilities." And when lower and higher races came into contact with one another, he insisted, there is no reason to believe they will improve significantly. "The same amount of educational effort does not yield equal results when applied to different stocks," he explained.

From there Giddings moved to apply the principle of the survival of the fittest biologically to different races. "There is no evidence that the now extinct Tasmanians had the ability to rise," he contended. "They were exterminated so easily that they evidentally [*sic*] had neither power of resistance nor adaptability. Another race with little capacity for improvement," he added, "is the surviving North American Indian. Though intellectually superior to the Negro, the Indian has shown less ability than the Negro to adapt himself to new conditions," he asserted, drawing on the Darwinian principle of adaptation. The Negro, on the other hand, yielded easily to what Giddings called environing influences, but even he, when deprived of "the support of stronger races, ... still relapses into savagery. . . ." Yet, so long as the Negro is left in contact with the superior whites, Giddings reassured his readers, "he readily takes the external impress of civilization, and there is reason to hope that he," unlike the Indian, will acquire a measure of the spirit of that civilization.[22]

Such views were pretty standard racist doctrine for the time, but in other places in the same book Giddings's ambivalence about the influence of race or biology is also plain. After spending a number of pages tracing the intermingling of races which went into the creation of the peoples of Europe, he concluded that the white race "is composite to the last degree." Whites were, in short, hardly the pure race some Americans liked to believe. He admitted that this was a hypothesis, rather than a proved fact, but, in his judgment, a view preferable to that which assumed a pure white race. He expressed confidence that future "research will demonstrate that the negro and yellow races, which evidently are destined to play an important role in future developments of the world's population are not primitive races, too simple in their biological composition to be capable of further evolution, but already highly composite races capable of progress."[23]

A similar ambivalence toward racial explanation by a prominent turn-of-the-century sociologist emerges in the writings and speeches of Edward A. Ross. Younger than Giddings, but destined for leadership in the profession by the time of the First World War, Ross was known for his liberal political opinions, of which he made no secret. Early in his career, for example, he lost his job at Stanford University for supporting Populist monetary principles; later, at the University of Wisconsin, he continued to speak out in behalf of liberal politics.

At the opening of the twentieth century he was also more outspoken against racial interpretations of behavior than Giddings. In a speech in 1901 before an audience of social scientists, Ross noted that there were two ways to account for differences between groups of people. "There is the equality fallacy inherited from the early thought of the last [eighteenth] century, which belittles race differences and has a robust faith in the power of intercourse and school instruction to lift up a backward folk to the level of the best." The counter-fallacy, which he saw growing up "since Darwin," "exaggerates the race factor and regards the actual differences of people as hereditary and fixed." He told his audience that at the present time "the more besetting fallacy" was that of race. For race was "the watchword of the vulgar," and therefore social scientists ought to be wary of it.

Writing four years later on the same subject Ross was still calling race into question and in sentiments that sound almost modern. "More and more the time-honored appeal to race is looked upon as a resource of ignorance or indolence," he began. "To the scholar the attributing of the mental and moral traits of a population to heredity is a confession of defeat, not to be thought of until he has wrung from every factor of life its last drop of explanation. 'Blood' is not a solvent of every problem in national psychology," he asserted, "and 'race' is no longer a juggler's hat from which to draw explanations for all manner of moral contrasts and peculiarities."

Having said that, Ross seemed to take it all back by remarks later in the same book. It was not easy to tell, he contended, whether behavioral traits in human beings arise from biological or social sources. At one time, he began, "we thought the laziness of the anaemic Georgia cracker came from a wrong idea of life. Now we charge it to the hook-worm and administer thymol instead of the proverbs of Poor Richard." It would be a mistake, he cautioned, to overlook biological (not to say racist) roots of human behavior. "The negro is not simply a black Anglo-Saxon deficient in school," he warned, "but a being who in strength of appetites and in power to control them differs considerably from the white man."[24]

By 1907, Ross's recognition of the role of biology intensified, and his racism became more explicit as well as more Darwinian. "The theory that races are virtually equal in capacity," he insisted, "leads to such monumental follies as lining the valleys of the South with the bones of half a million picked whites in order to improve the conditions of four million unpicked blacks." Color of skin, he continued, was not the only difference between the ruling and the ruled races. "I see no reason why races may not differ as much in moral and intellectual traits as obviously as they do in bodily traits."[25] As late as the 1920s, Ross was still accounting for invidious differences in behavior between blacks and whites on grounds of biology or heredity.

Charles Cooley, a Univeristy of Chicago sociologist, like Ross, was a Darwinian, but of a less competitive or racist stripe. Along with Darwin, Cooley admitted "differences of race capacity," but described them as "subtle" and therefore of minor importance. They did not differentiate one group from any other "in the generic impulses of human nature," he contended. "The more insight one gets into the life of savages, even those that are reckoned the lowest," he remarked, almost in Darwin's own words, "the more human, the more like ourselves they appear." He boldly deplored the legal separation of the races in the South because it implied a refusal to recognize the Negro as "fundamentally a man like the rest of us."[26]

For some other social scientists at the opening of the twentieth century, the idea of evolution through natural selection was the most persuasive reason for seeing race as a determinant of human behavior. As one physical anthropologist put the matter of 1896, the shape and size of the skull and pelvis of the various races determined which were higher and which lower. Measured by such criteria, "the Caucasian stands at the head of the racial scale and the Negro at its bottom." The scale, however, was not eternal, it had evolved over time. As a professor at Mercer University wrote in the *North American Review* in 1900, "the evolution of the race comes slowly—a part of each new element of strength being transmitted by the laws of heredity from father to son and on to the succeeding generations; and so, slowly and painfully a race advances. It is not a matter of decades, but of centuries."

The last phrase was intended to show that whatever improvement some people may have been observing in the achievements of blacks in America, the gap between the white and black races would not close for a long, long time, if ever. Or, as economist Joseph Tillinghast contended in 1902, "the Negro finds it surpassingly difficult to suppress the hereditary instincts that do not harmonize with American social organization. He is finding that two or three centuries are all too brief a period in which to compass almost the entire range of human development. There is nothing in this conclusion to surprise the student of evolutionary phenomena," he added.[27]

The references in the preceding paragraphs to the length of time required by evolution reveal more than a desire to justify the inferiority of the nonwhite races. They also contain sharp echoes of the idea that the behavior or actions of parents could be passed on biologically to their children. The echo is clearly heard in psychologist R. Meade Bache's suggestion in 1895 that the experience of slavery helped to account for differences in behavior between blacks on the one hand and whites and Amerindians on the other. Anyone who has seen the listless hoeing by slaves, he wrote, "must feel assured that the mental attitudes thereby betrayed could not fail in the course of generations to modify physical function."[28] There was nothing in Darwinian evolution that precluded changes in behavior from being inherited and thereby altering, in time, the racial character of a people, along the lines of Bache's remarks. Today that process is identified as a principle of Lamarckianism, a pre-Darwinian theory of evolution propounded by the early nineteenth-century French naturalist-philosopher Jean-Baptiste Lamarck.

According to Lamarck, evolutionary change occured as a consequence of an organism's effort to improve its situation in its habitat. In perhaps the best-known example from the theory the giraffe's long neck was explained by reference to the continual effort of the original short-necked and short-legged animal to reach for food at higher and higher levels. Implied in Lamarck's explanation for evolution is the assumption that behavior patterns of parents could be inherited by their offspring. This process came to be known as the principle of acquired characters. Darwin categorically rejected Lamarck's teleological explanation for evolutionary change. His own concept of natural selection had no place for an animal's or a human being's will in bringing about evolution; for him the process was, in effect, accidental. Darwin, though, raised no objection to the idea that the behavior of parents, under certain circumstances, might be inherited by their offspring. As he wrote in *The Descent of Man*, "habits . . . followed during many generations probably tend to be inherited."[29] Again and again in his *Expression of Emotions* he pointed to the inheritance of habits as the source of the muscular expression of emotions common to both animals and men. Like the great majority of natural scientists of the late nineteenth century, Darwin was ignorant of the principles of Mendelian genetics, even though Gregor Mendel's path-breaking work had been published soon after the appearance of *The Origin of Species*. (Darwin's own copy of Mendel's paper on the basic principles of genetics was found in his library after his death with its pages still uncut.)

That Darwin himself accepted the principle of acquired characters meant that social scientists who considered themselves Darwinians needed to feel no conflict between their commitment to Darwinism and their belief that behavior, or habits, to use Darwin's term, might be inherited. Although for some social scientists acquired characters explained why blacks were inferior to whites, for others the same principle could explain why observed differences between races need not be permanent. The idea of acquired characters, after all, worked in both directions; bad traits could certainly be passed on, but advantageous behavior patterns could also be inherited. In that way, education, for example, could diminish or remove in time an undesirable racial trait from a person or group. As the Mercer University professor observed, that was precisely how the white race achieved its pre-eminence. "The evolution of the race comes slowly," he had written, "a part of each new element of strength being transmitted by the laws of heredity from father to son and on to the succeeding generations; and so, slowly and painfully a race advances."

The Mercer University professor, however, did not extend his analysis to blacks or other allegedly inferior peoples. But some American social scientists with Darwinian principles did. The well-known sociologist Lester Frank Ward was just such a Darwinian. Rather than seeing nature or biology as limiting for individuals, he saw it as progressive as well as almost mystical. "Evolution," he told his fellow sociologists in 1913, is "an ascending series. This prolonged spontaneous upward movement of the entire organic world is the result of that form of the universal energy which inheres in the life-principle, which makes life a progressive agent. . . ." It was environment, he insisted, that "resists the upward pressure of the life-force and holds all nature down." Among human beings, the life force and the power of the human mind "press forward together toward some exalted goal. The environment," he was convinced, "lies across the path of both and obstructs their rise. The problem everywhere is how to unlock these prison doors and set free the innate forces of nature." Fortunately, he insisted, "there are many ways in which Nature strives to maintain a perfect race and even to improve it. I have grouped all these tendencies together under the phrase 'biological imperative' and it constitutes one of the most salutary principles of sociology." It was the "social imperative," by which he meant eugenics, that was the true danger to human progress because it tinkered with nature's own force for good. "Every plant and animal possesses potential qualities far higher than its environment will allow it to manifest," he maintained. The limiting factor is environment, not biology as the eugenicists maintained.[30]

Given his belief in the equality of people, and the role of environment in shaping the lives of people, it was not surprising that Ward never abandoned his belief in the Lamarckian principle of acquired characters. For if one were a persistent enough Lamarckian, evolution provided the surest way of improving the race. Change the social environment for the better and the good qualities in people that emerged would be genetically passed on to offspring and thus perpetuated. And because Ward was a social reformer before he was a social scientist, to his dying day he rejected all attempts to disprove scientifically the idea that behavioral characteristics could be inherited.

By the time Ward died in 1913, however, few social scientists could go along with him in affirming the correctness of the Lamarckian principle of acquired characters. The scientific disproof had been set forth as early as 1889 by the German embryologist and devoted Darwinian August Weismann. He had shown that, no matter what changes occurred in an animal's body or behavior in its lifetime, none of them appeared in its offspring. The classic experiment was the severing of the tails of mice over the course of dozens of generations without any alteration in the tails of the offspring. Weismann concluded that there was a sharp and impenetrable disjunction between heredity and environment; only changes in the germ plasm or heredity material could be passed on to subsequent generations. So long as the hereditary material was unaffected, no bodily or behavioral alterations, no matter how enduring, would be inherited. Most social scientists in America were not influenced by Weismann's conclusions until the opening years of the twentieth century, when, in conjunction with the acceptance of the rediscovered principles of Mendelian genetics, the role of heredity in human life gained a new prominence among biologists.

Lester Ward, if only because of his deep interest in biology, knew about and understood the implications of Weismann's work as early as 1903. One ironic result was Ward's insistence that Herbert Spencer, his longtime ideological enemy, had proved Weismann to be in error. Ward detested Spencer's defense of the principle of the survival of the fittest and the social Darwinism that often sprouted from it. Yet Ward eagerly accepted Spencer's conclusion that Weismann's refutation of acquired characters was erroneous. (Spencer's own theory of evolution depended on the ability of parents to pass on their habits to their offspring. Not surprisingly, therefore, in his debate with Weismann, Spencer insisted that "either there has been the inheritance of acquired characters or there has been no evolution."[31])

An important reason Spencer, Ward, and some other social scientists could persist so long in their acceptance of Lamarckianism was that many biologists in the first decade of the twentieth century were themselves not convinced that Darwinian natural selection adequately explained the evolution of man. For them, Neo-Lamarckian evolution, as it came to be called, provided a more satisfactory explanation.

The objections to Darwin's explanation were both scientific and philosophical. From the beginning, Darwin's theory of natural selection had lacked one major and necessary element: a source of the small variations in organisms of the same species upon which natural selection worked. The theory also suffered in the eyes of many scientists from a philosophical deficiency, as historian of science Peter Bowler has demonstrated. Darwin's whole theory of evolution was materialistic; it excluded any moral or religious force in accounting for the evolution of human beings. For people used to thinking of human beings as architects of their own lives under the guidance of a Supreme Being, the philosophical implications of Darwinism were difficult to assimilate when they were not downright repugnant. The underlying philosophy of Darwinism had no place for a divine order or purpose in the universe or for man's agency or will. On the other hand, Neo-Lamarckianism, by postulating will and purpose as the agencies bringing about evolutionary change, neatly supplied the elements lacking in Darwin's explanation of how evolution came about. As a result, during the first fifteen years of the twentieth century, objections from natural scientists to Darwinianism were common. "The fair truth is," wrote biologist Vernon L. Kellogg in 1907 in his survey *Darwinism Today,* "that the Darwinian selection theories . . . stand to-day seriously discredited in the biological world."[32]

Not even the growing acceptance of Mendelian genetics during the first decade of the new century provided much scientific support to the principle of natural selection. In fact, for a while, the new genetics provided an alternative explanation for natural selection in the form of the mutation theory enunciated by the Dutch botanist Hugo DeVries. DeVries argued that the small variations within organisms, upon which Darwin based natural selection, were not sufficient, in the light of the new Mendelian genetics, to account for the changes that resulted in the formation of a new species. Large mutations, DeVries maintained, were necessary. The leading young American geneticist, Thomas Hunt Morgan, followed DeVries for just that reason, and because, as he revealingly added, it gave him the opportunity to escape from that "dreadful calamity of nature, pictured as the battle for existence," that Darwin's natural selection theory postulated.[33]

The new genetics may have raised serious doubts among natural scientists about the validity of the concept of natural selection, but its effect upon the Lamarckian doctrine of acquired characters was nothing less than fatal. For Weismann's principle that no changes in the body or behavior of an organism could affect the germ plasm or hereditary material soon became accepted doctrine among geneticists and biologists. Some scientists almost immediately perceived a consequence of great social and moral effect. "If Weismann ... [is] right," commented Berkeley biologist Joseph Le Conte in 1891, "if natural selection be indeed the only factor used by nature in organic evolution and therefore available for use by Reason in human evolution, then, alas, for all our hopes of race improvement, whether physical, mental or moral! ... All our schemes of *education,* intellectual and moral, although certainly intended mainly for the improvement of the individual, are glorified by the hope that the race is also thereby gradually elevated." But if it is true that "selection of the fittest is the only method available" then it seems to follow that "if we are to have race-improvement at all, the dreadful law of *destruction of the weak and helpless* must with Spartan firmness be carried out voluntarily and deliberately. Against such a course all that is best in us revolts." He noted that "the use of the Lamarckian factors, on the contrary, is not attended with such revolting consequences." Le Conte clearly still hoped, as would others like Herbert Spencer and Lester Frank Ward even later, that "the Lamarckian factors are still operative, that changes in the individual, if in useful direction, are to some extent inherited and accumulated in the race."

The social implications of the end of acquired characters were not lost upon social scientists, either. As social worker Amos Warner pointed out as early as 1894, "if acquired characteristics be inherited, then we have a chance permanently to improve the race independently of selection, by seeing to it that individuals acquire characteristics that is desirable for them to transmit." But if acquired characters cannot be transmitted, then the only hope of social reformers like himself, "for the permanent improvement of the human stock would ... seem to be through exercising an influence upon the selective process."[35]

As Warner implied, and as the next chapter will show, the abandonment of the belief in acquired characters was a stimulus for a eugenics movement. The decline in the acceptance of acquired characters had the additional and immediate effect of hardening the concept of racism among some social scientists. By showing that environment could not change behavior based upon race or biology, the new genetics had given racism a scientific basis it had lacked so long as acquired characters was an accepted principle. Behavior that was thought to be derived from race could now, for the first time, be said to be permanent. For the obvious conclusion to be drawn from Weismann's demonstration was that no amount of education or improvement in the social environment over time could either eradicate anti-social behavior or foster socially desirable actions. Increasingly, the differences between blacks and whites, for example, would be seen as permanent rather than merely requiring a long time for them to be eliminated. "It is not in the conditions of life," concluded economist Frederick L. Hoffman in 1896, "but in race and heredity that we find the explanation of the fact to be observed in all parts of the globe, and in all times and among peoples, namely the superiority of one race over another, and of the Aryan race over all." Or, as psychologist George Ferguson concluded as late as 1921, the "psychological study of the Negro indicates that he will never be the mental equal of the white race," though he is capable of "great progress."[36]

The impact of biological ideas on social scientists in regard to race was aptly summarized by a University of Wisconsin professor in 1905 in the *American Journal of Sociology*. "The last few years have witnessed a great change of mind in matters of humanitarianism," he began. "The absolute unity of human life in all parts of the globe, as well as the idea of the practical equality of human individuals wherever they may be found, has been quite generally abandoned," he forthrightly explained. To treat all peoples "as if they were all alike, to subject them to the same methods of government, to force them into the same insitutions, was a mistake of the nineteenth century which has not been carried over into our own."[37]

Clearly, the idea of a unified human nature was not a central concern of the Wisconsin professor. Indeed, insofar as social scientists recognized differences among races, they were drifting away from a concern for what a common humanity might be. Nor was race the only division within the human family that social scientists at the turn of the century directed their attention to. Sex, for instance, seemed an obvious source of differences among human beings. Darwin played a prime role here, also. In fact, contrary to his approach to racial differences, he saw sexual differences as profound and important. After all, a central element in his theory of natural selection was reproduction, a process that depended upon individuals with different biological functions and therefore, presumably, with different natures. But there was more to Darwin's concern with sexual differences than that, as his principle of "sexual selection" made clear.

Darwin's admission that he could arrive at no explanation for the external racial differences among human beings compelled him to recognize that though natural selection could explain differences *between* species, it could not account for differences among individuals *within* species. Hence, in order to complete his theory of evolution, he devoted well over half the pages in the *Descent of Man* to providing a second mechanism for evolution, namely sexual selection. Its primary function for Darwin was to account for the often striking differences between males and females in a species. The essential element in the theory was that differences between the sexes could be accounted for by the competition for mates. Those characteristics that gave an advantage to maximizing reproduction would gradually become standard in the two sexes of a species.

On the opening pages of the *Descent of Man,* Darwin applied the theory to human beings. "Man differs from woman in size, bodily strength, hairiness, etc., as well as in mind," he observed, "in the same manner as do the two sexes of many mammals." As always, the connection with animals was crucial to him. "There can be no doubt," he continued, "that the greater size and strength of man, in comparison with woman . . . are due in chief part to inheritance from his half-human male ancestors. These characters would, however, have been preserved or even augmented," he believed, "during the long ages of man's savagery, by the success of the strongest and boldest men, both in the general struggle for life, and in their contests for wives." The process was clear: the success of such bold men would have been measured in "their leaving a more numerous progeny than their less favoured brethren."[38]

Darwin was well aware that his contemporaries John Stuart Mill and his wife Harriet Taylor, among others, doubted that women differed in mental abilities from men. Yet he thought it quite "probable that sexual selection has played a highly important part" in creating those differences. One reason for his thinking so was an analogy with "the lower animals which present other secondary sexual characteristics." No one disputes, he wrote, "that the bull differs in disposition from the cow, the wild-boar from the sow, the stallion from the mare." Similar differences, he thought, can be observed in the human sexes. "Man is the rival of other men; he delights in competition, and this leads to ambition which passes too easily into selfishness." Regretably, he acknowledged, "these latter qualities seem to be his natural and unfortunate birthright. It is generally admitted," on the other hand, "that with woman the power of intuition, of rapid perception, and perhaps of imitation, are more strongly marked than in man." Some of their qualities, he added in what today we can only designate as sexist and racist language, "are characteristics of the lower races, therefore of a past and lower state of civilization."[39]

As the foregoing implies, Darwin harbored a rather limited conception of the human female. "The chief distinction in the intellectual powers of the two sexes," he believed, "is shewn by man's attaining to a higher eminence, in whatever he takes up, than can woman—whether requiring deep thought, reason, or imagination, or merely the use of the senses and hands." The evidence for his conclusion seemed obvious to him. "If two lists were made of the most eminent men and women in poetry, painting, sculpture, music (inclusive both of composition and performance), history, science, and philosophy, with half-a-dozen names under each subject, the two lists would not bear comparison."[40]

And apparently time would not change the situation, for in Darwin's eyes the rivalry between men that in the course of evolution had made them superior in mind and body to women was still shaping their being. Darwin admitted that the purely physical contests between men over women no longer obtained. Yet even in the nineteenth century, he believed, men "generally undergo a severe struggle in order to maintain themselves and their families; and this will tend to keep up or even increase their mental powers, and, as a consequence the present inequality between the sexes."[41]

Darwin's sexist conception of woman did not arise from his theory of sexual selection; rather sexual selection was his way of accounting for what he thought he saw all around him. Other social observers of woman's place in society often arrived at similar conclusions, with or without the theory of sexual selection. In the United States the most frequent occasion when people were moved to draw upon biology to account for behavioral differences between the sexes was in regard to matters intellectual, such as education. Undoubtedly the best-known work in the middle years of nineteenth-century America to raise doubts about woman's ability to profit from higher education was Dr. Edward H. Clarke's book *Sex in Education,* published in 1873. Clarke was too modern a man to deny access to collegiate training to women out of hand. After all, by that time, several colleges had already opened their doors to women, including Vassar College, which had been founded over a decade earlier. Instead, Clarke raised questions about the long-range effects upon women of an education of the rigor that was provided for men at the leading colleges. His contention was that women's physique was seriously injured in the long run, if not in the short, when they undertook such training. He admitted that many young women had by then shown themselves to be scholars of high competence, but within a very short time after completion of their studies, he wrote, they almost invariably succumbed to physical and mental breakdowns. Women's nature, in short, was not suited for such high-powered mental activity.

Nor were all sexists men. Women such as M. A. Haraker, writing in the *Popular Science Monthly* in 1882, also drew upon Darwin. She concluded, for example, that women had less "reasoning power and creative imagination" than men, but they excelled men in intuition. And though not much study has been given to the nature and distribution of intuition between the sexes, "there is considerable evidence," she insisted, "that it is acquired by heredity, that it is closely akin to instinct, that some modification of it is the common possession of women, children, and the lower animals." Appealing to the ultimate authority, she noted that Darwin himself had identified observation, reason, imagination, and invention as male qualities, while he saw women as high in intuition, rapid perception, and, in his words, "perhaps of imitation." She then went into an analysis of women's physiology, from which she concluded that since they are smaller, have less speed of digestion, as well as expend more energy on reproduction than men, *"men will always think more than women."* [42]

Social scientists also called attention to the physiological differences between the sexes, sometimes adding that such differences ruled out certain activities by women. Writing in *The Psychological Review* in 1895, R. Meade Bache thought that it had been well established that women had faster reaction times than men, and that the difference was "in strict accordance with the fact that the brain development of men, as compared with that of women is greater, even when taking into account the relatively greater weight of normal individuals of the male sex as compared with that of normal individuals with the opposite one." Bache did not draw any policy conclusions from this difference, but psychologist John Dewey writing in *Popular Science Monthly* in 1886 implied that the responses of women to higher education raised some serious questions that did not arise when one looked at men. In reviewing a study of college women, Dewey observed that such training seemed to place limits on woman's ability to properly fulfill the roles of wife and mother. Only 26 percent of the college women in the study had married, he noted, and of those merely 63 percent had borne children. Furthermore, a quarter of the 12 percent of women who had died did so in childbirth. Without offering general statistics that would have put these figures into some comparative perspective, Dewey tersely observed that "these figures speak for themselves." [43]

Sociologist William I. Thomas, of the University of Chicago, took a different biological approach to the differences between the sexes. In a long article in the *American Journal of Sociology* of 1897, Thomas observed that natural scientists distinguished between "anabolic" and "katabolic" organisms. The former produced more energy than they needed; the latter consumed more energy than they created. Plants are anabolic, for they store energy; animals are generally katabolic, for they expend energy in movement; they make up for energy deficiencies by consuming the stored energy of plants. Women, Thomas contended, stand closer to plants, because women represent the constructive side of human nature. Following Darwin's lead, Thomas thought males were the disruptive or active sex. That women put on more fat than men, that is, stored more energy, had less lung capacity, produced less urine, and needed less sleep than the other sex, proved that women were anabolic in nature. The interpretation he gave to this finding clearly placed women beneath men. After noting that some lower animals could regenerate lost limbs or organs, he pointed out that "the lower human races, the lower classes of society, women and children, show something of the same quality in their superior tolerance of surgical disease." Women can tolerate pain and misery better than men, he asserted. And that helped to explain, he suggested, why "anthropologists regard women intermediate in development between the child and the man."[44]

Thomas, though clearly a Darwinian, did not accept the idea of sexual selection, but he fully accepted the Darwinian assertion that human males were more variable than females. Like other students of the time who delineated the differences between the sexes, Thomas believed that women's brains were smaller than men's, a difference that he thought helped to explain the greater variability of males. Or as he put it, among males there were more geniuses and idiots.

Women's "sensibility, feeling, emotionality, or affectability" Thomas linked to the "larger development of her abdominal zone, and the activity of the physiological changes located there in connection with the process of reproduction." Darwin, too, not surprisingly, had ascribed women's nature to her reproductive role. Woman differs from man in disposition, Darwin wrote in the *Descent of Man*, "chiefly in her greater tenderness and less selfishness." Owing to "her maternal instincts," woman displays "those qualities towards her infants in an eminent degree; therefore, it is likely that she would often extend them toward her fellow creatures." And that was Thomas's thought as well.

The result, in Thomas's mind, was a symbiotic relation between men and women that reached its acme in the Victorian nuclear family. "Man's katabolism predisposed him to activity and violence; woman's anabolism predisposed her to a stationary life. The first division of labor was, therefore, an expression of the characteristic contrast of the sexes.... This allotment of tasks," he argued, "is not made by the tyranny of men, but exists almost uniformly in primitive societies because it utilizes most advantageously the energies of both sexes."[45]

Nationally prominent psychologist G. Stanley Hall also discerned the anabolic nature of women: that is, her reproductive function was the key to her social character. "Our modern knowledge of woman represents her as having characteristic differences from man in every organ and tissue," he wrote in his influential study *Adolescence* in 1904. Woman is "conservative in body and mind, fulfilling the function of seeing to it that no acquired good be lost to mankind, as anabolic rather than katabolic.... Her whole soul, conscious and unconscious, is best conceived as a magnificent organ of heredity, and that to its laws all her psychic activities, if unperverted are true."[46]

For Hall, who was not only the president of Clark University but also a nationally known psychologist, the practical implications to be drawn from these supposedly scientific differences were that they should be reflected in social policy. Since nature seems to decree "that with advancing civilization the sexes shall not approximate but differentiate," Hall advised, "we shall probably be obliged to carry sex distinctions . . . into many if not most of the topics of higher education." This is so because boys and girls differ in their tastes and interests as shown by "history, anthropology, and sociology as well as home life. . . . This is normal and biological." He wondered, therefore, if the coeducational high school, with its identical training for both sexes, and which the United States has carried forward farther than any other country, "has not brought certain grave dangers, and whether it does not interfere with the natural differentiation seen everywhere else."[47]

Social scientists' doubts that women could profit from the higher education available to men continued to be heard through the first decade of the twentieth century. D. Collins Wells, of Dartmouth College, writing in *The American Sociological Review* in 1907, supposed that "we must submit to the higher education of women. It appears inevitable; but it seems to me not yet proved that this education should be the same in kind or amount as that afforded to men." James McKean Cattell of Columbia University's department of psychology also expressed serious reservations about higher education for women, and, as usual, those doubts stemmed from the biological differences between the sexes. "Girls are injured more than boys by school life," he wrote in 1909; "they take it more seriously, and at certain times at a certain age are far more subject to harm." He also worried about what he considered the high social price paid for allowing women to attend colleges and universities. Because college-educated women apparently bore fewer children, he calculated that "to the average cost of each girl's education must be added one unborn child." Furthermore, he warned, as women gain wider opportunities for employment outside the home, as was increasingly the case, the American family must suffer since women "can conveniently leave their husbands should it so suit their fancy."[48]

Cattell's student, the Teachers College pscyhologist Edward Thorndike, writing in 1914, discerned an enduring line between the capabilities of boys and girls. Even if "we should keep the environment of boys and girls absolutely similar," he predicted, the sexually rooted "instincts would produce sure and important differences between the mental and moral activities of boys and girls." As late as 1919, sociologist David Snedden was echoing Thorndike's views in the *American Journal of Sociology*. The whole question of the nature of woman's mind, he was convinced, needed additional study. He had few doubts "that most women, by instinct and as a result of custom inheritance, are peculiarly qualified for 'homemaking' as that has evolved through the ages."[49]

Race and sex as explanations for human differences and behavior were not the only ways in which biology captured the attention of American social scientists at the turn of the century. Darwin's emphasis upon the continuity between animal and human life encouraged students of human behavior to seek out fresh answers to why men and women behaved the way they did. The influence of Darwinian ideas extended beyond the thought and attitudes of scholars; it shaped social policy as well, as the next chapter seeks to show.

2

Beyond the Darwinian Imperative: Instinct, Eugenics, and Intelligence

It is impossible not to regret bitterly, but whether wisely is another question, the rate at which man tends to increase; for this leads in barbarous tribes to infanticide and many other evils, and in civilised nations to abject poverty, celibacy, and to the late marriages of the prudent. But as man suffers from the same physical evils as the lower animals, he has no right to expect an immunity from the evils consequent on the struggle for existence. Had he not been subjected during primeval times to natural selection, assuredly he would never have attained his present rank.

Charles Darwin, 1871

Over the centuries, men have been fascinated by what the life of animals could tell them about human nature. If human beings and animals shared much, they also differed greatly. Ever since Descartes in the seventeenth century, at least, one of the things that were believed to distinguish human beings from the rest of animate nature was that human beings acted out of their own volition, that they thought about, or willed action while animals acted out of instinct. To Descartes animals were close to machines.

After Darwin, however, the Cartesian gap narrowed considerably. Those who sought to understand the nature of human beings increasingly looked to nature, to life processes, to biology. They emphasized the animal in human beings and the correlative insight that the activities of animals might throw light on the sources of human behavior and the nature of man. One such insight was that human beings, too, had instincts. Darwin himself, as we have seen, spoke again and again about the social instincts in human beings and animals. He had no doubt, as he wrote in the *Descent of Man*, that "the instinct of sympathy," or the desire for praise from others, "was originally acquired, like the other social instincts, through natural selection."[1] In none of the social sciences was this Darwinian assumption of the relevance of animal behavior to the study of human beings more apparent than in the new science of psychology that was emerging in the last decades of the nineteenth century.

For all practical purposes, the study of human psychology in America began with the publication in 1890 of William James's *Principles of Psychology*. Although originally intended as a brief introduction to the subject for college students, the final product, which took James eleven years to complete, was a turning point in the history of the subject. One reason for the delay had been James's decision to build his book upon the Darwinian idea of the animal roots of human behavior. Instinct, James contended, drawing on Darwin, was common to both animals and human beings. He defined instinct simply "as the faculty of acting in such a way as to produce certain ends, without foresight of the end, or without previous education in the performance." In James's view instincts manifested themselves early in human life, in such things as holding a head erect, standing, and walking. "Climbing on trees, fences, furniture, bannisters" he also listed as a "well-marked instinctive propensity which ripens after the fourth year" of life. Among human adults, he thought emulation or rivalry was "a very intense instinct" along with pugnacity, anger, and resentment. "The hunting instinct," James told his readers, "has an equally remote origin in the evolution of the race"; fear, too, he thought to be a genuine instinct.[2]

When James came to discuss less obvious instincts, such as agoraphobia, he made even more pointed connections to Darwinian ideas. "This emotion has no utility in a civilized man," he wrote, "but when we notice the chronic agoraphobia of our domestic cats, and see the tenacious way in which many wild animals, especially rodents, cling to cover, and only venture on a dash across the open as a desperate measure—even then making for every stone or bunch of weeds which may give momentary shelter—when we see this," he suggested, "we are strongly tempted to ask whether such an odd kind of fear in us be not due to the accidental resurrection, through disease, of a sort of instinct, which may in some of our ancestors have had a permanent and on the whole useful part to play?"[3]

In delineating his conception of instinctive behavior, James did more than simply link his views to Darwin's evolutionary scheme. He deliberately chose to explain, as Darwin had done before him, why the then reigning explanation for human habits—associationism—was inadequate in the light of new biological knowledge. To believers in associationism, human behavior was explained by the association between one idea or activity with another, and as such it was a peculiarly human activity: animals could not associate ideas because they did not think. Instinct accounted for their behavior. It is not surprising, then, that a leading proponent of associationism, John Stuart Mill, minimized, when he did not deny outright, the role of race and sex in shaping human behavior. James contended that associationist explanations could not account for acquisitiveness or appropriation in human beings; to him, they were instincts. He thought that "constructiveness" was as "genuine and irresistible an instinct in man as in the bee or the beaver," as he revealingly phrased it. And though he did not think love or sex were instinctive since they were often conditioned by other forces, he pronounced jealousy to be "unquestionably instinctive."[4]

Other early twentieth-century social scientists similarly found the concept of instinct essential in seeking to account for human behavior. Sociologist Charles Ellwood, for example, writing in 1901 in the *American Journal of Sociology* faulted French psychologist Gabriel Tarde's and American psychologist J. Mark Baldwin's conception of imitation because they ignored instinct in trying to explain behavior. He was sure that human behavior was strongly influenced by innate forces such as the "economic instinct," "instinctive religion," and "instinctive morality," as well as instincts of race. Psychologist John Boodin of the University of Kansas thought that most of human activity depended upon instinct now that it was clear that there were no such things as acquired characters. All progress takes place, he wrote in Darwinian language, "through spontaneous variations and natural selection." He thought that "fundamental virtues which underlie social life, such as honesty, truthfulness, and kindness cannot be produced in people" by education or environment; they are present at birth and only await the proper time in the life cycle to emerge. "They are acquired no more than love is acquired . . . ," he insisted.[5]

When William James died in 1911, his fellow psychologists judged his emphasis upon instinct in accounting for human behavior to be a fundamental insight of the profession. An editorial in the *Psychological Review* described James as "following closely in the footsteps of Darwin." James, the editorial continued, "made instinct an essential part of the study of the human mind in a way no other psychologist had done. He took a large, flexible, dynamic view of instinct which gave it a place in the very forefront of human life and so of human psychology, instead of relegating it to the limbo of 'left-overs' from our animal ancestry for which apologies must be made or moralizing indulged."[6]

No practicing psychologist at the time of the First World War placed more emphasis upon instincts in accounting for human behavior than William McDougall. Instincts stood at the center of his argument in his highly influential book *Introduction to Social Psychology*, first published in the United States in 1909 while he was still teaching at Cambridge and Oxford in England. In his book McDougall compiled a list of human instincts, some of which came from James and some of which he himself identified. They ran the gamut from the parental instinct, through the gregarious instinct, to the Jamesian "Instincts of Acquisition and Construction."

A dozen years later, after he had become professor of psychology at Harvard, and perhaps the most widely known psychologist in America, McDougall was still contending that if "the facts of human experience and of human and animal behavior be impartially surveyed, we are fully justified in accepting the conception of instincts in the human species as innate tendencies to pursue by purposive actions certain biological ends," among which he listed mating, parenting, escape from danger, "purposive striving," dominance, and companionship. Indeed, virtually all textbooks in psychology, not only McDougall's work, followed James in drawing upon the concept of instinct in explaining human actions. Even some economists of the stature of Frank Taussig, Wesley C. Mitchell, and Thorstein Veblen accepted instincts in human beings. "Between 1900 and 1920," writes historian Hamilton Cravens, "at least six hundred books and articles published in America and England advanced the instinct theory."[7]

Anthropologists, too, found instinct an indispensable concept. Alfred Kroeber of the University of California baldly contended in 1910 that "there can be no doubt that the essential moral ideas of man spring from instinct." Opposition to murder, incest, and cannibalism were only the most obvious universal moral ideas that derived from man's evolutionary experience, Kroeber argued. Drawing a Darwinian connection between man and beast, Kroeber thought it easy to see how "these instincts have arisen. . . . In the animal world, they are, in the main, a necessity," he explained. "The species that consumes itself, habitually inbreeds, or neglects its offspring, perishes. The unreasoning conditions of nature therefore have impressed strong aversion to such practice, or have suppressed the instincts toward them, in virtually all higher animals. From our animal ancestors we no doubt derive the same feelings."[8]

In the early twentieth century, instinct was most frequently drawn upon by psychologists and other social scientists to explain normal human behavior. Nevertheless, there was nothing in the concept to prevent its being used in accounting for pathological or anti-social behavior as well. Before August Weismann had ruled out acquired behavioral traits as hereditary, a social reformer like Charles Loring Brace in his influential book *The Dangerous Classes of New York and Twenty Years' Work Among Them*, published in 1872, warned that the pathological character traits of drunken or criminal parents might well be inherited by their offspring. Medical men, Brace pointed out, were convinced that children could inherit insanity, criminal habits, and even a tendency toward prostitution.

By the early 1880s European works on criminality, which often stressed the hereditary origins of crimes, were laying a foundation for what American penologists and anthropologists later came to call "criminal anthropology." It began as a field of scholarly investigation in America in 1888 upon the publication of an article favorably summarizing the work of the Italian criminologist Cesare Lombroso. Lombroso's theory that criminals could be identified by their physical characteristics was destined to be widely known throughout the European world before the century was out. "The new school of Criminal Anthropology" was the subject of the presidential address of Robert Fletcher in 1891 upon his retirement from the Anthropological Society of Washington, D.C. After praising the work of Lombroso, Fletcher confidently told his audience that "the influence of heredity in the formation of criminal character has been long since admitted. . . ." A few years later Henry M. Boies, a prominent Pennsylvania authority on prisons, wrote that "the science of heredity" does more than simply explain the 50 or 75 percent of the cases of criminal depravity which have been successfully traced; it also shows that "nearly every case of it is due to a diseased or disordered organism or function of organs, produced by ancestral influences. Good seed," he sententiously concluded, "generates sound and healthy fruit, and imperfect parentage can only yield defective offspring."[9]

Sociologists, too, looked to instinct in accounting for anti-social behavior. Charles Ellwood of the University of Missouri, for instance, played down what he called the contagiousness of vice, crime, and general shiftlessness. It was thought by some, he said, that bad examples caused the young to go into crime and vice. More, however, was involved than imitation in accounting for the "'instinctive criminal'" or the "'hereditary pauper.'" They become so, he believed, "because inborn tendencies lead them to seek such models for imitation rather than others; because they naturally gravitate to a life of crime or pauperism." He then went on to assure his readers, who were sociologists, not penologist or criminologists, that "this is practically the unanimous conclusion of all experts engaged in the study of these classes."[10]

During the first decade of the twentieth century a new twist was given to the argument that criminality was biologically induced. The innovation was the introduction of the intelligence test to social science. Some psychologists had been interested for some time in devising a measure of individual mental differences. Soon after Alfred Binet, a French psychologist, developed a test for mentally deficient school children in 1905, American social scientists, particularly psychologists, began mental testing on a large scale. An early prominent figure in American testing was Herbert H. Goddard, director of research at Vineland Training School of New Jersey. Goddard had long been interested in ways to differentiate more precisely among the mentally deficient children at the Vineland school. Soon after he had become acquainted with the Binet test while on a European trip, he devised his own Americanized version and began to administer it.

By the opening of the second decade of the twentieth century, other officials at homes for delinquent children, along with prison authorities, were following Goddard's lead. In the course of the testings a strikingly large proportion of the delinquents and prisoners were found to be "feebleminded." The publication and use of Stanford psychologist Lewis Terman's revision of the Binet test in 1916 further confirmed the existence of a linkage between weak minds and weak morals. It was said that between a third and a half of the delinquents, prostitutes, and criminals were feebleminded, or, in the parlance of the testers: they were "morons," a term invented by Goddard to identify those who tested between a mental age of eight and twelve. (To round out the nomenclature: "imbeciles" stood below morons, while below them "idiots" were to be found.) As a result of these findings by experts, and the public attention given to them, the United States around the time of the First World War was beset by what came to be called the "menace of the feebleminded." Goddard made his contribution to the rising fear with the publication in 1912 of his book on the Kallikak family.

The Kallikaks were a real family; the name by which they went down into psychological history was invented by the relentlessly neologistic Goddard from two Greek roots meaning "beauty" and "bad." As a family, according to Goddard's report, the Kallikaks were an especially horrendous example of how feeblemindedness, simply because it was inherited, could mushroom over time into a gigantic social problem. Goddard's study of the results of feeblemindedness in a particular family down through the generations was not the first such examination of the inherited sources of crime and immorality. That distinction belonged to Robert L. Dugdale, whose book *The Jukes, a Study in Crime, Pauperism, Disease, and Heredity* was first published in 1877.

In the popular history of hereditarianism, the Kallikaks and the Jukes are usually portrayed as dual studies leading to the same conclusions. That portrayal is misleading because it obscures the way in which the overthrow of the Lamarckian concept of acquired characters intensified hereditarian concepts among social scientists. Dugdale's original study, in striking contrast to a later study of the Jukes, which Arthur Estabrook published in 1916, had not emphasized the role of heredity in accounting for the poverty, disease, and criminality he found among the family he called the Jukes. Instead, Dugdale described his study as dealing with the effects of both heredity and environment. "Heredity fixes the organic characteristic of the individual," while "environment," he contended in standard Lamarckian fashion, "affects modifications in that heredity."[11]

In his book, Dugdale frequently called attention to the improvement in delinquents that a changed environment could bring about. "The tendency of heredity is to produce an environment which perpetuates that heredity," he wrote. "Thus, the licentious parents make an example which greatly aids in fixing habits of debauchery in the child. The correction," he advised, "is change of environment." As a believer in acquired characters Dugdale could easily envision the alteration of heredity by environment. "For instance," he wrote in one place, "where hereditary kleptomania exists, if the environment should be such as to become an exciting cause, the individual will be an incorrigible thief; but if, on the contrary, he be protected from temptation, that individual may lead an honest life, with some chances in favor of the entailment [or hereditary force] stopping there." Or, as he wrote in another place: "Where the environment changes in youth the characteristics of heredity may measurably be altered. Hence the importance of education," he pointed out. To Dugdale, as to Lester Ward and many other later social scientists, Lamarckianism worked both ways: "environment tends to produce habits which may become hereditary, especially so in pauperism and licentiousness, if it should be sufficiently constant to produce modification of cerebral tissue." But since it did work both ways, Dugdale told his readers, "the whole question of the control of crime and pauperism become possible, within wide limits, if the necessary training can be made to reach over two or three generations." Changes in the environment, in sum, would do the job if persisted in.[12]

Dugdale could still be a Lamarckian in the 1870s, but students of social pathology working in the twentieth century could not. Indeed, by then the scientific successes of genetics were such that heredity was increasingly seen by biologists as the key to understanding human behavior. If the facts of family life seemed to lend themselves to a genetic explanation, the tendency was to seek out genetic or hereditarian causes. That is where H.H Goddard began when he examined the history of Martin Kallikak's descendants. Kallikak was a soldier during the American Revolution who made the mistake, as Goddard told the story, of having had a casual liaison with a mentally defective young woman he met in a tavern. From that union issued a series of mentally defective, criminally inclined, poverty-stricken descendants. Martin Kallikak, Jr., the offspring of that union, had 480 descendents, Goddard reported, of which 143 "we have conclusive proof, were or are feeble-minded, while only forty-six have been found normal. The rest are unknown or doubtful."[13] It was Goddard's conclusion that all those defective offspring derived from the feeblemindedness of Martin Jr.'s mother, the young woman in the tavern. Goddard was sure the defective children resulted from her genetic contribution because Martin Sr. later married a very fine and upstanding Quaker maiden; the desecendants of that union proved to be as morally upright as the mother, and as successful materially as Martin Kallikak, Sr., himself.

Goddard, of course, knew of Dugdale's study of the Jukes, but as a student of the new genetics, he recognized that it was not an examination of heredity's effect upon sociopathological behavior. "So far as the Jukes family is concerned," he correctly wrote in his 1912 book on the Kallikaks, "there is nothing that proves the hereditary character of any of the crime, pauperism, or prostitution that was found." The much discussed question of the place of heredity in the origins of crime and depravity may have gotten no answer from the Jukes family as studied by Dugdale, he admitted, "but in the light of present-day knowledge of the sciences of criminology and biology, there is every reason to conclude that criminals are made and not born. The best material out of which to make criminals, and perhaps the material from which they are most frequently made," he was convinced, "is feeblemindedness." That deficiency he saw as clearly inherited.

Goddard, differing markedly from Dugdale, ruled out changes in the environment as a remedy for feeblemindedness. The descendants of Martin Kallikak, Jr. (the offspring of the mentally defective tavern girl and Kallikak Sr.) "were feebleminded, and no amount of education or good environment," Goddard asserted, "can change a feebleminded individual into a normal one, any more than it can change a red-haired stock into a black-haired stock." This was carrying August Weismann beyond his own expectations. Goddard's conviction rested on what he considered incontrovertible evidence: "the striking fact of the enormous proportion of feeble-minded individuals in the descendants of Martin Kallikak, Jr., and the total absence of such in the descendants of his half brothers and half sisters is conclusive on this point," he asserted. "Clearly it is not environment that has made that good family."[14]

In his major work on feeblemindedness published in 1914, Goddard authoritatively announced that about half of all criminals were feebleminded and that this hereditary mental defect was the source of their law-breaking. Contrary to what may be thought of Goddard's rather obsessive attention to the alleged dangers of feeblemindedness, he was not the rigid hereditarian or elitist that Stephen Jay Gould in his *Mismeasure of Man* has made him out to be.[15] Contrary to Gould's description, Goddard never found it easy to accept the idea that intelligence or feeblemindedness, for that matter, was a single genetic character, or gene. "We do not know that feeblemindedness is a 'unit character,'" he wrote in his book on the Kallikaks. "Indeed, there are many reasons for thinking that it cannot be." Then, two years later in his big book on feeblemindedness, he thought new evidence strengthened the case for the unit character idea. It is from that work on which Gould depended for his characterization of Goddard's conclusion. Nevertheless, even then, contrary to Gould, Goddard had not yet resolved all his doubts. He "confesse[d] to being one of those psychologists who find it hard to accept the idea that intelligence ever *acts like a unit character*," but he found it hard to deny the conclusion to which his statistics seemed to be driving him.[16] Those were not the sentiments of a confirmed believer in a single character for feeblemindedness or intelligence.

In his single-minded pursuit of Goddard, Gould also makes a point of contrasting Goddard's approach to testing with the more flexible and sympathetic attitude taken by Binet. Yet, contrary to Gould, Goddard himself looked approvingly to Binet for guidance in interpreting test results. "A man may be intelligent in one environment and unintelligent in another," Goddard explained in his 1914 book. "It is this point which Binet has illustrated by saying 'A French peasant may be normal in a rural community but feeble-minded in Paris.' The peasant life is simple," Goddard elaborated; "the environment requires little adjustment. In Paris, it is different, all is complicated and requires the highest functioning of certain mental powers in order to enable one to adapt himself. That fact," Goddard advised, "should be borne in mind thruout this discussion."[17]

Goddard's caution and sophistication are better appreciated when it is recognized that among biologists and geneticists feeblemindedness as a "unit character" carried none of the doubts that still plagued Goddard. In 1922, for example, biologist Vernon Kellogg maintained that the contention that feeblemindedness was "a unit human trait following the general Mendelian order as regards its mode of inheritance . . . is hardly any longer open to doubt." Feeblemindedness, he told the readers of his popular book on biology, was as specific in its genetics as eye color and hair form; he even went so far as to see "feeblemindedness paired with full-mindedness" in human genetics. And as late as 1929, one of the leading geneticists of the time, Edward M. East, in his book *Heredity and Human Affairs,* described feeblemindedness, along with insanity and epilepsy, as a simple recessive trait.[18]

An even stronger and more explicit position was taken on the subject by the liberal biologist H. S. Jennings in his popular book *The Biological Basis of Human Nature,* published in 1930. More than once in the course of describing the way in which heredity worked, Jennings turned to the inheritance of feeblemindedness to illustrate how a single, recessive trait affected heredity. Early on in his book he made the point quite matter-of-factly: "It is known that feeblemindedness may result from a defect in one pair of genes." Then, at another time, he reported that feeblemindedness "is the simplest and least affected by the environment of any of the defects with which eugenic measures can deal." The reason geneticists were confident that feeblemindedness may be produced by the "alteration of a single gene," he explained, is that "feeblemindedness is often inherited according to the simple Mendelian system, which results when

father and mother differ in a single gene affecting the characteristic considered."
At one point he made clear that he believed, along with Kellogg, that a single
gene could cause feeblemindedness unless it were countermanded, as it were, by
the presence of a single, dominant gene that produced normal intelligence. In
the course of explaining how genes functioned he wrote that a "certain gene
may fail in laying a proper foundation for the brain; the result will be to produce a feeble-minded individual—unless there is also present, as its mate, a gene
that performs fully this function."[19]

Given the acceptance by biologists of the inheritability of feeblemindedness
well into the 1930s, it is not surprising that as late as the 1940s social scientists
were still seeing low intelligence as an inheritable characteristic. A reviewer in
the *American Sociological Review*, for example, in 1940 opposed reproduction
by feebleminded people not only on the ground that such behavior would add
"to the total number of feebleminded," in the country, but also that feebleminded families "are largely characterized by promiscuity, desertion, illegitimacy, crime, unhappiness, ill health, and other associated pathological
conditions."[20]

The recognition of heredity's role in spawning social pathologies, disease, and
immoral actions soon led to demands for measures to mitigate the danger. In
fact, early in the history of the new hereditarianism proposals for regulating
heredity were pressed upon the public. Only a concern for clarity of presentation has kept the eugenics movement, hereditarianism's best-known offspring,
out of these pages until now.

Both the term eugenics and the movement that stood behind it were the
brainchild of Francis Galton, an English statistician and life-long student of biology and heredity. He coined the word "eugenics" in 1883 from the Greek
words for "well-born." His writings over the years brought the idea of regulating heredity to the attention of Americans. Like many other natural and social
scientists of the time, Galton received much of his inspiration from Darwin,
who happened also to be his cousin. The essential ideas of eugenics, Galton liked
to think, had been laid down by his distinguished cousin. And, indeed, in more
than one place in *The Descent of Man*, Darwin had referred to the dangers
posed to good heredity by the well-intentioned humanitarianism of civilized
society. Of all animals, only human beings, Darwin recognized, were in a position to escape the controls generally imposed by natural selection. "We build
asylums for the imbecile, the maimed, and the sick," Darwin worried; "we institute poor-laws; and our medical men exert their utmost skill to save the life of
every one to the last moment. Thus the weak members of civilized societies
propagate their kind. No one who has attended to the breeding of domestic
animals," Darwin pointed out, "will doubt that this must be highly injurious to

the race of man." Darwin, however, was less sanguine than his cousin Galton that human beings could be induced to control their personal interests in behalf of social improvement through biology. "Both sexes ought to refrain from marriage, if they are in any marked degree inferior in body or mind," Darwin wrote. "but such hopes are Utopian and will never be even partially realised until the laws of inheritance are thoroughly known."[21]

Galton, however, did not want to wait that long. His investigations and writings on the power of inheritance in human affairs began in the 1870s and continued until his death in 1911. His movement for the adoption of social policies based on eugenic principles, however, did not catch on until the Lamarckian concept of acquired characters had been routed among biologists and social scientists. For under that principle human heredity could be improved simply by creating a better environment. Heredity was not fixed; even hereditary defects could be diminished or removed in time if the environment were bettered and kept so. Thus, for those reformers and social scientists interested in improving human nature, Galton's recommendation of breeding better people, a method that was slow at best, won few converts. But once the concept of acquired characters was no longer available, breeding became the only way to better the race. For those who were convinced that heredity played a central role in shaping human nature, and their numbers were growing rapidly, eugenics became a very attractive means by which to improve the social order.

Unlike social Darwinism, which sought to defend the status quo, eugenics was reformist in intention, a movement that sought to improve society through the application of the latest scientific knowledge. Again, contrary to social Darwinism, eugenics looked to the intervention of the state in society; in effect it repudiated laissez faire. In the name of its philosophy, it put society's good above that of the individual; it countered selfish individualism with social responsibility. Thus it is not accidental that eugenics did not come into its own as a movement in the United States until early in the twentieth century. By then, the work of Weismann and the new genetics based upon Mendelian science had made a deep impression upon the thinking of biological and social scientists. Even more important in accounting for the growth of the eugenics movement in those years was the advent of Progressivism, a national social reform movement that captured both major political parties for almost two decades, headed by Theodore Roosevelt and Woodrow Wilson.

Although eugenics was a reform in the midst of a reform era, at bottom its outlook was necessarily elitist. For the central purpose was to improve the masses, to bring them up to the standard exhibited by the few. Thus eugenics could hardly spawn grass-roots organizations, as, for example, causes like prohibition or even immigration restriction have. Yet simply because large numbers of Americans had been hearing much about hereditarianism and what it held in store for the nation, the eugenics movement came to exert a powerful and enduring influence, particularly on public policy.

The elitist composition of the eugenics movement was most apparent at its inception. The great preponderance of Americans who formed organizations or spoke out in favor of improving the race in the early twentieth century were biologists, some of whom had long been interested in better breeding of animals. In fact, the first organizational effort in behalf of eugenics was the creation of a committee of the American Breeders' Association in 1906 to "investigate and report on heredity in the human race." The committee was expected to make clear "the value of superior blood and the menace to society of inferior blood." The chairman of the committee was David Starr Jordan, then the president of Stanford University, and a biologist by training. Another member of the committee was Charles Henderson, who was a sociologist at the University of Chicago, but at the beginning of the movement social scientists were far outnumbered by biologists and geneticists. The prominent geneticist Charles Benedict Davenport was also a member of that founding committee. Before too long, Davenport became the nation's best-known advocate of eugenics, speaking out frequently on the subject from his post as the director of the Eugenics Record Office at Cold Spring Harbor, New York, the first formal eugenics organization in the United States.[22]

Professional sociologists were directly informed about the promise of eugenics as early as 1906 by an article in their professional journal entitled "The Biological Foundations of Sociology." Since heredity and environment both affect the race, the author maintained, human beings should try to improve both. "We should bear in mind, however," he warned, apparently aware that the theory of acquired characters was no longer available, "that, were eugenic breeding possible, we could improve the race to an unlimited extent; whereas our power of improving the individual by placing him under better conditions is strictly limited. We should remember," he added, echoing Darwin's admonition, "that an improved environment tends ultimately to degrade the race by causing an increased survival of the unfit."[23]

On the eve of the First World War, eugenics was a fashionable social reform movement on both sides of the Atlantic. The first International Congress of Eugenics, held in London in 1912, saw Winston Churchill as an English vice president, along with the American vice presidents: Gifford Pinchot, the well-known conservationist, and Charles W. Eliot, the president of Harvard University. Even socialists Beatrice and Sydney Webb and Harold Laski counted themselves eugenicists.

American social scientists, especially psychologists, were strongly attracted
to eugenics. "In intellect and morals, as in bodily structure and features," wrote
psychologist Edward Thorndike in 1913, "men differ, differ by original nature,
and differ by families. . . . Selective breeding can alter a man's capacity to learn
to keep sane, to cherish justice or to be happy," he contended. "There is no so
certain and economical a way to improve man's environment as to improve his
nature." H. C. McComas, writing in the *Psychological Bulletin* in 1914, thought
that recent writings on the inheritance of bad traits ought not to panic the
profession into taking rash steps to control heredity. "There can be no question,
however," he added, "that practical and effective measures for the isolation, or
sterilization, of the congenitally defective will be adopted in the future. Society
will not leave the matter to the defective, trusting . . . that the neuropathic
strain finds marriage uncongenial." An Oberlin professor, writing in *Popular
Science Monthly* that same year, predicted that "careful scrutiny" of the heredity of prospective marriage partners would soon be standard procedure before
marriage licenses were issued, and that the state would "debar from marriage"
those whose children might become a burden upon society. "The bearing of
children is, of course," he told his readers, "not an individual right, but a social
privilege, and in time it must come to be so recognized."[24]

Sociologists, too, found the principles of eugenics appealing. Leaders of the
profession like Edward A. Ross of the University of Wisconsin, Charles Ellwood
of the University of Missouri, and Frank Hankins of Smith College, all of whom
would become presidents of their professional organization, the American
Sociological Society, supported eugenics around the time of the First World
War. Ales Hrdlicka, perhaps the leading physical anthropologist in the country
at the time, thought in 1919 that the "growing science of eugenics will essentially become applied anthropology."[25]

Some social scientists found eugenic principles not only philosophically congenial, but also drew upon them in their own work. Robert Yerkes and Lewis
Terman, nationally recognized leaders of the psychological profession, were
active in eugenic organizations in the 1920s. And in 1921 William McDougall
of Harvard, perhaps the nation's best-known psychologist, explained that the
purpose of his Lowell lectures that year, published as *Is America Safe for
Democracy?*, was to provide a popular introduction to eugenics. He took as his
thesis the proposition *"that the great condition of the decline of any civilization
is the inadequacy of the qualities of the people who are the bearers of it."*[26]

Belief in eugenics, however, implied more than accepting the principle that
traits and habits could be inherited. It also laid an obligation on society to do
something about controlling heredity, an obligation that usually translated into
preventing the reproduction of mentally defective or criminally inclined people.

As one social scientist observed, society could not rely upon the self-restraint of "the neuropathic strain" in the population. Other ways of regulating reproduction would have to be pursued. The most common method resorted to, and also the most obvious sign that eugenics captured not only the attention but the approval of the public, as well as that of the social scientists, was the policy of rendering sterile those persons designated as unfit to reproduce.

Sterilization was not an invention of the eugenicists, though the idea was based on the same biological premise upon which eugenicists rested their cause: certain undesirable human behavioral patterns were inheritable. In modern America the systematic practice of involuntary sterilization began as a means of controlling crime. The first published recommendation of sterilization for criminal activity was made in 1887 by a superintendent of the Cincinnati Sanitarium. The practice was recommended as both a punishment and a way of helping individuals to control their criminal proclivities. And insofar as criminal tendencies were inherited, the practice was also justified on what would later be called eugenic grounds: it would prevent the propagation of the unfit.

In the case of male offenders, sterilization meant castration until 1899, when vasectomy was found to be practicable as well as having the advantage of not depriving the victim of his sexual powers. Apparently the first institutional use of sterilization was made at Indiana State Reformatory, where, in a single year, a Dr. Harry Sharp carried out vasectomies on several dozen boys in an effort to prevent masturbation. Later, Sharp reported "that it occurred to me that this would be a good method of preventing procreation in the defective and physically unfit."[27] So far as is known, Sharp ordered the vasectomies without legal authority.

By that time eugenicists and other reformers were already pressing for the necessary legal authorization of sterilization. In 1897, for example, a bill authorizing the sterilization of the so-called unfit came before the Michigan legislature, but it went down to defeat. The Pennsylvania legislature enacted such a bill in 1905, but it was lost by the governor's veto. The first law permitting sterilization was enacted in Indiana in 1907. It specified that "confirmed criminals, idiots, imbeciles, and rapists" could be involuntarily sterilized if, in the judgment of a committee of experts, procreation was considered inadvisable.[28] Other states soon followed with similar laws, until in 1915 thirteen states had empowered the government to render sterile certain criminals and mentally defective persons in public institutions. By 1930 some thirty states had enacted such laws.

Some of the laws authorizing involuntary sterilization made better provisions than others for the protection of the liberty of those being treated, but as the emphasis placed on heredity in accounting for human behavior intensified in the first two decades of the twentieth century, most people considered the laws reasonable as well as reformist. Nationally prominent sociologist Edward A. Ross, for example, personally testified before the Wisconsin legislature when it was considering a law authorizing sterilization. The objections to the bill, Ross curtly remarked, "are essentially sentimental, and will not bear inspection."[29] Ross was a leading progressive academic and the Wisconsin legislature that took his advice and enacted the law was then under the control of reformers and Progressives. The Socialist newspapers of the state opposed sterilization, it is true, yet it is revealing of the reformist color attaching to such laws that four of the five Socialists in the Wisconsin legislature voted for the sterilization bill.

That the socially conservative Roman Catholic church invariably opposed all sterilization laws wherever they might be proffered further convinced many reform-minded or Progressive Americans that permanently preventing the unfit from procreating was forward-looking as well as socially necessary. The reformist character of eugenics in general and sterilization in particular is also evident in that before the First World War not a single state legislature in the South had enacted laws for involuntary sterilization. The "advanced" states of the North and the West, wherein all the laws in fact had been been passed, might find such legislation in line with modern scientific thought, but the conservative South would have none of it. Only after the First World War was the march joined by any southern state. It is significant, too, that "progressive" North Carolina was the state in which, next to "frontier" California, most of the sterilizations have been performed. By the opening of the 1930s, some 12,000 sterilizations had been performed in the United States, of which over 7500 were carried out in California alone.

Eventually other countries besides the United States enacted laws permitting sterilization of so-called defectives, but the U.S. was far in advance of any other nation. The first sterilization law outside the United States was enacted in Canada, but it was twenty years after the Indiana law of 1907. The earliest in Europe was passed by Denmark in 1929, followed in rapid succession in the other Scandinavian nations. England, though largely Protestant, never enacted a law allowing involuntary sterilization, and Germany's was not passed until 1933. Its writing, it is worth noting, preceded Hitler's accession to power.

Social scientists, as the activities of Edward Ross in Wisconsin suggest, frequently accepted sterilization as a means of improving the heredity of the nation. A review of the psychological literature on eugenics and heredity that appeared in *The Psychological Bulletin* in 1913 reported that most of the writers whose works were surveyed favored sterilization. In summarizing the survey, the reviewer remarked that "the burden of supporting these people [that is, the mentally defective] must not rest any more heavily upon the normal race." Since the "unfit" were getting to be too numerous to be segregated—a practice once thought to be the proper solution—it followed that "the only thing to do is to sterilize them. With procreation stopped," the author concluded, "the matter would be practically under control in a generation." (Her solution, however, was much more drastic than that offered by most proponents of sterilization since she apparently contemplated sterilizing *all* defectives, not merely those in institutions. Her view was that unless a sweeping approach were taken, the problem could hardly be brought "under control in a generation." Too many fertile members of the "unfit" would still be free to procreate.[30])

The general belief that feeblemindedness and other deficiencies of behavior were inheritable and ought to be limited by preventing procreation reached a high point with the United States Supreme Court decision in 1927 in *Buck v. Bell.* The case concerned the constitutionality of a Virginia law enacted in 1924 which permitted the involuntary sterilization of persons in state institutions who were deemed to be feebleminded. In an unusually brief opinion of eight pages, Justice Oliver Wendell Holmes, Jr., spoke for the court. Justice Pierce Butler, the one Roman Catholic on the court at the time, was the only one to vote against the decision, but he offered no explanation.

Justice Holmes, in his opinion in behalf of the court, remarked that "experience has shown that heredity plays an important part in the transmission of insanity, imbecility, etc." From his treatment of the subject, it was clear that in his mind the question was not complicated. If a state may compel a young man to serve in the army in time of war, thereby putting his life in jeopardy, Holmes wrote, then it certainly ought to be able "to call upon those who already sap the strength of the state for . . . lesser sacrifices, often not felt to be such by those concerned, in order to prevent our being swamped with incompetence. . . . It is better for all the world," Holmes continued, "if instead of waiting to execute degenerate offspring for crime, or let them starve for their imbecility, society can prevent those who are manifestly unfit from continuing their kind. The principle that sustains compulsory vaccination," he concluded, "is broad enough to cover cutting the fallopian tubes." Then followed his best-known sentence in the opinion: "Three generations of imbeciles are enough."[31]

Holmes's private correspondence leaves no doubt that he personally favored the eugenic principles that stood behind the sterilization laws. Soon after announcing the decision of the court, he told Harold Laski that he had written an opinion "upholding the constitutionality of a state law for sterilizing imbeciles . . . and felt that I was getting near to the first principle of real reform." To another English correspondent he described his decision as one that "gave me pleasure."[32]

Holmes's opinion, it is worth noting, was supported by his liberal friend and fellow justice Louis D. Brandeis. (Although in the Virginia sterilization law case,

Brandeis offered no personal opinion, in another case that same year, in which he and Holmes dissented, Brandeis, in his dissenting opinion, drew upon Holmes's opinion in *Buck v. Bell* to support his position.) The principle enunciated by Holmes in *Buck v. Bell* has not yet been overturned by any Supreme Court decision, though the issue has been before the court at least once, in the case of *Skinner v. Oklahoma* in 1942.*

Behind the decision in *Buck v. Bell* stood the I.Q. test, which had, in fact, identified Carrie Buck and others like her as mentally deficient. During the 1920s the use of I.Q. tests spread beyond the identification of the mentally deficient. Increasingly they were drawn upon to compare the intelligence of the various ethnic and racial groups in the country. The practice was sparked by the enormous influx of immigrants into the United States during the previous fifteen years. (Between 1900 and 1914 more than 13 million European immigrants entered the United States.)

Undoubtedly the most influential publication on the issue, though by no means the only one, was *The Passing of the Great Race* (1916), authored by Madison Grant, a wealthy New York socialite and amateur zoologist. The message of Grant's book was simple: the immigrants then entering the United States in such great numbers from southern and eastern Europe, particularly Poles, Italians, and Russian Jews, were decidedly inferior mentally and morally, as well as physically, to the Irish, English, Germans, and Scandinavians who had entered the country in earlier years. In Grant's view the latter group constituted the Great Race of his title—the Nordics, as he called them—who were now on the verge of being inundated by the influx of their more numerous inferiors. Grant's general message was echoed by a number of social scientists.

In fact, even before Grant's book appeared, the nationally prominent sociologist Edward A. Ross, of the University of Wisconsin, put forth a similar argument in a series of articles he wrote for *Century Magazine* as early as 1912 and which were later collected into a book entitled *The Old World in the New*. Ross deplored the declining quality of the population of the United States as a result of the increasing number of immigrants from southern and eastern Europe. Edward Doll, writing in the *Journal of Applied Psychology* in 1919, seconded Ross's and Grant's apprehensions. It is "commonly recognized that the type of immigrant received in this country since 1900," he wrote, "is distinctly inferior

*Ironically enough, *Buck v. Bell* was cited by Justice Thurgood Marshall as the "initial decision," reaffirmed in *Roe v. Wade,* that the constitution provided no special protection for procreation. *San Antonio School District v. Rodriguez* 411 U.S. 1 (1972). Irony is piled upon irony when it is further recognized that Carrie Buck, the principal in *Buck v. Bell,* was not in fact an "imbecile" and that her child Vivian, whose birth provoked Carrie's sterilization, turned out to be normal. In short, there were no "three generations of imbeciles."

to the type of immigrant previously received." A comparable worry beset anthropologist Albert E. Jenks of the University of Minnesota. Many people think, he wrote in *Scientific Monthly,* that differences between ethnic groups are only "skin deep. But biologists know that ethnic groups differ beneath the skin," as experiments with animals have demonstrated. "The man who runs sees the outside differences between breeds of people," he pointed out, but "the anthropologist knows they begin inside in the seeds of the breeds." From those obvious physical differences between people emerge differences in "psychic characteristics," Jenks contended.[33]

Jenks did not specifically compare the quality of the different ethnic groups then entering the country, but Stanford University psychologist Lewis Terman had no doubt that "the immigrants who have recently come to us in such large numbers from Southern and Southeastern Europe are distinctly inferior mentally to the Nordic and Alpine strains which we received from Scandinavia, Germany, Great Britain, and France." He stressed the biological or genetic root of the problem. "No nation can afford to overlook the danger that the average quality of its germ plasm may gradually deteriorate as a result of unrestricted immigration."[34]

Kimball Young, a young psychologist at the University of Oregon and a former student of Terman, also concluded that "the rapid incursion of racial stock from Southern Europe has begun an inundation of our older populations which will result in a racial amalgamation of possibily serious consequences." It was not simply a matter of Latins intermarrying with native Americans, he emphasized, "but of those classes of South Europeans and Mexicans of less average ability than our own stock of the lower middle classes." Young, in short, attributed inferior native intelligence to the lower classes as well as to the lower races. In view of the inferiorities of these people, whether they stem from race or class, he maintained, our present psychological knowledge "must not be ignored by those who would have us admit all immigrants promiscuously in the pious hope that the future will wipe out present differences: economical, mental and social."[35]

Intelligence tests were increasingly used also to identify differences between races as well as between ethnic groups. As one reviewer of the psychological literature wrote with some pride in a professional journal in 1922, "though for a long time work in the field of race psychology seemed unprofitable . . . mental testing has caused an encouraging revival of interest in the field."[36] One consequence of the renewed interest was the discovery, one academic writer pointed out, that only about a quarter of blacks, on the average, scored equal to or above the average I.Q. for whites. In fact, virtually every tester who published his or her results in the early 1920s reported that blacks achieved scores significantly lower than those of whites. Lewis Terman, perhaps the leading tester in the country, emphasized the finding by drawing a comparison. After observing that blacks and Amerindians generally scored lower on mental tests than whites, he found it "interesting to note that intelligence tests of Chinese and Japanese in California indicate that these races are approximately the equals of Europeans in mental ability. Unselected Chinese children in San Francisco," he continued, "test almost as high as unselected California white children and enormously higher than the children of our Portuguese and South Italian immigrants." To Terman these comparisons demonstrated the inadequacy of cultural explanations and affirmed his belief that differences in intelligence were largely genetic.[37]

As early as 1916, Terman had ruled out class as an explanation for differences in intelligence. "The common opinion that the child from a cultured home does better in tests solely by reason of his superior home advantages is an entirely gratuitous assumption," Terman insisted. "Practically all of the investigations which have been made of the influence of nature and nurture on mental performance agree in attributing far more to original endowment than to environment," he maintained. "The children of successful and cultured parents," he assured his readers, "test higher than children from wretched and ignorant homes for the simple reason that their heredity is better."[38]

The nationally known psychologist William McDougall of Harvard University summed up in 1921 what his profession believed the tests of ethnic and racial groups revealed. "We have . . . pretty good evidence that capacity for intellectual growth is inborn in different degress, that is hereditary. . . . Further, we have good evidence," he continued, "that different races possess it in widely different degrees; that races differ in intellectual stature, just as they differ in physical stature."[39]

Undoubtedly the single event or circumstance that provided the greatest impetus to testing and to publicizing the conclusion of many social scientists that mental differences between races and ethnic groups were biologically based were the tests administered to the Army during the First World War. The immediate purpose for testing almost two million draftees was to help the Army in time of war to make more rational use in a short time of the enormous number of people suddenly put at its disposal. Army officials were never convinced that the tests would be of much use for their purposes, but there was little doubt among leading psychologists. As tester Robert Yerkes remarked, thanks to the Army tests the profession "achieved a position which will enable it to substantially help to the win the war and shorten the necessary period of conflict."[40]

Yerkes, Lewis Terman, and Herbert Goddard were the moving spirits in organizing the Army tests. Soon after the war ended they published a massive volume describing what they thought they had learned from administering the tests to so many subjects. In time, the Army experiment and the conclusions drawn from it would come in for extensive criticism. But the immediate reaction of those who conducted the tests bordered on euphoria. Soon after the report came out in 1921, Yerkes wrote privately to Lewis Terman, his colleague in the enterprise. "Well, old friend," he began, "we have the supreme satisfaction of knowing that our work won recognition of value and a place in the Army. Its standing is even better today than in November, 1918 and tends steadily to improve. Comparisons are invidious and unkind," he conceded, "but we have no cause to envy anybody!" As far as Yerkes was concerned, their work was so successful that "if war were to be declared tomorrow we should be hailed [*sic*] into the service immediately to classify recruits by intelligence."[41]

Among the works of social scientists supporting the idea of differential intelligence among ethnic and racial groups, none received more attention than Princeton psychologist Carl C. Brigham's *A Study of American Intelligence*, published in 1923. Brigham drew heavily upon the results of the Army tests. Robert Yerkes himself wrote a glowing introduction to the book, praising its method and its conclusions. Though written to be accessible to the general public, the book was also intended to be a scholarly analysis of the data gained from the Army tests, an intention supported by its scholarly footnotes and publication by Princeton University Press. The conclusions of the book, however, did not differ significantly from those advanced in Madison Grant's popular tract *The Passing of the Great Race* of seven years before, which had reached a wide audience.

At the very outset, Brigham announced that he was testing what he called Grant's "race hypothesis," by which he meant the contention that immigrants of Nordic ancestry were superior to those of Alpine or Mediterranean background. His test, however, revealed that Brigham disagreed with Grant in only one relatively minor particular. Grant had argued that Mediterranean immigrants were superior to Alpine, but Brigham's analysis reversed their standings. Both authors, however, were in complete agreement that Nordics were superior to all other groups. The anti-black worries of Grant also found support in Brigham's more scholarly analysis. "The most sinister development in the history of this continent" in regard to immigration, Brigham asserted, was "the importation of the negro."[42]

Brigham also gave full credence to the alleged biologic dangers to the nation as a result of the poor quality of the new immigration from southern and eastern Europe. "American intelligence is declining and will proceed with an accelerating rate as the racial admixture becomes more and more extensive," he predicted. Moreover, "the presence of the negro" will make that decline in intelligence even more precipitous. "These are the plain, if somewhat ugly facts that our study shows," he soberly concluded. Then, like some other social scientists, publicists, and politicians at the time, Brigham turned to urging the government to follow a policy severely limiting immigration of the undesirable groups.[43] And that was precisely what Congress enacted in 1924.

Several students of the effect of biological ideas on social thought and practices have pointed to the Immigration Act of 1924 as a prime instance of the undesirable effects of intelligence testing on social policy.[44] A close examination of the history of the legislation, however, suggests that the role of intelligence testing in the law's enactment was insignificant, and that social scientists interested in testing were similarly peripheral in influence. It is worth looking a little closer, therefore, at the law's history.

There can be no doubt that in its provisions the 1924 statute clearly expressed racial and ethnic biases. For example, the act excluded both Japanese and Chinese from immigrating into this country. But that admittedly racist policy was not new; it had been first put into practice in 1882 against the Chinese and later, in the early twentieth century, against the Japanese. Moreover, the new attention to mental testing could hardly have had anything to do with these restrictions, however racist its proponents may have been. As Terman himself had pointed out, Asians had generally been found to be equal to Caucasians in intelligence scores; their scores could not have been an excuse for the discrimination.

And as far as the restrictions on numbers allowed into the country, which the new law imposed, were concerned, that was neither a new idea nor one to be charged against the mental testers. The movement to limit immigration had been gathering strength ever since the 1890s, when labor unions and certain nativist groups began to agitate for the closing of the doors on immigration. The labor unions disliked the economic competition, and the nativist or patriotic groups disliked and feared the social character of the new immigration from southern and eastern Europe. And though there was undoubtedly racist hostility behind the movement to limit immigration after the First World War, the idea of restriction was not novel.

What was new was the obvious discrimination against the immigrants from southern and eastern Europe. Many Americans seemed to feel threatened by the unprecedentedly high number of immigrants who entered the country in the dozen or so years before the outbreak of war in Europe in 1914. In 1905, 1906, 1907, 1910, 1913, and 1914, more than a million immigrants entered the country each year. Numbers were important because these so-called new immigrants from southern and eastern Europe appeared to be strikingly different from not only old-line Americans but earlier immigrants as well. Generally, they were poor, Catholic, and Jewish, often illiterate, unskilled, and given to congregating in large cities, which were already seen as prone to crime, immorality, and violence. On balance, then, the more important motivation behind the successful effort to narrow the welcome to immigrants was the concern for social homogeneity, not mental fitness. In retrospect that was no more enlightened a reason than racial bias, but it was a different reason and that is the point being made here.[45]

Over and over again in the public discussions of the issue of immigrant restriction, whether on the floor of Congress or in the popular press, the dangers of national division and the consequent destruction of American social cohesion figured prominently. True, some biologists and eugenicists, like Charles B. Davenport and his assistant H. H. McLaughlin, were active in bringing their genetical arguments to the attention of Congress when it was considering immigration legislation in the early 1920s. But, significantly, very few social scientists participated in such moves. During 1921–22, psychologist Robert Yerkes wrote to the chairmen of the immigration committees in the House and in the Senate, calling their attention, as he liked to do, to the results of the Army tests and especially the published findings on the intelligence of the foreign-born. Yet, as the historian of psychology Franz Samelson has shown, by the time the legislation was actually being drawn up in 1923, Yerkes had begun to doubt that discrimination against certain ethnic groups was the proper road to take. Distinctions between so-called old and new immigrants, Yerkes was now saying, were "not scientifically based. We have not actually the facts which would enable us to evaluate the differences presented," he told the Commission on Migration that year. "And yet our government," Yerkes continued, "is on the verge of discrminating against certain European peoples." Although Yerkes had corresponded earlier with Congressman Albert Johnson, who headed the House committee on immigration, and who was an ardent proponent of selective immigration legislation, he did not testify at any of the hearings in 1923 or 1924.[46]

Even more indicative of the small part played by psychologists and other social scientists in shaping the content of the Immigration Act of 1924 is another finding reported by Samelson. When he searched the correspondence files of the House Committee on Immigration for 1922–24 he found hundreds of letters opposing any further immigration from eastern and southern Europe. Very few of them, however, made any reference to mental tests. He further notes that not one of the three reports submitted by the House Committee in support of the new immigration bill made any reference to intelligence tests or their use in ascertaining the comparative worth of different nationalities. The minority reports similarly lacked any reference to intelligence tests of immigrants.

In a later and independent examination of the congressional documents concerning the 1924 act, Mark Snyderman and R. H. Herrnstein fully corroborated Samelson's conclusions. Results from intelligence testing, they write, "are brought up only once in over 600 pages of congressional floor debate, where they are subjected to further criticism without rejoinder. None of the major comporary figures in testing—H. H. Goddard, Lewis Terman, Robert Yerkes, E. L. Thorndike, and so on—were called to testify, nor were any of their writings inserted into the legislative record," Snyderman and Herrnstein point out.[47]

In sum, ideas about the allegedly hereditarian roots of intelligence were not nearly as important in shaping the new immigration policy of the mid-twenties as were apprehensions about social cohesion and national unity (including anti-semitism and anti-Catholicism), which the massive immigration of the previous two decades clearly aroused. A member of the Immigration Restriction League may have offered the best clue as to why the new biological ideas about the distribution of intelligence among immigrant groups would not carry much weight with policy-makers, nor with the American public. "The country is somewhat fed up on high brow Nordic superiority stuff," he observed in 1924.[48]

Even those social scientists who believed that the new knowledge about intelligence ought to be drawn upon in shaping immigration policy objected to using it to exclude whole nationalities or races. As psychologist Robert Woodworth wrote as early as 1910, if immigration was to be restricted in order to improve the mental qualities of the population, then the better part of wisdom "would be to select the best individuals available from every source, rather than trusting to the illusory appearance of great racial differences in mental and moral traits, to make the selection in terms of races or nations."[49] Physical anthropologist Ales Hrdlicka and sociologist J. J. Spengler, who advocated the use of mental tests in making immigration more selective, similarly pressed for individual screenings, rather than placing limits on whole nationalities.

Finally, there is a much broader reason why it is historically dubious to ascribe the Immigration Act of 1924 to the influence of social scientists' acceptance of differential intelligence among immigrants. During the very time in which the immigration bill was being hammered out in Congress, an increasing number of social scientists were beginning to doubt that biology had anything to do with shaping human intelligence or behavior. Indeed, by the 1920s some social scientists were denying outright the validity of race and sex as useful explanations for differences between human groups. That profound alteration in outlook, which would eventually reshape and even dominate social science theory, was not achieved quickly or easily. Indeed, the process was at once complicated and lengthy. To unravel it requires a return to the nineteenth century when the intellectual bases for that fundamental transformation were first laid down.

II

The Sovereignty of Culture

3

Laying the Foundation

Human nature is . . . something more, on the one hand, than the mere instinct
that is born in us—though that enters into it—and something less, on the
other, than the more elaborate development of ideas and sentiments that make
up institutions. . . . Man does not give it birth; he cannot acquire it except
through fellowship, and it decays in isolation.

Charles Cooley, 1909

Most people know that Charles Darwin shared the discovery of natural selection as the motive force of evolution with his fellow British naturalist Alfred
Russel Wallace. It is less well known that, only a few years after their joint
discovery, the two men disagreed profoundly on the place of man in evolution.
Out of that disagreement emerged the alternative to race as an explanation for
differences in human social behavior.

As we have seen, Darwin waited until 1871 and the publication of the
Descent of Man to make public his long-held conviction that human beings
were included in evolution. Wallace was bolder and quicker. Seven years before
Darwin made his conclusions public, Wallace had announced in a prominent
English journal his belief that human beings in both mind and body were as
much the result of natural selection as other animals. As might be anticipated,
Darwin expressed immediately his pleasure at this first public assertion of the
implications of natural selection for humankind.

Darwin's pleasure, however, was short-lived. Within five years Wallace had
changed his mind. Before making public his departure from his own and Darwin's view, Wallace warned his fellow-discoverer. "I venture for the *first time*
on some limitation to the power of natural selection," he wrote Darwin in 1869.
Wallace was right on target in anticipating that Darwin and Thomas Huxley,
the well-known Darwinian "bulldog," would find his arguments "weak and
unphilosophical." He assured them, nonetheless, that his new view was the
result "of a deep conviction founded on evidence which I have not alluded to
in the article but which is to me absolutely unassailable." Darwin's reply was
immediate and laced with foreboding. He described himself as "intensely curi-

ous" to read the article, closing with the hope that "you have not murdered too completely your own and my child."

The "absolutely unassailable" considerations to which Wallace referred, but which he did not include in his article, was his conversion three years earlier to Spiritualism. For as he wrote later to Darwin, he himself only a few years before would have been as dismayed as Darwin by the alteration in his ideas. His views, however, "have been modified," he explained to Darwin, "solely by the considerations of a series of remarkable phenomena, physical and mental, which I have had every opportunity of fully testing, and which demonstrate the existence of forces and influences not yet recognized by science." The forces and influences were undoubtedly those of a supernatural nature. At the same time, Wallace had fashioned a scientific reason for his contention that human beings had escaped from the operation of natural selection. (Doubtless, the religious and scientific reasons were closely connected.) The substance of his scientific argument related directly to the issue of racial differences.

Wallace's scientific case rested on his conclusion that the human brain, including that of the most primitive peoples, was more powerful than was necessary for survival. For a large part of his early life Wallace had lived among primitive peoples in South America and Southeast Asia, an experience that convinced him that these people, simple as they may have appeared in mind and action, were equal in intelligence to Europeans. As the modern anthropologist Loren Eiseley remarked, Wallace displayed "scarcely a trace of the racial superiority so frequently manifested in nineteenth-century scientific circles," in which were included Darwin and Thomas Huxley. If human beings possessed brain capacities beyond what was needed for survival, Wallaced reasoned, then how could natural selection bring about its evolution? Where was the "survival value" of that capacity if that capacity was not fully used? After all, natural selection improved an organ only through its adaptation to the pressure of environment. In the case of the human brain, however, the capacity was greater than human beings really required or that the pressure of environment could account for. Wallace logically concluded on those grounds that "some higher intelligence directed the process by which the human race was developed."[2] Man was outside evolution.

Needless to say, Darwin refused to accept any "miraculous additions at any one stage of ascent" of man.[3] He and Wallace never reconciled their differences. Indeed, the differences only widened with time as Wallace went on to become a socialist and critic of the European social and economic system. Modern scholars like Loren Eiseley and contemporary thinkers like the social philosopher Prince Kropotkin found in Wallace the "human" side of evolution, a side that emphasized cooperation and community instead of competition and "survival of the fittest."*

What was centrally relevant to the question of race in the disagreement between Darwin and Wallace was their divergent conceptions of the mentality of primitive peoples. It will be recalled that Darwin could find no useful value in the physical (racial) differences among human groups. Thus he could not account for those differences through the operation of natural selection. He did, however, accept the common anthropological view of the time that the differences in levels of culture or civilization which occurred among the diverse peoples of the world derived from differences in their biological capacities. Some cultures were higher than others because the people in those societies were biologically superior. That was the opening in his theory of human evolution through which racism entered. It was that opening which Wallace closed with his conception of the intellectual equality and therefore the equal cultural capacity of all peoples.

As things turned out, Wallace looked to other ways and matters in his effort to make evolution less competitive and threatening. He did not develop any further his assertion of the mental equality of all peoples, or at least few took notice of its relevance. Yet that was the precise argument, elaborated and tirelessly defended, that undermined in time the concept of racism in America. Its elaboration and defense underpinned the concept of culture, an idea that in the twentieth century became not only an alternative to a racial explanation for human behavioral differences but also a central concept in social science. The theoretical substance of the concept was first laid down in the United States in the 1880s by a recent immigrant and fledgling anthropologist, lately a refugee from physics and geography, Franz Boas.

Boas's influence upon American social scientists in matters of race can hardly be exaggerated. At the same time that racial segregation was being imposed by law in the states of the American South, and eugenics was emerging as a hereditarian solution to social problems, Boas was embarking upon a life-long assault on the idea that race was a primary source of the differences to be found in the mental or social capabilities of human groups. He accomplished his mission largely through his ceaseless, almost relentless articulation of the concept of culture.

*Ironically enough, it was Wallace who persuaded Darwin to employ Herbert Spencer's phrase "survival of the fittest" as a shorthand for natural selection. "This term," he wrote Darwin in 1866, "is the plain expression of the fact, natural selection is a metaphorical expression of it, and to a certain indirect and incorrect sense, even personifying Nature, she does not so much select special variations as exterminate the most unfavourable ones." Quoted in Malcolm J. Kottler, "Charles Darwin and Alfred Russel Wallace: Two Decades of Debate over Natural Selection," in David Kohn, ed., *Darwinian Heritage* (Princeton: Princeton Univ. Press, 1985), 373–74.

Boas's most striking and certainly his best-known statement of his views on both race and culture appeared in 1911—the very same year, interestingly enough, in which Charles B. Davenport, the leading eugenicist in the country at the time, published his book *Heredity in Relation to Eugenics.* Boas's volume has a quite different, but equally suggestive title: *The Mind of Primitive Man.* Davenport's book was new; Boas's, significantly enough, was not since it consisted of the rearrangement and restatement of essays published over a span of years. It was, in short, a summary rather than a novel assertion of his ideas. That same year, Boas also published a report which was destined to be even more famous: *Changes in the Bodily Form of Descendants of Immigrants,* a part of a massive study on the effects of immigration, which had been authorized and financed by the federal government.

Both of Boas's works declared war on the idea that differences in culture were derived from differences in innate capacity. The essential message of *The Mind of Primitive Man* was that so-called savages did not differ in mental capabilities from civilized people, even if in their present state of existence they had not produced the artifacts or cultural achievements traditionally associated with civilized life. This assertion by Boas, which Alfred Wallace had set forth some years earlier, proved to be truly revolutionary for the development of anthropology and eventually of social science in general. It did nothing less than overturn the conception of social evolution that had reigned in anthropology even before Darwin had introduced natural selection as the driving force of biological evolution. The traditional view of social evolution held that human groups, or races, passed through a series of stages: from savagery to barbarism and culminating in civilization. The primitive peoples of the world, the traditional view maintained, were still in the earlier stages because they lacked the necessary biological wherewithal to reach the highest stage. These races were seen as socially and intellectually inferior to those which were deemed "civilized." It happened that these latter people were also white, while the others were usually colored.

Boas's rejection of the traditional view was truly radical; it simply denied the existence of any significant innate differences between savage, colored people and civilized, white people. The differences in physical appearance, in short, did not lead to any significant difference in mental or social function. That there were observable social differences was undeniable, but the explanation for those differences, Boas maintained, was that they were the product of different histories, not different biological experiences. Boas's view, as will be evident later, spread widely among social scientists in the course of the first decades of the twentieth century. Indeed, his introduction of history or culture as the cause of differences among peoples might be said to have been the sword that cut asunder evolution's Gordian knot in which nurture was tightly tied to nature. It also constructed a single human nature in place of one divided by biology into superior and inferior peoples.

The Mind of Primitive Man presented no new evidence or arguments in behalf of nurture over nature. The study on the descendants of immigrants did. Through it, Boas delivered a stunning empirical blow against those who doubted the power of the social environment on human beings. His research directly attacked the most widely accepted measure of the way biology resisted the impact of social or environmental influences. That measure was the so-called cephalic index, which was the ratio of the length to the width of the human head. The measure had long been used by physical anthropologists and other students of racial groups in Europe and America as the most reliable measure or sign for the identification or classification of human types or race. Thus all races could be identified by a percentage figure within a spectrum running from the wide-headed (brachycephalic) to the long-headed (dolichocephalic). The principal value of the index, aside from its ease of attainment, was that it was stable; it was apparently immune from environmental or social influences. From generation to generation, regardless of the environment in which the members of a racial group might live, the index could be counted on to identify them. That was the assumption and, apparently, the empirical findings of physical anthropologists, including Boas.

Boas's study was based on the patient measuring of several thousand immigrants and their children in New York City. The results were surprising even to him. For they showed that the head shapes of children of immigrants changed after the mother had been in the United States for a period of time. In fact, the changes in this alleged stable measure occurred within ten years after the mothers' arrival in the American environment. Since the genetic input was identifical—the mothers and fathers were the same—the social environment must have been the source of the changes. In his preliminary report to the public and later in his official report to the Immigration Commission, which sponsored the research, Boas emphasized one particular finding above all others. The shape of the head, he wrote in his final report, "undergoes far-reaching changes coincident with the transfer of the people from European to American soil. For instance, the east European Hebrew, who has a very round head, becomes long-headed; the south Italian, who in Italy has an exceedingly long head, becomes more short-headed; so that in this country both approach a uniform style, as far as the roundness of the head is concerned."[3] This was asserting environmental influences with a vengeance since it included not only a change in the supposedly unchangeable index figure, but an environmentally induced change in which two indices *converged* toward a common type.

Boas was well aware of the innovative character of his findings. Both publicly and privately he expressed his surprise at the results. The finding of convergence in the cephalic indices of the Jews and the Italians, for example, he thought "one of the most suggestive discovered in the investigation, because it shows that not even those characteristics of a race which have proved to be most permanent in their old home remain the same under the new surroundings." He thought, too, that more was involved than physical change. "We are compelled to conclude that when these features of the body change, the bodily and mental make-up of the immigrant may change."⁴ He found the results "so definite that . . . the evidence is now in favor of a great plasticity of human types." That plasticity may not be unlimited, he acknowledged, but from what we now know about "the plasticity of human types, we are necessarily led to grant also a great plasticity of the mental make-up of human types." He justified that conclusion on the ground that since head shape, which reaches its final form early in life, altered in a new environment, then it seemed to follow that those elements in human development that achieved maturity later would change even more under the influence of novel surroundings. Thus, he felt it reasonable to "conclude that the fundamental traits of mind" were also subject to change in a new environment. All this was inference, he admitted, but of sufficient strength to recast fundamentally the discussion over the role of race in shaping mind. For "if we have succeeded in proving changes in the form of the body, the burden of proof will rest on those who, notwithstanding these changes, continue to claim the absolute permanence of other forms and functions of the body."⁵

Boas was not only surprised by the results, he was also fearful that the very novelty of the results would call them into question. For that reason he was especially eager to account for the convergence. He worked, for instance, to rebut the possible objection that the fathers of many of the children born in the United States were really natives, rather than immigrants, that some of the children were in fact illegitimate. He himself thought the possibility of illegitimate conceptions "sociologically" unlikely, as he put it in a letter to his superior on the Commission, Professor Jeremiah W. Jenks of Cornell University. "But with an important conclusion of this kind," he added, "even an objection of this type had to be met." So he conducted a check-up study, from which emerged the fact that the children born in the United States resembled their fathers as much or more so than those born in Europe. In short, illegitimacy was not an explanation.⁶

Nor were changes in child-rearing practices an explanation, though that occurred to Boas as a possibility. As early as 1910 he opened a correspondence with an anthropologist in Italy concerning child rearing there, "particularly in relation to the question whether the style or bedding of the children changed by the Italians when they came to this country." In his official report, Boas frankly admitted that "the explanation of these remarkable phenomena is not easy." By then he had specifically ruled out child-rearing practices.[7]

From the outset Boas was convinced that he had hit upon a significant finding. Even before he published his results, he wrote about them to fellow scholars in other countries; he also reported at some length on his findings at international conferences. It appears that he sent out copies of his report to hundreds of persons, since he asked the Immigration Commission to provide him with a thousand copies. As he wrote in the report itself in 1911, "it is probably not too much to say that [the findings] indicate a discovery in anthropological science that is fundamental in importance."[8]

The public was also impressed and, as often happens with such matters, chose to interpret Boas's findings in ways not always in agreement with his. Among Boas's professional papers is a clipping from *Leslie's Weekly* for March 10, 1910, commenting on a preliminary report on Boas's research. *Leslie's* writer thought that "we are rapidly approaching a uniform fusion of the races that seek a home in this land of freedom" and that Boas had demonstrated physical and mental changes in the children "of alien races born in this country." The writer concluded "that the amalgamation of these races is producing a uniform type of offspring." Boas, of course, was arguing precisely the opposite: that environment, not amalgamation or intermarriage, was causing the convergence of types. A year later another journal, *The Outlook*, contended that Boas's work "gives definite psychical and physiological refutation to the saying, You cannot change human nature." The causes of these changes may be "climate, food, or the democratic spirit, or . . . some other more obscure cause," such as "prenatal influence exerted . . . through the mother."[9]

As far as his professional colleagues were concerned, Boas's findings aroused little critical appraisal. A young, unknown anthropologist took him to task in the august pages of the *American Anthropologist*, but he was easily squelched by Boas himself and his students and supporters. The editor of the journal, who was a former student of Boas's, admitted a little shamefacedly to Boas that he had felt compelled to publish the criticism because "two men occupying prominent positions" had endorsed it—one measure of the controversial nature, even at that late date, of Boas's environmental or cultural approach.[10]

Whatever serious criticism Boas's study received came much later and from Europeans. The most respected came in 1924 from the English biological statistician Karl Pearson and was cited in 1928 by sociologist Pitirim Sorokin as conclusive evidence, which it was not, against Boas's emphasis upon environmental influences. The fact of the matter was that changes in head indices were not difficult to accept; such changes had been discovered for other nationalities or types. It was the convergence of the indices of Jews and Sicilians in America that was hard to accept or explain. As a later student of the matter remarked in 1954, Boas had inferred "that the change in environment was greater for Sicilians and Jews, but the data brought to bear upon this point remained unconvincing."[11]

The validity of Boas's inference is, of course, not the important point. The importance, rather, resides in the emphasis he placed upon the role of social environment in bringing about physical and other changes. As he wrote a Swedish anthropologist in 1910, "I have always been so thoroughly convinced of the great stability of the cephalic index, that it has taken me a long time to get ready to accept the results of my own investigations."[12] Thereafter, as we have seen, he never missed an opportunity to press home his point.

Boas's study of head shapes, it needs to be stressed, was not needed to convince Boas of the power of nurture over nature, however effective the finding may have been with those who were less convinced than he of the irrelevance of racial influences. That particular research came very late in the evolution of Boas's ideas on the nature of human types, or race. In a sense that study, like *The Mind of Primitive Man,* was more a culmination, a proof and argument addressed to others, than a contribution to Boas's own thinking about race and culture. His conclusions had been arrived at long before. Thus if one wants to understand why biology or race as a shaper of human action came under attack in the early years of the twentieth century, one needs to go back much earlier in the life of Franz Boas. In those early years is to be found the answer to the question of how and why Boas pressed so early and so hard against racial explanations, and why his determined effort eventually won the allegiance not only of his fellow anthropologists but of social scientists in general.

To put the matter another way, despite his assigning a "fundamental" tag to his changing head shape study, there were no obvious turning points or crucial pieces of evidence in the development of Boas's thought on culture, race, and nature. His ideas on those subjects prior to 1911, however, can be broken down into their component elements and arguments. In that way we obtain a clearer idea of how he arrived at his 1911 conclusions and perhaps why so many others in time followed his lead. What, then, were Boas's reasons for coming to the conclusions he did on racial differences? How did he establish a way of thinking about race that Alfred Wallace had arrived at through personal experience?

The first and in many ways Boas's fundamental critical approach was historical and relativistic. As early as 1887, in a letter to the journal *Science,* he wrote that it cannot be said too frequently that the reasoning of anthropologists "is not an absolutely logical one, but that it is influenced by the reasoning of our predecessors and by our historical environment; therefore our conclusions and theories, particularly when referring to our own mind, which itself is affected by the same influences to which our reasoning is subject, cannot be but fallacious." What he meant by "fallacious" was that our ideas were relative to our environment and were not absolute across the spectrum of human experience. In order to avoid that "fallacious" outlook, he continued, we must "study the human mind in its various historical, and speaking more generally, ethnic environments." The whole purpose of ethnology, he added, was the "dissemination of the fact that civilization is not something absolute, but that it is relative, and that our ideas and conceptions are true only so far as our civilization goes."[13]

The second aim of ethnology also pointed in the direction of emphasizing social influences: to show "how far each and every civilization is the outcome of its geographical and historical surroundings," for we know that early events "leave their stamp on the present character of a people," He then reduced his view to a principle: "The physiological and psychological state of an organism at a certain moment is a function of its whole history." The character and the future development of any biological or ethnological phenomenon, he concluded, "is not expressed by its appearance, by the state in which it *is*, but by its whole history."[14]

Boas used history, too, to reply to those who rested their conclusion of the superiority of western civilization on differences in race. "What then is the difference between the civilization of the Old World and that of the New World?" he asked. "It is only a difference in time. One reached a certain stage three thousand or four thousand years sooner than the other. This difference does not justify us to assume that the race which developed more slowly was less gifted. Certainly the difference of a few thousand years is insignificant as compared to the age of the human race."[15]

A skeptic might ask why some modern colored peoples seemed unable to absorb the civilization of white Europe to the same extent as others had done earlier. Boas's response was that disease, competition from European factory-produced goods which drove out native crafts, and the large number of European invaders slowed the assimilation of European culture. In short, history, experience, and circumstances, not race, supplied the answer. In the past, Boas continued, circumstances and history had been more favorable to assimilation. Thus the Arabs in the Middle Ages were much more open to accepting strangers and novel ideas than were white Europeans of his own time. He admitted that the distribution of faculties among the races of the world was "far from being known"; nevertheless, he insisted that "the average faculty of the white race is found to the same degree in a large proportion of individuals of all other races." It may be true, he conceded, that the proportion of "great men" among some races may be less than among Europeans, but there "is no reason to suppose that they are unable to reach the level of civilization presented by the bulk of our own people." That conclusion of 1894 only became more firmly set in Boas's mind as time went on. Before the opening of the twentieth century, Boas was still prepared to admit that differences in structure, which he acknowledged to exist between races, "must be accompanied by differences of function, physiological as well as psychological, and . . . we must anticipate that differences in mental characteristics will be found," as well. "But they had not been proved yet." And that proof is hard to come by, he continued, because "social causes" for those mental differences have not been satisfactorily eliminated. For "as soon as we enter into a consideration of social factors," he emphasized, "we are unable to separate cause and effect or external and internal factors."[16]

What Boas called "social factors" we today would call cultural influences. This reliance upon historical rather than internal or biological forces to explain human behavior or action appears very early in Boas's thought and nowhere in a more striking fashion than in a brief article he published in 1889 entitled "On Alternating Sounds." George Stocking, the leading student of Boas's thought, has remarked that it is "impossible to exaggerate the significance of this article for the history of anthropological thought." For in that article Boas demonstrated quite concretely how he came to see culture or history as a cause for human behavior, a form of behavior that other observers explained by pointing to a physical peculiarity.

The problem he addressed was "sound-blindness," or the inability of certain people to hear particular sounds, just as color-blind persons were unable to perceive certain colors. For many anthropologists of the time, sound-blindness was a sign of racial and linguistic primitiveness. Boas rejected the analogy with color-blindness; he argued that the inability to hear certain sounds was to be found not in any physical or linguistic inadequacy but in perception, in the auditor's lack of familiarity with the sound. His proof was his own recollection of having heard a variety of sounds for the same Eskimo words. Only through many repetitions, he recalled, was he able to discover that what he thought were different sounds were really the same sound. His lack of familiarity with the sounds of the Eskimo language had made it difficult for him to hear the sound in its true form.

His conclusion reflected his consistent determination to reduce the differences between primitive and civilized peoples. The occurrence of alternating sounds, he reported to his fellow anthropologists, "is in no way a sign of primitiveness of the speech in which they are said to occur. . . ." The clincher to his case was the discovery that the speakers of languages that had been denominated "primitive" because they contained alternating sounds heard alternating sounds in modern, western "civilized" languages like English! Modern Scots-man, too, Boas pointed out, could not consistently hear certain sounds or words in German.[17] Intriguingly, that single brief article recapitulates Boas's profes-sional shift from first physics (sound), then to geography (Eskimos), then to eth-nology (language), and finally to an early conception of culture as the shaper of human perceptions.

Boas's broadest response, then, to skeptics who wondered how he could be so confident that all races could reach the high cultural level of Europeans was that all people were of equal intellectual and social potential, but that was not the only retort he readied. He also advanced specific responses to particular arguments. When a social evolutionist like Herbert Spencer contended that sav-ages lacked certain key mental attributes needed for acquiring modern civili-zation, Boas countered by questioning the ethnological evidence. Such anec-dotal reporting, he maintained in 1894 was "not a safe guide for our inquiry because causes and effects are so closely woven that it is impossible to separate them in a satisfactory manner, and . . . we are always liable to interpret as racial characters what is only an effect of social surroundings."[18] (At that date he had not yet found the word "culture" as the alternative to "race.") He then went on to point out how such anecdotal material might be interpreted from *his* outlook.

Contrary to Spencer's assertion that certain behavior demonstrated that primitive people were impulsive and lacked the self-control that Spencer claimed was indispensable for attaining a civilized society, Boas countered with the observation that impulsiveness was little more than the optimism of the modern businessman who also believed in the stability of existing conditions. "We may recognize a difference in the degree of improvidence caused by the difference" between the poor and the well-off, he admitted, "but not a specific difference between lower and higher types of man." The first stemmed from social conditions, and could be accepted, the other derived from biology and could not be.[19]

On another occasion, Boas called into question an authority cited by Spencer by drawing on his own experience in the field. To demonstrate that primitive people were less attentive than civilized persons, Spencer had quoted from a report by a European observer on certain Indian tribes. "I happen to know through personal contact the tribes mentioned," Boas was able to write. The questions asked of the Indians by the European, Boas was convinced, would be seen by them as trivial and thus they would, quite rightly, pay little attention. He, on the other hand, was quite prepared to testify "that the interest of these natives can easily be raised to a high pitch and that I have often been the one who was wearied out first."[20]

Boas also drew upon his own experiences among Eskimo to demonstrate that, contrary to the traditional conception of primitive peoples, "the mind of the native enjoys as well the beauties of nature as we do; that he expresses his grief in mournful songs, and appreciates humorous conceptions.... Though most explorers affirm that their music is nothing but a monotonous humming, the following tunes and texts, which were collected by me in Baffinland, will show that this is not true." These few examples, he concluded in 1887, "show that the mind of the 'savage' is sensible to the beauties of poetry and music, and that it is only the superficial observer to whom he appears stupid and unfeeling."[21]

In answer to those linguists who supported their belief in the inferiority of some primitive people by pointing to the absence in their language of numbers above ten, Boas once again turned to a social explanation. Such people merely lacked any need to count any higher, he responded. Once they had a need, as when white men entered their region, they quickly added to their numerals.

The third avenue of attack that Boas pursued was to ask for more convincing proof from those who looked to biology or race to explain differences between societies or groups of people. That is to say, he placed the burden of proof upon them. He did this, it is worth repeating, even though he was quite prepared to concede, as he did in 1894, that some colored races seemed to have brains that on the average were smaller than those of the white race. He further admitted that "differences in structure must be accompanied by differences in function, physiological as well as psychological, and, as we found clear evidence of differences in structure between the races, so we must anticipate that differences in mental characteristics will be found." But, he insisted, such mental differences "have not been proved yet."[22]

By 1911 he was emphasizing a more positive interpretation of the connection between mind and body. In the course of reporting on his finding about changes in the shape of the heads of New York immigrant boys, he noted that "if the bodily form undergoes far-reaching changes under a new environment, concomitant changes of the minds may be expected." Such alterations in mental qualities, he admitted, are difficult to prove by observation, "but it is evident that the burden of proof is shifted upon those who claim absolute stability of mental characteristics of the same type under all possible conditions in which it may be found."[23] And there he left his case in 1911, not a refutation of biological or racial explanations for the differences between races, but, as George Stocking has labeled it, an assertion of agnosticism.

Once again it is worth noting that throughout this sketch of Boas's ideas there is almost no development over the quarter of a century. Ideas he expressed in the 1880s appeared almost word for word in his 1911 book *The Mind of*

Primitive Man. The only serious exception to that generalization is his use of the word "culture," the concept that later would come to summarize the contribution that Boas made to social scientific thinking about differences in human action. Until the 1890s Boas used the term as his fellow anthropologists did: as another word for civilization or "high culture." It was, in short, a part of a hierarchical conception of social orders: some were better than others and the best was "culture." By the end of the 1890s, however, Boas was using culture in the plural, as we do today—every society exhibits a culture. At that point, though, Boas had still not abandoned the hierarchical component; now the term "civilization" was reserved for the highest cultural achievement. After an extensive examination of the social science literature published between 1890 and 1915, George Stocking reported that he found no plural form of "culture" used prior to 1895, except in Boas's writings. After 1910 the usage was common among anthropologists and other social scientists.[24]

Aside from the shift in the meaning of "culture," Boas's ideas on the concept of race changed not at all. As early as 1887 he was proclaiming the equality of the mental or intellectual potentiality of Eskimos and Europeans, a claim he would be making at the end of the period in regard to blacks and whites in the United States. So the question becomes: Why did Boas, unlike so many of his fellow social scientists of the time, repudiate race, while they, in the age of Darwin's triumph, continue to see it as an obvious way to account for human differences? An explanation is of more than biographical interest since Boas, almost single-handedly, developed in America the concept of culture, which, like a powerful solvent, would in time expunge race from the literature of social science. Indeed, even before the end of the second decade of the twentieth century, as will be seen in the next chapter, Boas's alternative to racial explanation had already pretty well conquered several fields of social science.

The roots of Boas's social outlook are sunk deep in the history of ideas concerning diversity. Even before Darwin and his concept of natural selection came on the scene, a debate over racial differences had divided students of racial types. A large part of the argumentation took place in the United States in the years before the Civil War as a direct outgrowth of the moral and intellectual conflict over black slavery. One of the prominent defenses of Southern slavery asserted that the black slaves were racially inferior, biologically suited to be "hewers of wood and drawers of water." Those who explained human diversity on grounds of race or biology were known as polygenists; those who denied that the differences derived from biology were denominated monogenists. As the name suggested, the "monos" believed that all peoples around the world, regardless of color or appearance, derived from a common source, which was usually, at least for religious people, Adam and Eve. There was, in short, a common humanity in the present just as there had been from the beginning. To the "polys," however, who were clearly the more "modern" of the two, the Biblical story was more myth than a scientific description of human ancestry. They believed that the diversity of humanity was a product of evolution. Whatever may have been the origins of human beings, by the nineteenth century they were divided into separate races with different innate capabilities. In short, even before Darwin appeared on the scientific scene, racial differences and evolution had become closely associated. Darwin himself, as we have seen, rejected the contention of the polygenesists on biological grounds. All human beings shared a common animal ancestry; for Darwin humankind was a single species (though not necessarily one in which all subgroups were equal in mental capacity).

In the Germany in which Boas grew up and was educated—he was born in Rhineland Germany in 1858—the monogenesist viewpoint was well represented. No one advanced it more ardently or fully than Theodor Waitz, a young University of Marburg professor. Boas never attended Marburg or studied with Waitz, who died prematurely in 1865. Nevertheless, throughout his life Boas pointed to Waitz's work as not only a source but a justification of his own commitment to cultural explanations of human social behavior. What, then, was Waitz's contribution to Boas's thinking?

For a man who died at age forty-four, the volume as well as the content of Waitz's work are impressive. Before Darwin published *The Origin of Species* and in the same year as Boas's birth—1858—Waitz published the first volume of what would be a work of six volumes. In the light of the monogenesist and polygenesist controversy, of which it was clearly a part, the title of that first volume—*On the Unity of the Human Species and the Natural Condition of Man*—made clear where Waitz stood. Personally, and in his writings, Waitz identified with those figures of the Enlightenment who believed that all people, black or white, high or low in cultural achievement, were "equally destined for liberty." Whatever differences there might be between peoples, Waitz stressed in his first volume, derived not from something peculiar to them but from "something acquired in the course of their development, which, under favorable circumstances, might have been equally acquired by peoples who appear at present less capable of civilization." Here, of course, was Boas's later emphasis upon history as the source of human differences.[25]

Waitz made no secret of his fervent opposition to what he called the "American School" of anthropology, which had been spreading polygenesist ideas in the United States during the 1840s and 1850s. As Waitz saw it, these Americans contended that "the higher races were destined to displace the lower," one consequence of which was the justification of "the right of the white American to destroy the red man."[26]

In Boas's treatment of race over the years, no other authority achieved the prominence accorded Theodor Waitz. Sometimes he quoted directly from Waitz, as in his path-breaking critique in 1894 on racial explanations; at other times he simply referred to "Waitz's great work," as he wrote in an article in 1910. As late as 1934 he was still reminding his readers that his own view of culture had been "expressed by Waitz as early as 1858 and is the basis of all serious studies of culture."[27]

Important as Waitz's influence undoubtedly was for Boas, it does not encompass the full answer to the question of the sources of Boas's ideas. We still need to ask why Waitz's ideas found so congenial a lodging and such an enduring presence in Boas's mind. Again, a good part of the answer emerges from his background in Germany before he emigrated to the United States in 1885. Margaret Mead, undoubtedly the most famous of Boas's students, recounts a story of her relation with Boas that brings into focus the role of his personal background. When, in the early 1920s, she was trying to overcome Boas's refusal to let her pursue anthropological field work in the South Seas, which would be the basis of her *Coming of Age in Samoa*, she hit upon a sure way of getting him to change his mind. "I knew there was one thing that mattered more to Boas than the direction taken by anthropological research," she wrote in her autobiography, *Blackberry Winter.* "This was that he should behave like a liberal, democratic, modern man, not like a Prussian autocrat."[28] The ploy worked for Mead because she had indeed uncovered the heart of his personal values.

In later life, Boas himself offered some clues to the origins of those values. "The background of my early thinking," he recalled, "was a German home in which the ideals of the Revolution of 1848 were a living force." His father, he remembered, was "liberal, but not active in public affairs; my mother idealistic, with a lively interest in public matters." Both parents were religious skeptics, thus sparing him, as he put it, "the struggle against religious dogma that besets the lives of so many young people."[29] Indeed, as these recollections and his whole life testified, freedom of inquiry and freedom from dogma were at the center of his personal philosophy, a commitment that in its application, he insisted, should include all people. In early, as well as in later life, he was always ready to fight for his ideals, sometimes literally. Throughout his adult life Boas's face bore the scars from a duel entered into during his university days to punish an antisemitic slur.

Until the rise of Hitler, Boas exhibited a lively attachment to what was liberal and idealistic in German culture, yet it was the rising anti-semitism in the Germany of the 1870s and 1880s that helped shape his decision to emigrate to the United States. As he explained in 1882 to his uncle, Dr. Abraham Jacobi, then a practicing physician in the United States, Germany did not promise many career opportunities, "and Jewish teachers have great difficulty getting an appointment." The United States, on the other hand, seemed a more open society, in which opportunities for a geographer or budding ethnologist were already wide, and expanding. By the time Boas decided to emigrate, his interests had shifted from physics to geography, and after spending some time in Baffinland among Eskimos, he made the final transition to ethnology. In describing his "life's work" to his uncle, Boas's words can stand as a prediction of the place he would finally occupy in the development of the concept of culture. His aim, he told his uncle, was to see how far one can get in "determining the relationship between the life of a people and environment."[30]

In sum, Boas came to the United States with an outlook that emphasized equality of opportunity, freedom of inquiry, and openness toward people who were different and socially excluded. Racial explanations, as far he was concerned, only closed off opportunity and acceptance. Many scholars might see people of different appearance as different also in mind. But, as Boas wrote in 1894, he was still waiting for the proof. In short, Boas approached the question of race with a defined ideological position that shaped his answer.

That ideological commitment comes through in the justifications he offered for the research projects he pursued. For example, his deeply held belief in the value of primitive peoples is plainly evident in a recommendation he made in 1902 to Nicholas Murray Butler, then the president to Columbia University, where Boas taught. He urged Butler to bring Columbia into cooperation with the American Museum in collecting and exhibiting the handicrafts and art works "from cultures different from our own." Such collections are important, he told the president, because "we should try to counteract [the] tendency towards a one-sided valuation of one point of view to the exclusion of all others."[31] Boas's professional correspondence similarly reveals that an important motive behind his famous head-measuring project in 1910 was his strong personal interest in keeping America diverse in population and open in opportunities for all.

Even before the project was approved by the Immigration Commission, Boas had called the attention of Commission member Jeremiah Jenks to the important social questions raised by the changing character of immigrants arriving in the United States. "Instead of the tall blond northwestern type of Europe, masses of people belonging to the east, central, and south European types are pouring into our country," he wrote. This has raised the question "whether this change of physical type will influence the marvellous power of amalgamation that our nation has exhibited for so long a time." What was little more than a problem to investigate in March by September had become a solution. For by then Boas was reporting to Jenks that his preliminary findings "are in the direction of a decided assimilation of the foreign types" and that those who do assimilate are "individuals who live in favorable social surroundings."[32] Turned around, that statement meant that if immigrants did *not* assimilate, the cause should be attributed not to race but to unfavorable social circumstances.

By 1909 Boas was convinced that his research on immigrants was providing an answer to the question of the proper social policy to follow. That research had shown that "all fear of an unfavorable influence of South European immigration upon the body of our people should be dissmissed." The following year he justified further research projects to a head of a Jewish orphanage in New York on the ground that his work so far had shown that "the assimilation of the foreign-born is a very rapid one and that the fears based upon the divergence of immigrant types may be considered as unfounded."[33]

The idealogical underpinning of Boas's conviction that all ethnic groups were of roughly equal mental potentiality is especially apparent in some remarks he addressed to Jenks in 1910. He began by noting that some recent studies had demonstrated that native children did better in New York schools than foreign-born. The possibility that this comparison might be held against the immigrant children apparently troubled him. For he then went on to say that it is "of the greatest importance" to take into consideration the differences in the *physical* development of the foreign-born and the native children. Since the foreign-born children were the better developed. Boas interpreted that finding to show "convincingly that the slowness of their advance in school life is due not to a lack of ability, but the fact that so much energy is taken up by their social and linguistic assimilation, that physiologically speaking, they are older than the native-born children, who do not need to struggle against these difficulties."[34] Why Boas chose to advance such an ad hoc interpretation is hard to understand until one recognizes his desire to explain in a favorable way the apparent mental backwardness of the immigrant children.

Nowhere does Boas's commitment to the ideology of equal opportunity and the recognition of the worth of oppressed or ignored people become more evident than in his relation to Afro-Americans, a people whose life patterns had long been allegedly "accounted for" by race. During the early years of the new century, at precisely the time the Southern states were imposing legal segregation upon black people, Boas was pressing for new studies of the Negroes' situation in American society. In 1905, for example, while admitting that the subject was controversial, he nonetheless sought to inform the general public about his profession's reaction to the "arguments based on the tacit assumption of the physical and mental inferiority of the Negro." About the same time he was seeking to interest Andrew Carnegie in contributing half a million dollars to support a new Museum on the Negro and the African past, which he intended to establish. He frankly told Carnegie that the purpose of the museum was to counter the rising racism in the country and what he saw as a lack of appreciation of American blacks. The subject was of such importance, he told the philanthropist, that he had assigned it first priority among his research goals. Undoubtedly hoping to anticipate at least one objection from Carnegie, Boas characterized as "unfair" any judgement of American blacks based merely upon their situation in the United States. Their "cultural achievement in Africa," he informed Carnegie, demonstrates that their inventiveness, political organization, and steadiness of purpose "equal or even excel those of other races at similar stages of culture." Earlier that same year he complained to the editor of the *Century Magazine* about some articles on Negroes that had appeared there. The articles "continue to worry me," Boas wrote to Richard Gilder, the editor, "and I think you owe to the race a presentation of a more favorable aspect of their achievements." Meanwhile, he was teaching courses at Columbia on "The Negro Question" and "The Race Problem in the United States."[35]

His frequent public efforts in behalf of Afro-Americans naturally brought him to the attention of blacks. W. E. B. DuBois, destined to be the leading black social scientist of the early twentieth century, then teaching at black Atlanta University, invited Boas in 1905 to attend a university conference on the Negro Physique. Knowing Boas's penchant for measuring people, DuBois promised to provided him with many black students to measure if Columbia could cover the travel cost. Boas agreed not only to attend the conference without fee, but also to give the commencement address that year at Atlanta University. The burden of his remarks, on that occasion, as might be anticipated, was the greatness of the African past. You have a history, he told the students, not only of which they could be proud, but also one which would serve well as a proper response to those who dared to portray them as inferior. "If you will remember the teachings of history," he advised, the work of the future would become "a task full of joy." DuBois himself could well have been speaking for the students, too, when he recalled that Boas's presentation left him "too astonished to speak. All of this I have never heard."[36]

Over and over again in print Boas sought to meet any and all of the arguments advanced by proponents of the inferiority of African-Americans. In response to those who directed attention to the smaller average size of the Negro brain, he contended that nothing is really known about differences in structure that might result from differences in size. Furthermore, "the inference in regard to ability is ... based only on analogy and we must remember ... that we find a considerable number of great men with slight brain weight." There may be less average ability in blacks than in whites, he was quite prepared to admit, but we can "expect these differences to be small as compared to the total range of variations found in the human species." Finally, the brain weights of Europeans vary so much "that a great many occur that are below the average of the Negro brain," while many Negro brains are heavier than European brains. Consequently, "we may expect a similar distribution of ability."[37]

To those who contended that biologically the Negro was not capable of fitting into the complex social order of the United States, Boas responded with a litany of qualities that apparently consituted for him the very center of the American character. Anthropology, he confidently asserted, "can give the decided answer that the traits of African culture as observed in the aboriginal home of the Negro are those of a healthy primitive people with a considerable degree of personal initiative, with a talent for organization, with a considerable imaginative power; with technical skill and thrift." And to those who thought the Negro cowardly or timid, Boas retorted that "neither is a warlike spirit absent in the race, as is proved by the mightly conquerors who overthrew states and founded new empires, and by the courage of the armies that follow the bidding of their leader." In short, there was everything to prove that the laziness, licentiousness, and improvidence that were popularly believed about blacks at the time were really "the result of social conditions rather than of hereditary traits." Boas made a similar defense of Negroes to his fellow social and natural scientists in the course of his vice presidential address in 1909 at the meeting of the American Association for the Advancement of Science. And in 1911 in his introduction to Mary Ovington's *Half a Man: The Status of the Negro in New York* he praised the book for being "a refutation of the claims that the Negro had equal opportunities with the whites, and that his failure to advance more rapidly than he has, is due to innate inability."[38]

In considering Boas's obvious and persistent concern about the status of blacks in American society, it is at once relevant and puzzling to reflect that he never lavished the same degree of attention upon the status of Amerindians despite his deep involvement with them in his own research and that of his students. During the late nineteenth and early twentieth century the question of the well-being of the aboriginal population was both a highly visible reform and a political cause. The reservation system of the time was a travesty of humane treatment of a dependent people, and the decline in the Amerindian population was a vital topic of public discussion. Yet Boas's public writings and his professional correspondence are almost devoid of expressions of concern or interest in the status of Amerindians compared with those directed toward the question of African-Americans in American society.

No explanation for the difference in concern is to be found in any divergence in the vision he held for the two disadvantaged groups. George Stocking has described that vision a "melting pot" approach. If that term means the creation

of a new culture through the introduction of fresh elements, that does not seem
to be quite what Boas had in mind. A more accurate, if less colorful description
of his solution to the question of cultural diversity within one country would
seem to be "cultural or ethnic disappearance or integration." For what Boas
advocated was the submergence of both peoples in the general American
population.

In 1903 a correspondent, Natalie Curtis, asked Boas what the culture of a
particular Indian tribe would be like if its members could be completely isolated
from white intrusion or cultural contamination. Curtis had obviously picked
up the idea of cultural relativism and carried it to a high level. The idea, however, did not appeal to Boas, who suggested instead that what the tribe in question needed were teachers with better ethnological training. The purpose of that
training, he made clear, was the ultimate integration of the Amerindians into
the general American population. "I firmly believe," he wrote, "that if nothing
else could be accomplished," attention to the culture of the Indians would make
"the transition from the old life to civilized life . . . much easier, and that the
young generation, instead of being of very doubtful moral value, could become
useful members of society, and would introduce into our community such parts
of their own culture as were worth preserving." He recognized that the assimilation of Amerindians, however desirable it might be, would not be easy. He
suggested that policy ought to draw a distinction between "our treatment of
half-bloods, who are likely to merge into the white community," and that
which should be accorded "full-blooded Indians, who will always remain distinct from the white community."[39]

At about the same time he was expressing his views on the future of Amerindians to Curtis, he was advocating a similar future for them in a paper on the
scholarly direction he thought the Bureau of American Ethnology should move.
"A thorough inquiry . . . by competent anthropologists" into Indian history and
culture, he thought, "would enable us to encourage or discourage mixture of
the two races, to teach the children in such a way that they would become
useful citizens, and to mitigate the hardships involved in the transition from
savage to civilized life."[40]

The future Boas envisioned for Afro-Americans left them no more chance for
racial or cultural survival than his future for Amerindians. He wrote Jeremiah
Jenks in 1909 in connection with his study for the Immigration Commission
that the biological amalgamation of blacks was best for them and for American
society. He predicted that, as more and more Southern Europeans entered the
country, intermarriage between blacks and whites would likely increase. Consequently, he advised, "we ought to know what social and hygienic significance
the recent laws of Southern states, forbidding intermarriage between the two

races, will have." In Boas's mind, the prohibition of intermarriage between blacks and whites was likely to lead to undesirable social consequences. "Broadly speaking," he explained to Jenks, "the question before us is that of whether it is better for us to keep an industrially and socially [but, significantly, not biologically] inferior large black population, or whether we should fare better by encouraging the gradual process of lightening up this large body of people by the influx of white blood." And in the event that Jenks did not perceive the implications of that remark, he rephrased the question: Might it be "of advantage to accelerate the infusion of white blood" among the Negroes?[41]

When Boas considered "The Problem of the American Negro" for the *Yale Review* in 1921, intermarriage between blacks and whites was the heart of his message. Only by eliminating the differences in the appearances of African-Americans, he argued, could racial prejudice be eliminated. To his way of thinking, differential treatment of blacks and whites was patently unfair. Whatever physical differences that might exist between whites and blacks, he contended, were not significant, and especially because the range of differences within one race overlapped so much that of the other. In sum, in his mind, actual differences between the races were not the true source of racial prejudice. Prejudice resulted rather from the "tendency of the human mind to merge the individual in the class to which he belongs, and to ascribe to him all the characteristics of his class." There does not have to be a marked or noticeable difference for prejudice to arise, he stressed, for it occurs in the case of anti-semitism or in conflicts between capital and labor. The source of the problem is not consciousness of economic class or color, but "consciousness of the outsider." The essentially psychological nature of Boas's formulation became plain with his specific denial that competition between economic groups had much to do with the origin of racial prejudice. The identification of the outsider, he insisted, has to precede the economic rivalry. For until group identification is fixed, competition for economic gains cannot occur. Economic competition could exacerbate and deepen prejudice, he conceded, but it would be "an error to seek in these sources the fundamental cause for the antagonism, for the economic conflict . . . presupposes the social recognition of the classes."[42] By that he meant, of course, the well-recognized visible indicia of race, sex, and ethnicity.

Since Boas's explanation for racial prejudice rested on social awareness of differences, it followed that his remedy for prejudice would be to dilute or diminish those differences as much as possible. Thus intermarriage between the races became his sovereign solution to racial prejudice. He identified intermarriage as "the greatest hope for the immediate future." Once it became difficult to tell if a person were white or black, he predicted, "the consciousness of race would necessarily be weakened. In a race of octoroons, living among whites, the color question would probably disappear." Therefore he advocated that the law should "permit rather than restrain" marriages between whites and blacks.

His solution, of course, looked forward to the disappearance of Afro-Americans, just as his solution for the differentness of the Amerindians was their ultimate submergence or integration into the general population. He was, in short, no cultural pluralist. Apparently, he even contemplated the disappearance of his own ethnicity. He closed his *Yale Review* article with the observation that the Negro problem in America would not disappear "until the negro blood has been so much diluted that it will no longer be recognized just as antisemitism will not disappear until the last vestige of the Jew as a Jew has disappeared."[43]

As the foregoing suggests, it would be a mistake to see Boas as a cultural relativist, that is, someone who saw all cultures as equal, or who refused to recognize a hierarchy among societies. His willingness to see Amerindians and African-Americans integrated into the general American population is only the most obvious evidence of his rejection of cultural relativism. There is, to be sure, no doubt that he repeatedly stressed the necessity of recognizing the value of cultures other than his own, regardless of the degree of differences. Yet, as George Stocking has pointed out, Boas never abandoned the idea that behind all cultures stood a common system of values, especially apparent in the culture of Europeans. Indeed, his basic defense of primitive peoples implied a hierarchy since his invariable point in comparing cultures was that each could potentially achieve the highest culture, which always was Europe's.[44]

Boas never specified the social and moral values he considered essential for civilization, but they do emerge from his writings on cultures and civilization. Among them were individual freedom, human fellowship, and scientific knowledge, all of which were realized in modern civilization, and which Boas revealingly referred to as our "own" culture or civilization. He saw a number of those values, too, being acquired or advanced in certain cultures through historical experience. A culture that reflected those values was obviously better than one that failed to, another sign, if it is needed, of his continuing belief in a loose hierarchy of cultures. Indeed, in 1941, a year before his death, Boas explicitly denied holding a "general relativistic attitude."[45]

If Boas was not a relativist, neither was he prepared to reject completely the role of heredity or biology in shaping human behavior. For contrary to what some later commentators have written about Boas's ideas, he never denied that heredity was highly influential in determining the lives of individuals. On some occasions he even seems to have placed it above environment. Once, for example, in delineating a defense of physical anthropology, he specifically disagreed with those authorities who considered the environment to be more important than heredity. In his mind that was unsound, for "biological consideration makes it very probable that the influence of heredity should prevail, and thus far," he himself had "failed to find conclusive proof to the contrary." And even in his great paean to cultural explanation, *The Mind of Primitive Man,* he advanced a powerful defense of the importance of heredity. After a lengthy exposition of the way the environment shaped the behavior of racial types, he went on to say that "these influences are of quite secondary importance when compared to the far-reaching influence of heredity. Even granting the greatest possible amount of influence to environment," he warned, "it is readily seen that all the essential traits of man are due primarily to heredity. The descendants of the negro will always be negro; the descendants of the whites, whites; and we may go even considerably further, and may recognize that the essential detailed characters of a type will always be reproduced in the descendants, although they may be modified to a considerable extent by the influence of environment."[46]

Although Boas seems to be talking about races in the previously quoted passage, he is actually referring to individuals as the inheritors of traits. This distinction by Boas between group or race and individual becomes clearer in a remark he made considerably later, in 1931, when writing about the relation between biological form and physiological and psychological functions. He had no doubt that "there is a very definite association" between the two. "The claim that only social and other environmental conditions determine the reactions of the individual disregards the most elementary observations, like differences in heartbeat, basal metabolism, or gland development; and mental differences in their relation to extreme anatomical disturbance of the nervous system." In short, he concluded, "There are organic reasons why individuals differ in their mental behavior."[47] In fact, in Boas's view, it was this variety among individuals within any given racial type that called into question the concept of race itself. Individual variation within each race, as he often said, was so wide as to overlap to a large extent the range of variations in every other race.

Boas's emphasis upon the role of heredity in the individual should make crystal clear that at no time was he an extreme environmentalist. Certainly not of the variety exemplified by his contemporary, behavioralist psychologist John Watson, who boasted that, if given a dozen healthy infants, he would "guarantee to take anyone at random and train him" to be able to enter any occupation regardless "of his talents, penchants, tendencies, abilities, vocations, and race of his ancestors."[48] Boas, after all, began his professional career as a physical anthropologist, and that knowledge and experience never disappeared from his conception of the nature of human beings. He may have gone a long way in his own work and life to demonstrate how important culture and history were, but he never forgot that human beings, like all other animals, were products of their physical inheritance as well.

One consequence of Boas's recognition of the role of heredity in shaping an individual was that it reinforced his emphasis upon the individual. Because the heredity of each person (except for identical twins, of course) is unique, Boas deemed it improper to ascribe automatically to any individual the culturally determined attributes of the race or social group to which that individual belonged. He made that point in a typical manner in 1915 in a farewell talk to a Barnard College class in anthropology that he was unable to complete because of illness. "The concept of race" is artificial "when applied to individuals," he emphasized. "Nothing justifies the branding of an individual as inferior or regarding him as superior because he happens to belong to one or the other's race. He closed his remarks with a plea that was deeply consistent with his many efforts to extirpate racial interpretations from social relations. If his course had convinced the students of "the correctness of our views," he characteristically told them, "then it is your duty to overcome racial prejudice."[49]

Throughout this explication of Boas's conception of culture and his opposition to a racial interpretation of human behavior, the central point has been that Boas did not arrive at that position from a disinterested, scientific inquiry into a vexed if controversial question. Instead, his idea derived from an ideological commitment that began in his early life and academic experiences in Europe and continued in America to shape his professional outlook. To assert that point is not to say that he fudged or manufactured his evidence against the racial interpretation—for there is no sign of that. But, by the same token, there is no doubt that he had a deep interest in collecting evidence and designing arguments that would rebut or refute an ideological outlook—racism—which he considered restrictive upon individuals and undesirable for society.

That he may have had other motives as well, motives related directly to his professional interest in providing a secure intellectual base for his newly emerging field of anthropology, is an interpretation suggested by several scholars. Certainly the culture concept, by denying the influence or power of biology in the social sciences, offered to a new discipline like anthropology a clearly defined intellectual tool that could identify and thereby justify the discipline in competition with the better established and more highly regarded biological sciences of the time. Boas was unquestionably a powerful political figure in the development of American anthropology as he was surely the profession's most influential figure as well as its founding father in America. Yet in Boas's voluminous writings, both private and public, little direct evidence of that professionally oriented motive appears. Much evidence does come to light in that correspondence to suggest a persistent interest in pressing his social values upon the profession and the public. His correspondence, to be sure, is replete with details of the profession's growth and definition. Yet in that large body of writing, references to a need to counter threats from biology are few, indeed rare when compared with the concern that Boas's student Alfred Kroeber displayed between 1915 and 1917, when he forthrightly proclaimed the necessity of the total separation, in both method and substance, of social science from biology. As late as 1919, Kroeber wrote Boas bitterly complaining about natural and physical scientists' lack of understanding of the concept of culture. "Consequently, the sense always crops up in their minds that we are doing something vain and unscientific, and that if only they could have our job they could do our work for us much better."[50]

Kroeber's own writings, as we shall see in the next chapter expanded and defined further the concept of culture, which Boas had been formulating. But even before Kroeber arrived on the scene, Boas's criticisms of race as a category of analysis of human differences were winning appreciative responses from social scientists outside anthropology.

4

In the Wake of Boas

Heredity cannot be allowed to have acted any part in history.
Alfred L. Kroeber, 1915

Franz Boas's exposition of the concept of culture as an alternative to race in accounting for differences among human groups was a genuine seminal contribution. The idea not only conquered his own field of anthropolgy but spread as well to other social science disciplines. The diffusion was remarkably rapid. Again and again in the early years of the new century American social scientists outside of anthropology testified to the influence of his writings on their conception of human nature and the place of biology or race within it. One sociologist, for example, in 1914 specifically mentioned Boas as perceiving "no essential difference between the negro and white races." Two years later a University of Pennsylvania sociologist quoted at length from Boas's study of changes in bodily shapes in support of his own contention that physical types cannot be seen as stable.[1]

More significant than the recognition accorded Boas's opposition to race were signs that his writings had not only informed social scientists but had actually overturned the thinking of some of them. Sociologist Charles W. Ellwood of the University of Missouri passed through such a transformation. In 1901 Ellwood had published an article in the *American Journal of Sociology* in which he contended that "innate tendencies" account for behavior rather than imitation or environmental influences. After all, he pointed out, if environment were a sufficient cause, children of one race, who were reared in the home of another, could be expected to "develop the same general mental and moral characteristics." But that is not the case, he informed his readers, since "the negro child, even when reared in a white family under the most favorable conditions, fails to take on the mental and moral characteristics of the Caucasion race."[2]

Within five years, however, Ellwood had changed his mind. In reviewing a recent book that proclaimed the inferiority of blacks and declared them incapable of improvement, Ellwood specifically referred to the work of Boas when

84

calling the author to account. There is some "scientific authority," Ellwood conceded, to support the idea of "the natural inferiority of the negro race. But even here it is to be noted that a large and growing school of anthropologists and race-psychologists finds the explanation of the mental and moral differences between races, not in innate qualities or capabilities, but in differences in their social equipment or machinery."[13]

Carl Kelsey, later to become a nationally known sociologist at the University of Pennsylvania, passed through an even more dramatic and fundamental transformation than Ellwood, again thanks to Boas's work. In 1903, Kelsey was only beginning his career at the University of Pennsylvania, having published his doctoral dissertation that year on "The Negro Farmer" in the South. He described his book as a contribution to the understanding of the growing social problem of the races in America. Both North and South in their own ways, he wrote, had been historically prejudiced about slavery and blacks. "The South, perceiving the benefits of slavery, was blind to its fundamental weaknesses," while the North, "unacquainted with Negro character, held to the natural equality of all men." As a result of such distorted views, he contended, no sound studies of blacks or slavery had been undertaken. Consequently, after the emancipation of the slaves, many mistakes were made regarding the future of Afro-Americans. "Any thoughtful observer must agree," he confidently wrote, "that as a race [Negroes] were not prepared for popular government at the time of their liberation." The mistakes that had created Southern Reconstruction, he was glad to observe, were now behind Americans. In 1903, he reported "the North is slowly learning that the Negro is not a dark-skinned Yankee, and that thousands of generations in Africa have produced a being very different from him whose ancestors lived an equal time in Europe." Thus the differences between blacks and whites in modern America did not originate in slavery; "their origins lie farther back" in Africa.[4]

Kelsey's reference to "thousands of generations" implies that in his view the African experience had become embedded over time in the heredity of Afro-Americans. Kelsey, in sum, was a Lamarckian, a believer in acquired characters. That supposition is reinforced by a remark he made four years later. Within that interval he had learned "pretty definitely . . . that acquired characteristics are not passed on from generation to generation. This fact," he continued a little ominously, "is reacting powerfully upon our social theories." Very likely he was referring to social scientists like Lester Frank Ward and other Lamarckians—including himself, presumably—who believed that improvements in races or groups resulting from a better social environment could be made permanent through inheritance. That theory, as we saw earlier, had provided some comfort to those social scientists who wanted to improve the lot of blacks and other disadvantaged Americans. Once the theory of acquired characters ceased to be acceptable, those reform-minded social scientists were confronted by a choice between biology—which no longer could be seen as experientially cumulative—and culture, which was.

At that juncture—and together with other social scientists—Kelsey chose culture. Some recent researches, he remarked at a professional meeting in 1907, "weaken the belief in superior or inferior races. It now seems very probable that there is an approximate equality of mental ability among the various races, and that race differences are the result of different environments."[5]

Once he had grasped the distinction between acquired and inherited characteristics, Kelsey became an outspoken proponent of the power of environment or culture to shape human actions. He doubted, for example, in 1909 that it could "be shown that during all historic time the human race has made any material changes via the road of heredity." Rather, increasingly it becomes evident that the social order is the generator of the problems that plague it. Some people may worry about "racial degeneration," he admitted, "but is it not possible that the trouble lies in our own social institutions?" Could it not be that "the serious problems of immigrants," upon which much contemporary discussion concentrated, were really "due to social differences rather than to inherited physical differences?"

By shifting the explanation of social behavior from heredity to culture, from nature to nurture, Kelsey became even more optimistic than he had been as a Lamarckian. If the social environment is the primary cause of social differences, then the "situation is hopeful," he thought. For now that the Lamarckian doctrine of acquired characteristics has been repudiated, we can be confident that "no matter how bad the environment of this generation" may be, "the next is not injured provided that it be given favorable conditions." To recognize such facts, he thought, "is surely to have an optimistic view" of the future of society. "In a word," he concluded, "we create the evils as well as the good. Nature is impersonal. To an increasing degree man determines." For a reform-minded social scientist like Kelsey, environmentalism or culture had enormous appeal; social improvement could be carried out much faster than under even the Lamarckian dispensation.[6]

It therefore comes as no surprise that when Kelsey came to review Boas's *The Mind of Primitive Man* in a scholarly journal in 1913 he had almost nothing but praise for the book, denominating it a "genuine service" to scholarship by providing an "admirable survey of existing information" on human types. It is significant that Kelsey summarized Boas's conclusion that "racial differences are largely superficial" to be "the prevailing belief" among social scientists.[7]

Kelsey repeated his Boasian views in a general text published in 1916. The fact that the book was drawn from his teaching at the University of Pennsylvania suggests he had been disseminating those views in other ways as well. At one point in the text he noted that the dominant civilizations of today are largely among the lighter colored groups. It is easy to assume," he admitted, "that there must be some causal connection between these facts." But the more carefully the apparent connection is examined, the more doubtful it becomes. He then launched into a typical Boasian examination of the history of high civilizations among people of color in the past: Egyptians, Babylonians, and Chinese. "In the light of history it may be that other factors than color of skin have caused the lighter groups to play such a prominent role just now," Kelsey archly suggested. "At all events," he concluded in words echoing Boas's well-known agnosticism on the subject, "until some one is able to put his finger upon some physical difference which can be shown to have some connection with the degree of culture or the possibility thereof, we have no right to assume that one group of human beings is either superior or inferior to any other." Kelsey, however, could not go all the way with Boas; intermarriage between whites and blacks, he belived "no thoughtful person . . . can today advocate," though he opposed laws that prohibited interracial marriage.[8]

Contemporaneously with Kelsey's transformation, another sociologist destined for national prominence, Howard Odum, made an even more striking crossover from racism to Boasian environmentalsim and the unity of human nature. The transformation was especially dramatic because Odum was a native white Southerner, then teaching at the University of Georgia. In 1910 Odum published his doctoral dissertation in sociology, entitled *Social and Mental Traits of the Negro,* which he had completed at Columbia under Franklin Giddings. A large part of the material had been drawn from Odum's earlier accumulation of African-American folklore collected in the South, especially in the course of his work with a professor of psychology and education at the University of Mississippi. The portrayal of blacks in Odum's book was largely unfavorable inasmuch as they were described as shiftless, dirty, and irresponsible, along with other negative descriptions, though the study was intended to improve whites' understanding of the plight of Negroes. In the end, Odum admitted, there was some truth to the view that not much can be done to make a Negro anything but a "negro. But he may be assisted to be a good negro," Odum remarked, "and that is the highest privilege that can be given him."

If blacks were to be educated, Odum advised, then of first importance was the need to lead the black child "toward unquestioning acceptance of the fact that his is a different race from the white, and properly so; that it always has been and always will be." No hint appears here of Lamarckian acquired characters qualifying the permanence of Negro inferiority. He then proceded to justify his principle by recalling a black teacher who told a black child not to regret her skin color or kinky hair. " 'If you were a little pale faced, yellow haired girl,'" the teacher was quoted as advising the student, " 'and all the rich, well-educated people about you had brown skin; if those who rode in carriages and autos had kinky hair; if the dominant, cultured, successful race were negroes, you would long to be a Negro also, brown skin, kinky hair and all.'" It was all a matter of education, money, and morality, the teacher explained; " 'and just as soon as the majority of negroes acquire these, the question of color will begin to drop out.'" This rather advanced conception of the socialization of prejudice Odum indignantly denounced as "monstrously wrong." Contrary to what the black teacher assumed, he emphasized, "the question of color will not drop out." And just because it would not, "textbooks are needed which are especially adapted to the negro *mind.*"[9]

Three years after the appearance of his *Social and Mental Traits of the Negro,* Odum contributed to a national publication an essay on Negro children in the public schools of Philadephia. This time his conception of blacks was quite different. In the course of the intervening years, he had learned of Boas's study on changes in bodily forms and he had read *The Mind of Primitive Man.* In his paper on Philadelphia Negro children, Odum quoted a passage from Boas's book on the "plasticity" of the human mind, remarking that "it would clearly be impossible for the Negro chidren to show the same manifestations of mental traits as white children, after having been under the influence of entirely different environments for many generations." He thought "injustice would be done to Negro children if harsh judgment be passed upon them because they do not maintain the standard of the white children."

His explanation now for the mental deficiencies of the black children was virtually the reverse of his earlier one. Again and again he pointed to social barriers that impeded the educational and social progress of the black children. Recognition of the social background within which Negroes lived "lend support," he contended, "to the conclusion that the failure and defects of Negro children may be due only to environment which is unfavorable to their highest development. There is thus far," he added, "no evidence to contradict such a conclusion, while there is much evidence to show that the environment under which Negro children have grown is unfavorable to the development of the mental abilities commonly accepted as superior." It is true, he admitted, that changes in the environment cannot make mediocre minds into brilliant ones, but the identification or measurement of mediocrity in children is quite unreliable. As a consequence, "it is absolutely impossible to say how much and of what sort are the innate differences between white and Negro children," he concluded.[10] Boas himself could have written that conclusion.

Not all social scientists at that time needed to pass through the kind of ideological turnabout experienced by Carl Kelsey and Howard Odum. William I. Thomas, a rising sociologist at the University of Chicago, for example, does not seem ever to have held strong racist views about people of color. He claimed in a letter to Boas in 1907, for instance, that for the previous ten years he had been teaching a course "on the mental traits of the lower races" in which he had been arguing that "differences in mind are environmental in origin rather than innate." Indeed, as early as 1904, he had publicly written that race prejudice was grounded only on differences in appearances and therefore should be considered "a superficial matter." Prejudice often expressed itself with high emotion and intensity, he recognized; nevertheless, it is "easily dissipated or converted into its opposite by association, or a slight modification of stimulus." In short, Thomas would have nothing to do with those social scientists who saw hostility between the races as innate or natural because of inherent differences in mind as well as body. Indeed, race appeared sufficiently superficial to Thomas for him to predict that, as education spread, race differences would be no more socially significant than differences between occupations.[11]

If, in 1904, Thomas was discerning some superficial differences between the races, by 1907 even those traces of difference had disappeared. "it is probable," he wrote, "that brain efficiency (speaking from the biological standpoint) has been, on the average, approximately the same in all races and in both sexes since Nature first made up a goodworking model." Whatever differences there may be "in intellectual expression are mainly social rather than biological," he was convinced. Following a lead he probably picked up from Boas, he then illustrated the similarities in mental activities among civilized and primitive peoples, by citing proverbs from English and African languages that carried the same message: "one tree does not make a forest," the Africans say; "one swallow does not make a summer," was the English equivalent. Only when the environmental or social circumstances within which Negroes live are genuinely equal to those of whites, he now insisted, shall we "be in a position to judge of the mental efficiency of . . . the lower races." At the present time, he asserted, "we seem justified in inferring that whatever differences there may be between the higher and lower races . . . [they] are no greater than they should be in view of the existing differences in opportunity." Boas probably would not have referred to "lower races" but he certainly could have endorsed Thomas's acknowledgment of the environmental or cultural sources of the mental differences between civilized and primitive people, a view, incidentally, that Thomas described to Boas in his 1907 letter as "common now to a school of thinkers."[12]

Thomas's influence upon other social scientists was not on a level with that of Boas, who not only preceded him in prominence but also outdistanced him in intellectual power. Nevertheless, Thomas's role in convincing social scientists that race was an outmoded, or, more accurate, a misleading means of explaining differences in the behavior of social groups was both important and effective. (His influence was important, too, in the contemporary critique of sex differences, a subject to be addressed in the next chapter.) Thomas's theoretical contribution to the discussion of race differences, however, is not equal in power and subtlety to that of the University of California anthropologist Alfred L. Kroeber, who, not coincidentally, had received his training under Boas at Columbia.

In effect, Kroeber picked up where Boas had left off in making culture the alternative to a racial explanation. Whereas Boas's attack on race was intimately connected with his personal and ideological commitment to opportunities for blacks in American society, Kroeber's interest in the concept of culture was almost entirely theoretical and professional. Neither his private nor his public writings reflect the attention to public policy questions regarding blacks or the general question of race in American life that are so conspicuous in Boas's professional correspondence and publications. Kroeber rejected race as an analytical category as forthrightly and thoroughly as Boas, but he reached that position primarily through theory rather than ideology.

Essentially, Kroeber's contribution to the evolution of the idea of culture was to insist that culture must be totally free from any connection or dependence upon biology, something that Boas, with his experience in physical anthropology, never thought necessary. In a series of articles published between 1910 and 1917 in the principal journal of his profession, The *American Anthropologist,* Kroeber explained why that independence from biology was indispensable for understanding the meaning and use of the concept of culture, or what he preferred to call "history." The culminating essay in that formulation, "The Superorganic," published in 1917, became the classic expression of why biology had no place in anthropology. The term Kroeber took from Herbert Spencer; the meaning he gave it was his own.

Kroeber's determination to separate biology from history began with Boas's insistence on the equality of all societies in mental potentialities. Contrary to the social evolutionary school of anthropology, which had built the discipline largely around Darwinian evolution, Kroeber, along with Boas, denied that certain societies were lower than others in cultural potentiality, whatever they might be in present achievement. He began, appropriately enough, by identifying Darwin as the culprit who had confused the issue. Again and again in Darwin's writings, Kroeber complained in 1910, and quite accurately, one finds the argument that savages are halfway between animals and men. Unfortunately, he continued, that view has been widely disseminated by "appliers of the doctrine of biological evolution to ethnology, sociology, and history." But the truth is that "with all the breadth and acuity of his mind, Darwin was not an ethnologist." He may have been pre-eminent in natural history, but he revealed "little indication of any deeper understanding of human history." Darwin's distinction between savages and civilized men "cannot be rejected too insistently," Kroeber maintained. "Nothing is more erroneous than the widespread idea and oft-repeated statement that the savage is only a child."

The important thing for Kroeber was that "men are men and essentially alike wherever born and however reared." In knowledge and perhaps even in intellect the savage "may not rank above the children of civilization," he recognized, "but in character, in emotions, and in morals, he is essentially and absolutely a man." (By "morals" Kroeber meant what today would be called values, or those things which are deemed "good.") Whatever differences there are "between the morality of savages and ourselves," he explained, "is therefore not really in the morality but in the civilization," by which he meant that specific good and evil would differ from society to society, but no society would fail to identify ideas or actions that were evil or good. Those who do not recognize this, he held, "fail entirely to understand the people of whom they speak."

He was particularly insistent to establish the universality of morality or values among all human societies; it was the possession of values that set human beings apart from animals. Possessing and acting on values defined humanity. Morality, he contended, is "an inherent element of the human mind, it is psychologically unexplainable and finds its justification only in itself. As an integral constituent of man," he continued," it is common to all races in identical or virtually identical form"[13]

At that date, however, for all of his criticism of Darwin and the social evolutionists who had drawn upon Darwinian ideas, Kroeber still acknowledged a link between culture and biology. A central purpose of his 1910 article had been to show that the moral sense in human beings was, as he put it, "instinctive," something, he admitted, that derived from Darwinian evolution. "If we believe in evolution from animals," Kroeber pointed out, "we must find the source of human morality as of human sense and emotions, in animal life." Some people, he recognized, might find this origin "devoid of hopefulness and even depressing," but if the moral impulse is instinctive, as he contended, then "it is correspondingly universal and ineradicable; if inherited, it is permanent; and if incapable of racial improvement, it is equally incapable of deterioration as long as men are men."[14] In short, biology provided a reason for seeing all men as equal in morality, that is, in the capacity to make judgments about good and evil.

Morality was a characteristic of the human species, a distinguishing element of human nature, not something some human beings exhibited and others lacked. (Kroeber's identification of an animal or biological origin of morality, as will appear later in this book, would return seventy years later in a different but related form in the writings of certain sociobiologists. It appeared first, as Kroeber well recognized, in Darwin's *Descent of Man*.)

Kroeber, however, did not require a biological basis for his assumption that all human beings were equal in potentialities. As with Boas, that was the basic assumption with which he began. Unlike Boas, however, Kroeber frankly recognized that it was unproved assumption. (Boas took the more conservative or safer line: those who believed in race had made an unproved assumption.) As Kroeber wrote in 1915, "the absolute equality and identity of all human races and strains . . . has not been proved nor has it been disproved. It remains to be established, or to be limited" by continuing investigation, perhaps even by experiments. Biologists, he continued, believed that differences in heredity or race explained why some societies behaved differently from others, but "until such differences are established and exactly defined," the historian must assume their non-existence. "If he does not base his studies on this assumption," Kroeber warned, "his work becomes a vitiated mixture of history and biology."[15]

Kroeber, nonetheless, did not hesitate to draw upon recent findings in biology to achieve the separation he sought between biology and history. His earlier contention that the moral impulse was instinctive contained the relatively optimistic conclusion that if morality is "incapable of racial improvement, it is equally incapable of deterioration as long as men are men." With that phrase Kroeber revealed his awareness that the idea of acquired characters—Lamarckianism—was scientifically dead. As noted already, other social scientists about the same time had been compelled to reshape their views about race because Lamarckianism had been disproved. Kroeber did more than react; in 1916 he took August Weismann's assertion of the untenability of acquired characteristics and made that the centerpiece of his defense of the concept of culture.

The key to his case, as he pointedly phrased it, was that Weismann had proved "the Lamarckian structure . . . to be absolutely hollow. . . . His basic idea, that the hereditary idea is totally distinct from the organic body, that therefore the fate of the individual cannot affect the race," Kroeber emphasized, "has found the strongest corroboration" in the new field of Mendelian genetics. The new geneticists, Kroeber admitted, may not agree with Weismann's admiration for Darwin and natural selection, but that very difference in opinion gave additional credence to their agreement with Weismann that a wall exists between "gamete and zygote," or the unfertilized sex cell and the fertilized cell.[16]

Kroeber's point was that the wall between gamete and zygote made clear why biology could not explain the achievements or lack of achievements of any given human society. For if achievements of one generation could not be passed on to the next through inheritance, then advanced or higher races could not claim a biological basis for their standing, nor, by the same token, could so-called lower races be charged with having created their own inferiority by their lack of effort to improve themselves. Instead, each race was creating its own culture. "The accomplishments of a group, relative to other groups," he wrote in 1917, "are little or not influenced by heredity."[17]

Unfortunately, in Kroeber's view, historians and other students of human experience had been led astray by the biologists, who, ever since Darwin, had insisted upon a connection between organic and social evolution. A belief in Lamarckianism allowed, and perhaps even encouraged, them and social scientists to make just that connection. For with its emphasis upon use-inheritance, Lamarckianism made it possible to encompass social and organic evolution within a single theory. Biologists now well recognize, Kroeber added, the disproof of Lamarckianism, but they have been reluctant to act on that knowledge because that would deprive them of the opportunity to explain both organic *and* social evolution.

Even a biologist as enlightened as Weismann, Kroeber regretted to admit, failed to recognize the implications of his own disproof of acquired characteristics. Weismann was quite correct, in Kroeber's opinion, to have pointed out that Mozart's genius could not have flourished among the "Australian blackfellows," but he was quite wrong in the reasons he advanced. No Mozart grew up among the Australian primitives, Weismann explained, because of the absence of the necessary mental faculties among individual Australians. Since Weismann was a biologist, rather than an anthropologist, Kroeber pointed out, he had ignored the role of culture. To Kroeber, what was lacking among the Australians was not a person who was individually capable of becoming a Mozart, but a history or civilization that would foster or provide the cultural context within which a Mozart could emerge. In Kroeber's view any population of substantial size contained a range of individuals, among whom one or more was capable of becoming a Mozart, *providing* his social or historical environment was capable of realizing that potentiality. After all, he contended, "groups of men average substantially alike and the same in qualities." The true source of human genius, in short, was history or culture, not an appropriately talented individual, as Weismann had argued.[18]

Biologists, in Kroeber's mind, were not the only students of human evolution to have misunderstood the role of biology in accounting for human experience. "Many historians, especially sociolgists, anthropologists, and theorists," Kroeber complained, had been "tempted to imitate" biologists if only because of Darwin's success in explaining organic evolution. Lester Frank Ward, the putative father of American sociology, was a well-known example, Kroeber quite accurately remarked, because he believed that mental qualities "must be accumulated in man by acquirements and fixed by heredity," something which Weismann had shown to be erroneous. Ward's persistence in holding to that misconception, Kroeber objected, "reveals the tenacity, the insistence, with which many conscientious intellects of the day will not and can not see the social except through the glass of the organic." Other social scientists tried to show through experimental psychology how organic influences shape behavior, Kroeber, said. But there, too, the results are dubious since all "such investigations have ... revealed ... that social agencies are so tremendously influential on every one of us that it is difficult to find any test" that would reveal what was inborn or to what degree. Since, "in the present state of our knowledge," the actual role of race in accounting for human behavior is "unprovable," it is "really also not arguable."[19]

Then came his clincher, "What is possible, however, is to realize that a complete and consistent explanation can be given for so-called racial differences, on the basis of purely civilizational and non-organic causes." There is, in short, he explained, "another evolution ... in which use modification [that is, Lamarckianism] *is* permanent and transmittal of the acquired exists ... This non-organic process of evolution is that of civilization, of human accomplishment," or what we now call culture. The argument that most people believed race or biology accounted for differences between societies "carried the least weight of all" for Kroeber. It was merely another example of the objection raised against Copernicus's contention that the earth revolved around the sun when anyone could verify with his own eyes that just the reverse was true. Ethnologists are convinced, he reported, that "the overwhelming mass of historical and miscalled racial facts that are now attributed to obscure organic causes, or at most are in dispute, will ultimately be viewed by everyone as social and as best intelligible in their social relations." For the historian as well—"him who wishes to understand any sort of social phenomena—it is an unavoidable necessity today to disregard the organic as such and to deal only with the social."[20]

In sum, Kroeber was advocating more than a mere change in assumptions as Boas had done; he was insisting upon a new mode of explanation for human behavior. "Civilization and heredity are two things that operate in separate ways; ... therefore any outright substitution of one for the other in the explanation of human group phenomena is crass; and the refusal to recognize at least the possibility of an explanation of human achievement totally different from the prevailing tendency toward biological explanation is an act of illiberality." Indeed, Kroeber insisted, the difference was so profound and fundamental that "if the process of civilization seems the worth-while end of knowledge of civilization," then it must be pursued "as a process distinct from that of mechanical causality, or the result will be a reintegration that is not history."[21]

The fundamental difference between science and history in Kroeber's thought was that there were no "laws in history similar to the laws of physico-chemical science." All the so-called laws of civilization or history "are at most tendencies, which however determinable, are not permanent quantitative expressions, as are the laws of physics." History deals "with conditions . . . not with causes," he stressed.[22] Nor is history like biology. The relation between the phenomena of history is sequential, not causal in the way "the principles of mechanical causality, emanating from the underlying biological sciences, are applicable to individual and collective psychology." History, or culture, in brief, has no laws of behavior; it consists merely of behavior that shapes the behavior that follows.

Even a non-social circumstance like geography, Kroeber contended, does not shape history or culture. "Georgraphy does not act on civilization," he pointed out; it lacks causal power. On the contrary, civilization or human activity adjusts to geographic circumstances, and may even alter them. As a concrete explication Krober instanced farming. "Agriculture," he observed, "presupposes a climate able to sustain agriculture and modifies itself according to climatic conditions. It is not caused by climate"; human beings instead reshape their farming habits to fit the climate.[23]

As one might anticipate from the foregoing, Kroeber specifically denied that any Darwinian influences such as natural selection affected culture. Civilization introduces a factor, he insisted, "practically or entirely lacking in the existence of animals and plants"; consequently it is untouched by natural selection. "Prehistoric archeology shows with certainty," he added, "that civilization has changed profoundly without accompanying material alterations in the human organism." Animals adjust to environmental change through physical alterations—that is, through natural selection—but human beings do so through culture, or history. "The distinction between animal and man which counts," he emphasized, "is not that of the physical or mental," for they are matters of degree, "but that of the organic and social, which is one of kind. The beast has mentality, we have bodies," he noted; "but in civilization man has something that no animal has." That was the central element in human nature. Unlike the situation with animals, the direction and movement of history "involves the absolute conditioning of historical events by other historical events." The process that is history "is as completely unknown and unused," he was convinced, "as chemical causality was a thousand and physical causality three thousand years ago." Since the sources of animal and human behavior differ in kind, it follows that "the dawn of the social . . . is not a link in any chain, not a step in a path, but a leap to another plane." Its accomplishment Kroeber likened "to the first occurrence of life in the hitherto lifeless universe," an event that decreed "from this moment on there should be two worlds in place of one." For Kroeber, as George Stocking has pointed out, man was not so much a rational as a *rationalizing* being.[24]

The human sciences in Kroeber's view differed from both art and science. "If scientific methods give science, and artistic exercise yields literature," Kroeber suggested, then "it is at least conceivable that there may be a third activity, neither science nor art . . . but history, the understanding of the social . . . whose justification must be sought in its own results and not by the standard of any other activity." As one anthropologist was later to point out, Kroeber was the first American social scientist to assert "the complete disparity of biological and cultural evolution."[25]

Kroeber's Superorganic or culture was not only distinct from biology, it was separate from individual human beings as well. As members of a group or society human beings had created it, but as individuals they neither exerted influence over it nor were shaped by it. Kroeber was fully aware that in identifying this dichotomy his conception of history's relation to the individual differed sharply from that of his contemporaries. Their conception he described as resting on "the endlessly recurring, but obviously illogical assumption that because without individuals civilization could not exist, civilization therefore is only a sum total of the psychic operation of a mass of individuals. His own view was that "a thousand individuals do not make a society. They are the potential basis of a society; but they do not themselves cause it." It is true, he conceded, that mental activity, like motor behavior, depends on the organic in human beings, and therefore is derived from heredity or biology. But the *product* of that organically based "mental activity," he argued, has little to do with "civilization in that civilization is not mental action but a body or stream of products of mental exercise. . . . Mentality relates to the individual. The social or cultural, on the other hand, is in its essence non-individual. Civilization as such," he emphasized "begins only where the individuals ends. . . ."[26] Or put another way: culture is the social interaction of individuals, contemporaneous and across generations integrated by language. Social groups, whether they are what today are called races, or in Kroeber's time were called racial types, are not biological units at all, but socially integrated aggregates, the cultural or intellectual products of which are the sole result of human perception and experience. Biology or heredity played no direct part in their production, and therefore has no part in explaining a social group's actions or culture.

Kroeber's conception of the relation between the individual and culture is strikingly similar to one advanced much earlier by the French sociologist Emile Durkheim. "The group thinks, feels, and acts quite differently from the way in which members would were they isolated," Durkheim asserted. "If we begin with the individual in seeking to explain phenomena, we shall be able to understand nothing of what takes place in the group." Or in another place he wrote: "We must . . . seek the explanation of social life in the nature of society itself." Nor can a group of individuals be seen as a society. "A whole," Durkheim contended, "is not identical with the sum of its parts. It is something different, and its properties differ from those of its component parts."[27]

Durkheim anticipated another basic concept of Kroeber. Compare Kroeber's principle that "the only antecedents of historical phenomena are historical phenomena" with Durkheim's classic formulation: "The determining cause of a social fact should be sought among the social facts preceding it."[29]*

Durkheim's book from which those sentences are taken appeared in 1895, twenty years before Kroeber arrived at the same position. Since no reference to Durkheim appears in Kroeber's early writings in which his position is set forth, the relation between the ideas of the two men is at best unclear.** Given Kroeber's commitment to that position, it seems likely that, if he had derived his conclusion from Durkheim, he would have acknowledged rather than concealed that association since to do so would offer support for his own position. After all, he was a newcomer to social science; Durkheim was a leading figure. Perhaps nothing more is involved than the familiar story of a member of one discipline not knowing what a member of another is thinking (or writing), especially when the authors in question are from different countries.

The hard line that Kroeber drew between civilization and biology was so absolute and novel that even some of his fellow Boasians found his concept difficult to entertain. His friend Edward Sapir, the linguistic anthropologist and Boas student, for example, almost immediately after the publication of "The Superorganic" repudiated Kroeber's rigid differentiation between natural sci-

---

*Even more striking is that in elucidating his meaning of the phrase "the cause of a social fact should be sought among social facts," Durkheim was, like Kroeber, dealing with race. No social phenomenon "is known which can be placed in indisputable dependence on race," he wrote. "The most diverse forms of organization are found in societies of the same race, while striking similarities are observed between societies of different races. . . . If these things are true, it is because the psychological fact is too general to predetermine the course of social phenomena. Since it does not call for one social form rather than another, it cannot explain any of them." Emile Durkheim, *The Rules of Sociological Method* (8th ed., Chicago: Univer. of Chicago Press, 1938), 108.

**In 1925, Kroeber's friend and fellow Boasian, Alexander Goldenweiser, noted the similarity between Kroeber's position in "The Superorganic" and what Goldenweiser referred to as Durkheim's "institutionalism," but offerd no clue as to the relation between the two men who shared a similar idea. See Alexander Goldenweiser, "Cultural Anthropology," in Harry Elmer Barnes, *The History and Prospects of the Social Sciences* (New York: Aflred A. Knopf, 1925), 249n.

ence and social science. Sapir agreed that many sciences conceptualized and history did not, but there were sciences like geology, for example, that were as particularistic as history. Geology, in fact, he noted, "is a species of history, only the history moves entirely in the organic sphere." Geology, like the social sciences, "values the unique or individual, not the universal."[30] (Sapir might have added that Darwin's proof of his theory of evolution by natural selection was also historical in method rather than scientific in the way physics or chemistry are.)

Sapir also took Kroeber to task for denying any role for the individual, contending that certain individuals at least—he instanced Napoleon—clearly made a difference in the evolution of European history and culture. When Alexander Goldenweiser, another fellow student of Boas, objected to Kroeber's removal of individual influence from the development of culture, Kroeber attempted to explain his purpose. Only by removing the individual from a role in history, Kroeber wrote Goldenweiser privately, can we begin to understand history. If we begin with biography, or the individual, as Sapir and Goldenweiser seemed to want, Kroeber argued, then we will never be able to differentiate between "super-individual culture" and that which derives from the individual. Only after we have distilled a "pure history, a science of super-individual culture," he explained, can we "profitably begin to connect its findings with the findings of any study that brings in the individual." If we include biography or the individual, "we never emerge from that circle and never to what I call the one pure culture history." In short, as he wrote Goldenweiser, his aim was methodological, not substantive.[31] Kroeber's intention may have been methodological at the time, but throughout the remainder of his distinguished career he continued to write and teach what he called "culture history" in which the individual played little or no role.

Many years later, Kroeber admitted that, despite the character and language of his essay "The Superorganic," the biologists had never constituted a threat to social science or history. The threat against which he was reacting and against which the essay was directed was the arguments of those social scientists like Lester Frank Ward and Herbert Spencer, who insisted on combining culture and biology. "What the essay really protests," Kroeber wrote in 1952, "is the blind and bland shuttling back and forth between the equivocal 'race' and an equivocal 'civilization.'"[32]

Throughout Kroeber's writings and efforts in those early years, his professional purposes clearly dominate. The aim was to arrive at a sharply focused definition of the nature and methods of the social sciences, one that would set the study of human life apart from science in general and biology in particular. Therefore, he called upon social scientists "to press this great truth at every opening, and every turn, to brand each error and confusion as fast as it raises its head, to stigmatize all half-hearted evasion, to meet argument with argument, and if necessary, assumption and assertion with counter assumption and assertion."[33]

Even as he was asserting, in almost fanatical language, the independence of history from biology and the professional autonomy of the emerging social sciences, Kroeber still looked to biology in accounting for *individual* development. He was quite prepared, for example, to accept Francis Galton's conclusions regarding the inheritance of intelligence first set forth in *Hereditary Genius* in 1869. Galton's showing that mental qualities were just as likely to be inherited as physical traits, Kroeber found "reasonable as well as convincing." An irascible temper and musical aptitude, he thought, were probably as inheritable as blue eyes and red hair. There is no reason to rule out the real possibility, he concluded, "that characters of mind are subject to heredity much like traits of the body." He also accepted the implications of a report by the experimental psychologist Edward Thorndike that the future of an individual's life was "settled when the parental germ cells unite, and [is] already long closed when the child emerges from the womb." Or, as he rephrased the point, "nothing is further from the path of a just prosecution of the understanding of history" than to deny "differences of degree of the faculties of individual men."[34] Kroeber, in short, like Boas before him, was no believer in the infinite or even the equal potentialities of individual human beings.

Where Kroeber departed profoundly from both Galton and Thorndike, both of whom were committed eugenicists, was with their belief that altering the heredity of individuals through eugenics could improve society. Heredity might explain almost all the differences between the accomplishments of one individual as against another, Kroeber recognized, but "the accomplishments of a group, relative to other groups, are little or not influenced by heredity because sufficiently large groups average much alike in organic makeup." And that is why civilization or history was not shaped by great men or women; they, rather, as in his example of Mozart, were merely products of society. "The difference between the accomplishments of one group of men and those of another group is therefore of another order from the differences between the faculties of one person and another," he maintained. "It is through this distinction," he stressed once again, "that one of the essential qualities of the nature of the social is to be found."[35] The superorganic or culture was above and beyond any individual.

By the same reasoning Kroeber maintained that the individual, in turn, was uninfluenced by culture. "Because culture rests on the specific [i.e., individuals] human faculty," he reinterated, "it does not follow that this faculty . . . is of social determination." Social circumstances, he conceded, may expand or limit individual opportunities, but that "does not prove that the individual is wholly the product of circumstances outside himself, any more than the opposite is true that a civilization is only the sum total of the products of a group of organically shaped minds." To believe that the individual is "the result of his moulding by the society that encompasses him," Kroeber warned, is an assumption, and an extreme one at that, and quite at variance with observation. After all, individual human beings, like animals, are organic and to that extent are beyond cultural influence. Kroeber, it is clear, was no environmental determinist. Modern conception of socialization, which perceive individuals as products of culture alone would not have been any more acceptable to Kroeber than to Boas.[36]

And just because Kroeber did see an important role for biology in individual development, he did not rule out an eventual and legitimate place for biology in accounting for the development of society. Once biology and history have each worked out "its independent destiny justly and intelligently," he predicted in 1916, it is quite possible that the future will give them "new contacts and new bonds."[37]

As this book seeks to show, Kroeber's vague prediction of a place for biology in accounting for human social behavior did eventually come about. But, as he also predicted, that did not occur until the idea of culture had been developed to its fullest meaning, a meaning in which biology was allowed to play no role at all. That fulfillment was given popular exposition and expression by another Boas student, Robert Lowie. In the same year that Kroeber's "The Superorganic" appeared, Lowie published his public lectures, "Culture and Ethnology," which he had delivered to a lay audience at the American Museum of Natural History in New York City.

In the very first lecture Lowie drew a sharp distinction between culture and psychology, because the latter, he stressed, "deals on principle exclusively with *innate* traits of the individual." Therefore it cannot explain those aspects of culture that are peculiar to a people, as opposed to all human beings. It is true, he acknowledged, that differences in sense perception between societies have been identified, "but up to date we can simply say that experimental psychological methods have revealed no far-reaching differences in the mental processes of the several races." Rather, what strikes the ethnologist is the diversity of cultures among people of the same race or at different times in their history. Even if it should be shown sometime in the future that some races are not quite equal to others in potentialities, Lowie contended, "such information could not solve the . . . specific problem of why the same people a few hundred years earlier were a horde of barbarians and a few hundred years later formed a highly civilized community." For those reasons, he reminded his audience, his insistence on the autonomy of culture is not simply "adolescent self-assertiveness."[38]

Biology and psychology, Lowie contended, cannot explain the phenomena that ethnologists study. "Psychological laws no more account for cultural phenomena than the law of gravitation accounts for evolution." The customs and arts of a society may "appear trivial to the biologist," he rather defensively noted, "but they are, for good or ill, the subject of ethnology." Then, continuing to follow Kroeber, he posed the fundamental choice: recognize the need for new methods of inquiry into human nature or let the whole subject be abandoned. If the natural sciences can explain these social phenomena, he boldly proclaimed, then "ethnology is superfluous," but if they cannot, then "new methods are required, in which case ethnology is indispensable."[39]

If psychology or race cannot account for the differences among people, then the "inference is obvious. Culture is a thing *sui generis*, which can be explained only in terms of itself. This is not mysticism," he assured his listeners, "but sound scientific method." Deliberately, he drew an analogy from biology to capture the uniqueness of culture and its method. Just as the biologist insists that every cell is derived from another cell, so the "ethnologist will do well to postulate the principle, *omnia cultura ex cultura,*" which meant, he explained, that the ethnologist will "account for a given cultural fact by merging it in a group of cultural facts or by demonstrating some other cultural fact out of which it has developed."[40]

With Kroeber and Lowie, the revolution in thought about race and human nature toward which Boas had been working for over a quarter of a century had reached its maturity. Biology or race had been swept away as explanations for differences in the social behavior of peoples. All human societies exhibited a common nature; whatever differences there were between one people and another derived from culture or history. All people were capable of learning or adopting another culture, no matter how complex or "high" it might be. In the name of a special, even unique science of human experience, the anthropologists had issued a ringing declaration of independence from race, biology, or any science that depended upon universal laws or principles. Human society, whether called civilization, history, or culture, was unique; it stood apart from the remainder of nature. It study, therefore, required methods different from those employed in the study of nature.

The uniqueness of culture received a striking exposition in anthropologist Berthold Laufer's review of Lowie's *Culture and Ethnology* in 1918. Laufer's words nicely epitomized the triumph of the concept of culture among anthropologists. "Culture is a thing *sui generis* which can be explained only in terms of itself," Laufer summarized. In contrast, he pronounced "the theory of cultural evolution—the old concept that societies evolved through stages—to be the "most inane, sterile, and pernicious theory ever conceived in the history of science (a cheap toy for the amusement of big children)." Culture, on the other hand, "cannot be forced into the straitjacket of any theory whatever it might be, nor can it be reduced to chemical or mathematical formulas. As nature has no laws, so culture has none," Laufer insisted. "it is as vast and as free as the ocean, throwing its waves and currents in all directions. . . . Our present and our future lie in our past," he concluded with a flourish.[41]

By the time Laufer's review appeared, many sociologists, too, had adopted the concept that had reshaped the thinking of anthropologists about the role of race and biology in human affairs. One sociologists, for example, writing in 1917, deplored racial discrimination against Afro-Americans in the South on the ground that it was possible "to affirm that among scholars competent to render an authoritative judgement, the ancient doctrine that some races are by nature inferior has been rejected. Every argument advanced in its support has been tested and found wanting," he maintained. The evidence seemed so convincing to him that the burden of proof now rested on those who denied its truth. "Every year brings stronger support for the new doctrine of the potential equality of all races," he declared. Echoing the anthropologists, he contended that "peoples differ in their planes of cultural development, not in their inherent capacity for development." He even went so far as to bring the proposition directly home, for he found "no sound reasons for believing that the negro does not share in this equal potential chance for civilization."[42]

That same year, sociologist Edward B. Reuter, soon to be recognized as a major authority on race questions, also testified to the triumph of the cultural interpretation of race. It would be "a simple matter," Reuter confidently asserted, "to multiply authorities who hold that in inherent capacity there is an essential mental equality among races and that whatever differences are manifested are explainable solely on the grounds of unequal opportunity." The following year, a leading sociologist, Ellsworth Faris, agreed that the works of Boas and William I. Thomas, among others, had discredited, once and for all, the idea that there were any real or important distinctions between the civilized and the savage mind. "Instead of the concept of different . . . degrees of mentality, we find it easier," he said, speaking for his professional colleagues, "to think of the human mind as being, in its capacity, about the same everywhere."[43]

Unlike Reuter, however, Faris was also prepared to acknowledge the tentative character of the Boasian conception of the equality of mental potentialities among human groups. The whole concept, Faris reminded his audience, as Kroeber himself had said earlier, "is only a hypothesis. It has not been proved."

He observed that it is possible, for example, that behavioral differences between peoples might still be explained by differences in sizes of the brain. Only insufficient or inadequate tests and measurements have been made so far, Faris observed. The most recent he could cite were those completed at the St. Louis Exposition in 1904, and they were narrowly conceived and dubious in their import. He suggested, therefore, that an expedition of trained professionals travel to Africa, armed with the new knowledge of physical and mental measurements and familarity with African languages, to administer tests to a 1000 or 1500 "properly distributed individuals." Such a scheme had been planned in 1914, he recalled, but it had been interrupted by the war. With the return of peace, he hoped that the plan would be revived. "If so," he concluded, "it will be possible to write with more certainty concerning the mind of primitive man."[44]

In Faris's doubts are exposed once again the ideological roots of the triumph of the Boasian substitution of culture for race. For the St. Louis experiments of 1904 had been about the only experimental evidence that Boas had before him when he published *The Mind of Primitive Man*. Nor was any further information available along the specific and controlled lines Faris envisioned. Yet the anti-racist interpretation continued to win converts. In the absence of proof—as Boas and Lowie phrased it—a racial explanation for human differences was unacceptable. Instead, culture or the human environment became the primary source of human social behavior. As the sociologist Charles W. Ellwood emphasized in 1918, instinct could not account for civilization. Using the word "culture" in the traditional, rather than in the newly emergent anthropological sense, he noted that "it has been always the learned activities rather than the non-learned activities which have counted for the development of culture among all people." Human beings may have instincts, like the other animals, he conceded, but instincts *"could not give rise to a new type of social life and of social evolution."* In his concluding remarks he revealed his ideological interest in the new approach to human life. "If we follow the clue which modern anthropologists are giving us," he observed with obvious approval, "we shall reach a 'human'—one might almost say a 'humanistic'—sociology rather than the biological or mechanistic one which obtains among some social thinkers." Four years later Ellwood was even more convinced of the virtues of the new view of the human sciences and the promise it held for a better world. "This view is in accord with the general position of social science," he confidently announced; "for modern sociology and modern anthropology are one in saying that the substance of culture, or civilization, is social tradition and that this social tradition is indefinitely modifiable by further learning on the part of men for happier and better ways of living together. . . . Thus the scientific study of institutions awakens faith in the possibility of remaking both human nature and human social life."[45]

With the acceptance of the concept of culture the idea of a hierarchy of human societies based on innate differences was no longer tenable; human nature was now a unity, however diverse its expression may have been in the myriad of cultures around the world. Yet a physiological division within humanity still remained, one that was as eye-catching as any provided by race or color. It, too, had spawned assertions of fundamental differences in character and quality. The divider was sex. Darwin himself had called attention to the differences that seemed to flow from it, and many had followed his lead. By the end of the nineteenth century, however, some American social scientists were beginning to find reasons to doubt that the division carried serious meaning for human nature. The practical fruits of those doubts are the subject of the next chapter.

5

Does Sex Tell Us Anything?

Few, if any, of the so-called "sex differences" are due solely to sex. Individual differences often are greater than differences determined on the basis of sex. . . . The social training of the two sexes is and always has been, different, producing differential selective factors, interests, standards, etc.

Psychologist Chauncey N. Allen, 1927

As some social scientists at the turn of the century remarked, race and sex had much in common as biological signs of different natures. Indeed, in making the point the social scientists often felt they were identifying the obvious. After all, what white American could ignore the different physiognomy of the African-American, or the anatomical differences between males and females? One might try to overlook them, to be sure, but neither history nor the social climate of opinion at the end of the nineteenth century suggested that they ever had been ignored, or even that they ought to be. Throughout the nineteenth century, books, articles, and lectures proclaimed the inborn differences between sexes as frequently as they delineated differences between races. And just because of that similarity, social scientists' intellectual shift from seeing sex as a determinant of behavior repeated some of the steps taken in the abandonment of racial explanations. Yet the differences in the social perception of sex and race—that is, of women and blacks—made it inevitable that the refutation of sexism would not be simply a rerun of racism's extinction.

For one thing, the question of the nature of woman, unlike a similar inquiry regarding blacks, was complicated by greater familiarity and greater complexity. Most male social scientists, like the public at large, had close and enduring contact with women and usually almost none with blacks. How one perceived blacks or Indians, or immigrants or women, for that matter, depended on one's relation to them, for that shaped one's expectations. As early twentieth-century social scientists rightly pointed out, for a dominant group to classify or identify a subordinate group, whether a sex or a race, was, ipso facto, to exert some control over it. The very idea of definition or classification set limits to acceptable behavior or, more generally, the nature or character of the group. Certainly that was what the idea of race imposed upon blacks; the same was accomplished for women by sex. The purpose of social control, of course, is to enhance the power of the dominant group, and that, too, occurred in regard to both blacks and women. For blacks the dominant group was white men and women; for women, it was males. Here, however, the similarity ended with important differentiating consequences for the ways by which an implicitly biological explanation was called into question.

Males and females, after all, were intimately associated as husbands and wives, mothers and sons, fathers and daughters, brothers and sisters. And just because of the familiarity and diverse social contexts those relations inevitably provided, the traditional explanations for differences between the sexes might well be tested and found wanting. The absence of any comparable association between blacks and whites, except within the subordinated roles to which blacks were confined, provided few or no comparable opportunities for a member of the dominant group to question the traditional definition. Here daily observation and racial explanations reinforced rather than contradicted one another. As Franz Boas frequently pointed out in his effort to overturn racial thinking, it was necessary that whites see blacks in circumstances that were not traditional; hence his numerous efforts to bring African history and ethnology before the white public. (As we shall see in a later chapter, when blacks were seen in a diversified social context the racial perceptions of white social scientists did indeed begin to alter.)

Opportunities to see "the other" in a fresh context were always present for men and women. A man might well call into question a traditional conception of woman because his observations of his sister, wife, or mother contradicted it. Fathers could recognize in their daughters a sharp deviation from the traditional conception of female, and have a personal interest in encouraging it as well. Similarly, a son might learn from his mother, or a husband from his wife, that women were more varied and capable than the traditional stereotypes portrayed them. Rosalind Rosenberg in her excellent study of early twentieth-century women social scientists has illustrated this possibility. She pointed out that one reason for the open-mindedness toward women exhibited by John Dewey, George Herbert Mead, and James R. Angell of the University of Chicago's psychology department was that each of them was married to an activist and well-educated wife.

Nevertheless, simply because all human relations involve power to some degree, the mere close or even intimate association of males and females would not of itself automatically redefine the standard definition of woman that biology was said to have decreed: that her place was in the home and nursery. Too many men, like too many whites, profited from a narrow and subordinating stereotyping of "the other." Nor did the intimate association between men and women necessarily weaken the conviction that woman was designed by nature for a function only she could perform. In an age when control over fertility was unreliable at best, and immoral at worst, that conviction was highly resistant to change. In such circumstances, it was more than plausible that biology might well play a role in the lives and future of women that was indeed controlling. After all, it was possible to believe that color of skin, or kind of hair, or general appearance was superficial and thus not important in shaping behavior. But if one believed there was a connection between a natural function and behavior, the specially designed body of woman could easily be perceived as shaping her outlook and behavior.

Certainly that was the way Charles Darwin saw the matter. Indeed, he had built his whole theory of natural selection on the assumption that body shape and behavior went hand in hand. More than that, he had even devised a theory—sexual selection—which related the female's particular part in reproduction to the evolution of the bodily shape of the male. It is not surprising, therefore, that he saw a connection between the reproductive function of human females and their character. Indeed, at one point in *The Descent of Man*, as we saw earlier, he specifically differed from his few liberal contemporaries who denied that the human sexes differed in their mental powers. Since sexual differences were observable in the lower animals he thought it probable that they existed among human beings as well. No one doubts that sows differ in disposition from boars, bulls from cows, stallions from mares, he pointed out. "Woman seems to differ from man in mental disposition, chiefly in her greater tenderness and less selfishness," traits also observable in savages, he contended. Men are competitive and ambitious and therefore selfish. "These latter qualities," he ruefully noted, "seem to be his natural and unfortunate birthright. The chief distinction in the intellectual powers of the two sexes," he continued, "is shewn by man's attaining to a higher eminence, in whatever he takes up, than can woman—whether requiring deep thought, reason, or imagination, or merely the uses of the senses and the hands."[1]

The idea that the sexual or biological function of woman shaped her mind and behavior was socially epitomized in the nineteenth-century idea of "separate spheres" for men and women. Women's sphere was the home and children; men's was the world of work and power. Within the home woman was supreme; she was the inculcator of morals and values to children and husband. Inasmuch as she occupied so important a role, it was appropriate for her to be educated. Thus the doctrine of separate spheres, ironically enough, opened a path which, in time, would lead to the undermining of the doctrine. For if we want to understand how a biological basis for differences between men and women was brought into serious doubt, the effects flowing from the entrance of women into higher education can hardly be exaggerated.

Until late in the nineteenth century, the question of woman's ability to profit from or even survive higher education was controversial. Vassar College, the first of the major women's colleges, when it opened in 1865, was often described by its first president, John Raymond, as "an experiment." And as late as 1871, Dr. Edward H. Clarke of Harvard Medical School published a widely read book on the dangers of higher education for girls. He did not question their ability to do collegiate work; he merely doubted that the achievement was worth the cost exacted from their psyche and procreative powers.

One consequence of permitting young women into institutions of higher learning was that Clarke's fears and those of others like him were soon shown to be erroneous. As the president of the University of Michigan, James B. Angell, noted in 1881, "the solicitude concerning the health of the women" at Michigan "has not proved well-founded. On the contrary," Angell remarked, "I am convinced that a young woman, coming here in fair health . . . is quite as likely to be in good health at the time of her graduation as she would be if she had remained home." And for those who doubted the intellectual abilities of women, Angell's answer was equally positive. "There is no branch of study pursued in any of our schools in which some women have not done superior work. It was found," he added, "that in those studies which are thought to make the most strenuous demand of the intellect, some of the women took equal rank with the men. They have desired and have received no favors," he emphasized.[2]

In universities and colleges across the country, it soon became evident that women were often doing better than men, capturing more prizes and admissions to honor societies. Stanford University, for example, became so concerned about the possibility of being overwhelmed by female students because of their high intellectual quality that Mrs. Jane Stanford, one of the founders, ordered the number of young women in attendance to be limited to no more than five hundred.

Undoubtedly the most significant consequence of this increasing participation of women in higher education was that the subjects of the "woman question" were themselves becoming the investigating scholars. That was another striking contrast to the situation that prevailed in the comparable debate over the reality and impact of race. As historian Rosalind Rosenberg has shown, the presence of women within the ranks of the emerging social sciences, especially psychology, changed or catalyzed thinking on the subject in two ways. Male professors and male graduate students observed these "new women" in their psychological laboratories and in their classes, competed with them in studies, and learned at first hand from them about women's capabilities and nature. Even more pertinent in accounting for the intensification of the criticism of the traditional view of woman's nature was that women social scientists themselves initiated or carried out investigations of the assumptions upon which the conventional view depended.

A striking measure of the influence of Darwinian evolution and of women's own role in raising objections to assertions of women's inferiority was that some of them were constructed by women from *biological* concepts, particularly from Darwinian evolutionary ideas. One of the earliest defenses of women's abilities was written by Eliza Burt Gamble, a professional writer and feminist. The very title of her book—*The Evolution of Woman*—suggested her debt to Darwin, while the subtitle emphatically revealed the uses she intended to make of that debt: *An Inquiry into the Dogma of Her Inferiority to Man.* After reading Darwin's *Descent of Man* she "became impressed with the belief that the theory of evolution . . . furnishes much evidence going to show that the female among all the orders of life, man included, represents a higher stage of development than the male."

In the book she recounted the many examples from nature which demonstrated that "the female is the primary unit of creation, and that the male functions are simply supplemental or complementary." Moreover, Darwin's concept of sexual selection, she contended, demonstrated that it was the female who determined the shape and actions of the male. In species after species, she insisted, "the female represents a higher development than the male." The human male may be larger than the female, she conceded, but "he is still shorter lived, has less endurance, is more predisposed to organic diseases, and is more given to reversion to former types, facts which show that his greater size is not the result of higher development." Darwin himself, she added, "has proved by seemingly well established facts that the female organization is freer from imperfections than the male, and therefore that it is less liable to derangements."[3]

The establishment of the biological or physical superiority of woman was not Gamble's principal aim. Her primary goal was to show that morally, as well as physically, woman was superior to man. In effect, she stood Darwin on his head by noting that the very qualities that Darwin assigned to woman, and which he implied identified her as inferior, were really the source of social bonding, that force so essential for human society. "Motherhood was the primary bond by which society was bound together," she declared. Darwin failed to recognize this fact, she regretted to have to point out, but the logic of his argument nevertheless shows that "the maternal instinct is the root whence sympathy has sprung and that is the source whence the cohesive quality in the tribe origi-

nated." The qualities of perception and intuition, which are usually attributed
to woman, Gamble thought, will make woman equal in civilizing power to man
once she has equal opportunity to express them.

Paternal affection, on the other hand, she noted, is unknown among lower
animals, so in order for a human social and moral order to emerge, feelings of
sympathy and affection must develop. Only in very recent times, as evolution
goes, have kinship and rights of succession been through fathers, she observed.
For most of human history, those fundamental characteristics of a true human
social order were "reckoned through the mother. In other words," she concluded, "motherhood was the primary bond by which society was bound
together."[4]

Gamble was not the only feminist thinker to draw upon biology in general
and Darwinian evolution in particular to assert the equality of woman with
man. As early as 1883, Lester Frank Ward, soon to become a prominent figure
in the emerging discipline of sociology, advanced what in later publications he
would call a "gynecocentric" conception of human evolution.[5] He developed
the idea at length in 1903 in his book *Pure Sociology*. Like Gamble, Ward began
by noting the "primary" character of the female in nature. Males emerged later,
when nature discovered, as it were, the advantage of "organic progress through
the crossing of strains," by which he meant sexual reproduction. As a convinced
proponent of the positive role of biology in human advancement, Ward recognized the advantage of sex in reproduction while simultaneously noting the
limited role that males played in continuing the species. Animals increased their
variation and thereby enhanced their ability to adjust to environmental changes
when two parents, as opposed to one, contributed to an offspring's heredity.
Males, however, contributed only sperm, while the female was burdened with
the work of bearing and rearing offspring.[6]

Ward recognized, as Gamble had not, that the designation of the female as
primary in nature left something to be explained. How did it come about, he
asked, that the "male, primarily and normally an inconspicuous and insignificant afterthought of nature, has in most existing organisms attained a higher
stage of development" than the female? What is most surprising, he continued,
is not that the female "is usually superior to the male, but that the male should
have at all advanced beyond its primitive estate" as a mere fertilizing agent. The
principle that explains it, he answered, is the male's insistence on reproduction,
regardless of risk or even death, of "perishing at his post" if necessary. That
very insistence provided the force behind the Darwinian principle of sexual
selection, Ward emphasized. Darwin, Ward agreed, had made "an unanswera-
ble case in favor of his principle of sexual selection." In the course of it, the
female decides, for the sake of her offspring, or the species, to mate with the

best males. Thus she chooses the mate who is stronger and bigger than the others, as well as those with moral attributes like courage and beauty.[7]

Contrary to what some Darwinians might conclude from the functioning of sexual selection, Ward saw in it no justification for woman's subordination to man. Even when females had created through sexual selection males larger than themselves, they still exercised their power of choice. Ward heaped scorn on those who dared to suggest that the larger size of the male in nature resulted from his obligation to feed and protect the female and her young. That function, he stressed, is the female's. "It is she that has the real courage, courage to attack the enemies of the species. Many wild animals will flee from man," he noted, "the only exception being the female with her young. She alone is dangerous." Just approach some chicks, he advised his readers, and see how the mother hen reacts. "The cock is never with her. His business is with other hens that have no chicks to distract their attention from him."[8]

For Ward the feminist, the functioning of sexual selection in the prehistory of human beings was filled with irony. By selecting men with larger bodies and bigger brains, women provided the weapons that man, with "his egoistic reason, unfettered by any such sentiment as sympathy, and therefore wholly devoid of moral conceptions of any kind," employed to exact "from woman whatever satisfaction she could yield him. The first blow that he struck in this direction," Ward sadly recounted, "wrought the whole transformation." Women were deprived of choice, and though that was not accomplished all at once, "still it was accomplished very early, and for the mother of mankind all was lost." Because "woman was smaller, weaker, less shrewd and cunning than [man], and at the same time could be made to contribute to his pleasure and his wants . . . he proceeded to appropriate her accordingly."[9]

Force alone, Ward contended, not mental or moral inferiority, was the source of woman's subordination. Only because of that complete dominance of woman by man has the real truth—that woman is "really the superior sex"—been perceived as incredible. Only those who understand the new biology "are capable of realizing" the true nature of woman and the necessity, as well as the desirability, of bringing about a new "gynandrocentric" stage of human experience in which "both man and woman shall be free to rule themselves."[10]

The feminist and social thinker Charlotte Perkins Gilman was so impressed by Ward's use of Darwinian evolution to account for the historical subordination of women that she based upon his arguments much of her writing in support of women's freedom to participate in the world of work. Thus she looked to biology to assert the equality of the female. Her best-known book, *Women and Economics* (1898), is filled with analogies to animal behavior by which she exposed the illogicalities of the subordination of woman to man. The subtitle of another book of hers, *The Man-Made World; Or Our Androcentric Culture* (1911), is obviously dependent upon Wardian terminology as well as Wardian theory. Along with Gamble and Ward, Gilman accepted certain qualities as peculiarly female, though only rarely characterizing any of them as less valued than those she associated with males. ("Art, in the extreme sense," she wrote in *The Man-Made World*, "will perhaps always belong most to men. It would seem as if that ceaseless urge to expression was, at least originally, most congenial to the male."[11]) Gilman, again in agreement with Gamble and Ward, stressed the social and moral qualities of women that made civilization possible and which, in the modern world, were needed more than ever now that brute strength and personal power no longer counted as heavily as in the past. Gilman, in short, saw Darwinian evolution as a positive force that would advance women's freedom.

Those qualities peculiar to women, like their cooperative and unaggressive nature, were also emphasized by Lydia Commander, writing in a sociological professional journal in 1909. Following in the steps of Gilman, she justified women's participation in the world of work on the ground that modern society needed the qualities peculiar to women. "Woman is the working human creature," Commander asserted. "To work is an inherent tendency of woman's nature; with man it is an acquired characteristic. Woman works from instinct; man from habit," she concluded. Because of his combative nature, Commander pointed out, man works as he fights—competitively—while woman works cooperatively, and that is what the modern industrial world sorely needs. Woman's obligation is "to change the basis of industry from war to cooperation, to people before property, and life before labor."[12]

The theories and arguments of Ward, Gamble, Gilman, and Commander make clear that a belief in the role of biology in shaping human behavior was not necessarily socially conservative. All of these feminists were clearly interested in changing the world in which they lived. Social Darwinism, to be sure, was a defense of the social status quo, but the linkage it forged between biology and society was neither inherent in the evolutionary thought of the time nor prominent in the thinking of social scientists, most of whom, as we have seen, exhibited a strong reformist outlook regarding the social order.

Social Darwinists might rephrase Darwinism to suit their defense of the present, but reformers could not help but find Darwinian evolution congenial since it clearly proclaimed the ubiquity and persistence of change. The natural world in Darwin's depiction was in steady flux; what was true today was surely to alter as time passed, just as had occurred in the past. Moreover, when Darwin described a world in which organisms adapted to changing circumstances, he provided a powerful justification for social reform, which asserted nothing more than the need for society to recognize that its order, like that of nature, changed and required adjustment. If conservative social Darwinists interpreted Darwin's message to be that those who are today on top arrived there because they had defeated their less competent rivals, to reformers Darwin's message was that no one stayed on top because change and adjustment were the order of nature. That outlook gained its earliest and perhaps strongest expression at the new University of Chicago.

Among the social scientists at the University of Chicago in its opening years were John Dewey, James Angell, William I. Thomas, and George Herbert Mead, all of whom were Darwinians of the reformist breed. Their Darwinian outlook encouraged them to view the woman question with an openness that was rare in academia. Dewey and Thomas had also begun to appreciate what Ward and Gamble had identified as the cooperative and caring nature of women. They saw in women's nature an alternative to the competitive and aggressive traits usually associated with men, traits which, in the new age of cities and factories, seemed less socially useful. For these and other reasons, a number of young "new" women of the time found in the University of Chicago a more congenial atmosphere than what then prevailed at the more traditional and certainly older universities like Harvard or Yale. Columbia, with Franz Boas in anthropology, provided a similarly encouraging ambience for women. Indeed, probably more women anthropologists were trained at Columbia under Boas during the first twenty years of the new century than at any other university in the country.

Until the 1920s, however, the leaders of the attack on the question of woman's nature did not come from anthropology. They originated in the new women's colleges, and in the University of Chicago, especially in its departments of psychology and sociology. Published criticisms of the traditionalist view appeared even before the century opened. In 1895, a young woman student of psychology at Wellesley College criticized in a scholarly journal a study made by a leading male psychologist in which he had experimentally established women's limited mental and attitudinal outlook. His research consisted of comparing word associations by male and female college students at the University of Wisconsin. The conclusion Joseph Jastrow reached in his research was that women tended to mention words that were drawn from concrete things, immediate surroundings, decorations, and individuals; the words on men's lists referred to remote, useful, general, and abstract things. Thus, Jastrow concluded, men varied more than women—they were more variable, as the jargon of the time went. And variability was good because it produced a wider range of personalities and achievers as well as a greater number of geniuses. The Wellesley student, Cordelia C. Nevers, rejected his conclusion after she had administered a similar word test to students at her women's college. Her research did not identify the imitativeness and lack of variability among women that Jastrow had discovered in his research. Nevers's recommendation was that such discrepancy in results ought to constitute "a needed warning concerning the dangers of such comparative study of the mental processes of men and women."[13]

Given the disparity between Nevers's and Jastrow's professional standings, it might be anticipated that she was not acting entirely on her own. Her teacher at Wellesley was Mary W. Calkins, who had completed all the work for the doctor's degree in psychology at Harvard, but had been denied the degree because of her sex. The year after Nevers's article appeared, Calkins herself entered the lists against Jastrow, who, in the interim, had denied that Nevers's evidence had weakened his case. Calkins admitted that Nevers's evidence was inconclusive, but shifted ground to the role of environment in making such comparisons. It is "futile and impossible," she contended, "to eliminate the effect of environment" in administering such tests. The differences in training and behavior of men and women "begin with the earliest months of infancy and continue through life," she informed him. Consequently, the words that young women wrote on their lists were actually determined by their socially limited experience. Hence, nothing could be learned about the quality of woman's mind by comparing such lists with those from men. The experiences of the two were just too different.[14]

The debate was then entered by a graduate student at the University of Chicago, Amy Tanner. She took even higher ground than Calkins, condemning both Calkins and Jastrow for failing to make sure that the habits and background of the two sets of students were as identical as could be arranged. She dismissed the tests as showing nothing more than "a fact which it is entirely unnecessary to prove: that habit determines the association of ideas." She then enunciated the principle that would echo through the investigations of the next two decades into the differences between the male and female minds. The diversity of occupations, indeed of life experiences of men and women, Tanner insisted, "prove certain *particular* differences," but the existence of inherent differences, which was, of course, the central issue, "can not be demonstrated until men and women are not only nominally but actually free to enter any profession." Even now, she correctly pointed out, many are closed to women, especially married women. As a result, Tanner concluded, "the real tendencies of women cannot be known until they are free to choose, any more than those of a tied-up dog can be."[15]

Tanner's prescription for a proper investigation ruled out for the foreseeable future any further experimentation, the favorite mode of psychological argument and proof at the time. It was not likely that the high standard she had specified would be realized soon; indeed, it has not been fully realized at the end of the twentieth century.

Other students in psychology at Chicago, however, were prepared to continue to investigate the ancient question, albeit with somewhat lower standards of investigation and proof. Frances Kellor, for example, a Chicago student of criminal behavior, drew attention to the way in which expectations of the behavior of the two sexes affected judgments as to the relative degradation of their criminal behavior. Most people think women criminals are more degraded or abandoned than men, Kellor observed, but her investigations brought home to her how that judgment "seems due rather to the difference in the standards which we set for the two sexes. We say woman is worse, but we judge her so by comparison with the ideal of women, not with a common ideal." She pointed out that in identifying bad habits among women, she herself in her own research had included swearing and smoking among them, "but among men we should not consider them in the same light." It is true, she conceded, that licentiousness in behavior and conversation and uncleanliness in habits and person "do exist to a high degree; but the men and women come from the same classes, have the same standards, and know the same life." From that perspective, she concluded, "the woman is not more degraded than the man."[16] In short, the argument from social circumstances could work both ways: it explained away the allegedly superiority of men over women, and it could eliminate the supposedly greater degradation of women criminals.

Kellor, however, was no more a total environmentalist or cultural determinist than Ward or Gamble. She, too, thought biology shaped women's view of the world. The crimes of women, she concluded from her extensive study of the female criminal, "are more closely associated with immorality, because biologically she inclines to this rather than to crimes of force. . . . Where public safety is threatened, as in homicides," she suggested, "emotional conditions in woman, as contrasted with motives of gain in man, are often at work."[17]

The most important woman social scientist trained at the University of Chicago in those early years was the psychologist Helen Bradford Thompson. (John Watson, the founder of behaviorism, who was a classmate of Thompson's at Chicago, later recalled that he was told by his professors that he had done excellently on his doctoral examination, but that he had not equaled Thompson's performance; she was awarded a "Summa" to his "Magna.") By the end of her investigations into the differences between men and women, Thompson, in contradistinction to Kellor, would leave no place for innate "feminine" characteristics. That, however, was where Thompson ended, not where she began her research. Her initial effort was a straightforward, quite traditional psychological investigation as defined by the new psychological laboratories at the German universities. That meant testing the reactions of individuals in as objective a manner as possible. And that is how Thompson began her examination of sex differences, the results of which appeared in her book *The Mental Traits of Sex,* published in 1903.

The body of the book reported the results of tests she administered to fifty young women and men for motor ability, skin and muscle sensitivity, taste, smell, hearing, vision, and intellectual and affective characteristics. As might be expected, she found some differences between the sexes but the results were not consistent with sex. Sometimes females were superior to males and vice versa.

At the end of the book, she turned to the question that had gripped her from the beginning: How well did her findings fit the standard conception of the differences between women and men? The conventional view, of course, had found a striking congruence between biology and the life roles of the two sexes. The female egg, she summarized the traditional view, "is large and immobile. It represents stored nutrition. The male cell is small and agile. It represents expenditure of energy." From these differences, the social and psychological divergences had been deduced. Women's characteristics "are continuity, patience, and stability. . . . She is skilled in particular ideas and in the application of generalizations already obtained, but not in abstraction or the formation of new concepts," was the way Thompson paraphrased the standard argument. Males, she continued, were supposed to be more variable, and "everywhere we find the male sex adventurous and inventive. Its variety of ideas and sentiments is greater. . . . Men are more capable of intense and prolonged concentration of attention than women. They are less influenced by feeling than women. They have greater powers of abstraction and generalization."[18]

After that lengthy recitation of the traditional conception of sex differences, Thompson admitted that "on the surface, at least, the results at which we have arrived accord very well with this theory."[19] There was indeed a correspondence between woman's biology and her behavior, and man's and his. She did not say, but she might have, that it was inevitable that her results should be consistent with the standard view. After all, by using the standard tests, she had accepted the terms of the debate set by the traditionalists. She measured what they measured. Only if they had mismeasured could her findings differ in the end from their's.

At the close of her study, then, she moved to alter the terms, to probe beneath the assumptions of the new psychology with its laboratory tools, to ask if the questions were really reaching the fundamentals of the issue. First she questioned the use of analogies between sex cells and the actual behavior of men and women. "Women are said to represent concentration, patience, and stability in emotional life," from which it might be thought "that prolonged concentration of attention and unbiased generalization would be their intellectual characteristcs," she observed. But, she pointed out, that was not the interpretation arrived at, for those last "are the very characteristcs assigned to men."[20] And so she went down the list to spell out how the analogies had been shaped not by evidence or logic but by preconceptions about the differences between the sexes.

She crowned her assault upon the use of analogies with a bit of ironic humor. Suppose, she suggested, the nature of the sex cells had been reversed. Then it would be quite easy for the biologically oriented students of sex differences "to derive the characteristics of sex with which they finally come out. In that case, the female cell, smaller and more agile than the male, would represent woman with her smaller size, her excitable nervous system, and her incapacity for sustained effort of attention." Since, under Thompson's supposition, the male sex now contained the large, immobile sex cell, it "would image," she slyly suggested, "the size and strength, the impartial reason, and the easy concentration of attention of men."[21]

Having disposed of analogies, she then turned to the old question of the variability of males. Here she relied in part on new research, which had not found any reliable scientific basis for accepting that description. But even if new evidence did, in the end, establish the principle, she suggested, that might mean that there were more geniuses among males, but by the same token it also meant that there would be more idiots, so where was the gain? Her conclusion was that "the biological theory of psychological differences of sex is not in a condition to compel assent."[22] In sum, the methods of psychology could not settle the question. A fresh approach had to be taken. Her recommendation was much like that which Franz Boas and her teacher at Chicago, W. I. Thomas, had already begun to emphasize: look to the role of environment in shaping behavior.

To do that, Thompson moved intellectually from psychology to sociology, from innateness to environment. She now pointed to experiments with children "from the criminal classes" who had been "placed in good surroundings" and who then "usually develop into good members of society." Modern practical sociological theory, she stressed, is based "on the firm conviction that an individual is very vitally molded by his surroundings, and that even slight modifications may produce important changes in character." Some students of sex differences, she admitted, have derided the idea that differences in environment can explain the differences, but she thought the approach "at least worthy of unbiased consideration." In support of her position, Thompson proceeded to describe the different ways in which boys and girls were reared, educated, and perceived in society, a social fact that she though might well explain most if not all of the sex differences identified by the various psychological tests she and her colleagues had developed and applied.[23]

As a final confirmation of her new approach she raised a question the biologically oriented students of sex differences often asked: Is it not likely that the different rearing practices and education of girls and boys arose in the first place from the simple fact that "their natural characteristics are different?" Boys, for example, being more active than girls, are naturally given tools rather than dolls. Not at all, she responded. She then proceeded to reveal her fundamental assumption that few if any innate differences existed between the sexes. "There are many indications," she began, "that these very interests are socially stimulated. A small boy with an older sister and no brothers is very sure to display an ambition to have dolls," she imagined. "It is in most cases quenched early by ridicule, but it is evident that a boy must be taught what occupations are suited to boys." Girls with only brothers, she added, are known to be saddened when told they cannot play the games and engage in the sports of boys. If there were fundamental, instinctive differences between the sexes, she pointedly asked, would it be necessary "to spend so much effort in making boys and girls follow the lines of conduct proper to their sex?" No, she concluded, it was not biology, but "the necessities of social organization" that created a division of labor between the sexes, and upon that division social ideals have been erected and still persist.[24]

More cautious than Tanner, Thompson was not yet ready to ask if that social necessity was still defensible. She was prepared, nonetheless, to suggest that no one should overlook the changes in the roles of the sexes that could be observed all around. "The question of the future development of the intellectual life of women is one of the social necessities and ideals rather than of the inborn psychological characteristics of the sex."[25] The future of woman, in short, was unlimited by biology, provided society lifted the present limits on her activities. Thompson, together with Tanner and Calkins, had discovered social conditioning or socialization as the antidote of biology. No new body of evidence had been brought forth, simply a change in assumption. Whereas the traditionalists had assumed, along with Darwin, that biology differentiated the behavior of the sexes, those who rejected a biological explanation now substituted social environment or culture as the true source of differential behavior patterns.

None of the professors with whom Thompson studies at Chicago was as persistent and effective in questioning the traditional biological justifications for the subordination of women as William I. Thomas. Ironically enough, he had begun his career as a sociologist, as we have seen, with a doctoral dissertation that set forth the very concepts that Thompson rejected at the conclusion of her own dissertation and book *The Mental Traits of Sex.* He had based his work on the then fashionable, though clearly sexist concept of metabolic differences between the sexes. That sexist beginning, however, did not prevent Thomas from following Ward and Gamble in seeing women's role in reproduction as the fundamental source of human sociality, and man's superior strength as the means, though not a justification, for his domination of woman. Thus in 1898 Thomas argued that "women became an unfree class, precisely as slaves became an unfree class—because neither class showed a superior fitness on the motor side," that is, evolution had left them behind physically. But, he quickly added, both classes—blacks and women—are "regaining freedom because the race is substituting other forms of decision for violence."[26] It is not surprising, then, despite his earlier acceptance of metabolic differences, that by 1899 Thomas was denying that sex differences were "inherent in the male and female disposition." What differences there were, he suggested, seem to be "partly a matter of habit and attention."

The reasons behind Thomas's change of heart—and that change would deepen and accelerate in the first decade of the new century—are not difficult to uncover. As we have seen already, his questioning of race was well under way during the 1890s, and his wide reading in ethnology, or what today would be called anthropology, made it increasingly difficult to find katabolic and anabolic patterns among the great diversity of female and male activities around the world. William James in his *Principles of Psychology* may have determined that modesty was an instinctive form of behavior, but already in 1899 Thomas found that hard to accept in the light of his knowledge of different societies. Too many of them simply did not define the concept in the same way as did Americans or Europeans. All societies have a sense of modesty, Thomas conceded, but among some nakedness is perceived as modest, not immodest, as James had implied.[27]

Thomas's skepticism, which caused him to subject the standard generalizations about women to the test of diverse experience, whether in ethnology or in everyday life, soon brought him to a rejection of the traditional descriptions. With his classes at Chicago now half occupied by bright, eager women students, he had ample opportunities for fresh insights into the nature of the sexes. Similarly, the new world of women in the workplace offered critical tests of the old conceptions. How could women be described as clumsy or lacking in intellectual confidence, Thomas wondered in 1907, now that the fastest typist is a woman, and the rapidity with which women in department stores make change so obviously belies the ascription to them of lacking manual dexterity? And he could not help noticing, now that women were in the professional schools, that "with the enjoyment of greater liberty," the American woman has "made an approach toward the standards of professional scholarship, and some individuals stand at the very top in their university studies and examinations."[28]

Thus, he concluded in 1907, along with Helen Thompson, that until women enjoy full equality with men no fair comparison can be made. And his standard for that comparison was as high as Thompson's, and as unlikely as hers to be met in the immediate future. "The world of modern intellectual life is in reality a white man's world," he pointedly observed, bringing the two principal disadvantaged groups under a common rubric. "Few women and perhaps no blacks have ever entered this world in the fullest sense. To enter it in the fullest sense," he explained, "would be to be in it at every moment from the time of birth to the time of death, and to absorb it unconsciously and consciously, as the child absorbs language." Until that happens, he insisted, we shall not "be in a position to judge the mental efficiency of women and the lower races." He felt justified, therefore, in concluding that "the differences in mental expression between . . . men and women are no greater than they should be in view of the existing differences in opportunity." On second thought, he continued, it is possible that women's "capacity for intellectual work is . . . under equal conditions greater than in men." And the reason he thought so was their superiority to men in "endurance" and "cunning," which he defined as "the analogue of constructive thought—an indirect, mediated, and intelligent approach to a problem."[29]

Other male social scientists around the turn of the century also came to accept a social explanation for whatever behavioral differences may have been identified between the sexes. Not surprisingly, Franz Boas was among the earliest. In 1894, for instance, he asserted that, despite smaller brains, "the faculty of woman is undoubtedly just as high as that of man." Sociologist Edward A. Ross, in 1905, indicated his acceptance of social conditioning by extravagant praise for Charlotte Gilman's assertion in *Women and Economics* that when women became economically independent they would shed some of the characteristics that traditionally had been ascribed to sex. What "a broad clearing in the jungle" of sex differences she had opened, he exclaimed. We now know that "'sex' like 'race' is the recourse of the lazy," he concluded. Mrs. Gilman, he noted with glee, had shown that so-called sex differences are really rooted "in the surface soil of modifiable social conditions," not in the sex cells. Gilman had demonstrated, he exulted, "that the woman question is for the sociologist as well as the biologist."[30]

The connection between the concepts of race and sex, which Thomas and Ross drew, appeared with understandably increasing frequency in the writings of social scientists as the new century wore on. Both racism and sexism were vulnerable to criticism from the same argument of social conditioning; thus an opposition to racism easily transformed itself into a critique of sexism. Mary Coolidge, a sociologist who wrote one of the earliest scholarly studies of the Chinese in America, made an effortless transition from her knowledge about the Chinese to a comparable way of thinking about the character of woman. "In some aspects the woman questions are analogous to race questions," she wrote in 1912. "Surely, if in so short a time the 'Heathen Chinese' can rise to be a progressive human being in our estimation, it is not impossible that women may become social entities, whose acquired 'femininity' may be modifying faster than the carefully digested ideas of scientific observers." The implication, of course, was that freedom of opportunity would expand women's activities. If the "semi-feudal, almost unreasoning peasant of Eighteenth century Europe" can reach the present level of Americans "under the stimulus of wider economic opportunities," one can barely foresee "what womankind might be with an equal liberation and as strong an impetus," she suggested. Or, to phrase the issue more directly: "tradition and convention have operated with much more force upon women than upon men."[31]

Coolidge enthusiastically singled out Lester Frank Ward, William Thomas, and another Chicago professor, John Dewey, as notable warriors in the struggle against traditional ideas about women's capacities. "It is scientifically demonstrable," she quoted Dewey as asserting, "that the average difference between men and women is much less than the *individual* differences among either men or women themselves."[32] That argument, too, had been advanced to dispute race as an explanation of human differences.

A close companion to the socialization argument in attacks upon sexist thinking was the charge of bias—usually male—in the interpretation of evidence. Indeed, Helen Thompson, who as early as 1903 had rung the changes on that argument, returned to it in 1914 in the course of reviewing the recent literature on sex differences for a professional journal of psychology. She noted that in many studies "girls have stood better than boys in measures of general intelligence." That finding, however, she observed with not a little sarcasm, has so far not caused anyone to draw "the conclusion that girls have great native ability than boys. One is tempted to indulge in idle speculation," she continued in the same vein, "as to whether this admirable restraint from hasty generalization would have been equally marked had the sex findings been reversed." Following the critical approach she had perfected over a decade earlier in her book, she commented that those writers who say girls "are more docile and industrious than boys" also see them as more emotional and volatile in mood. "They seem to find no contradiction in the fact that the sex which is most dominated by emotions and moods is also the one which has the greatest capacity for plugging away at a task whether it is interesting or not."[33]

Some male social scientists, too, were drawn to irony and sarcasm when they encountered defenses of innate sexual differences. For example, Robert Lowie, the anthropologist, found it "amusing to note how every sex difference that has been discovered or alleged has been interpreted to show the superiority of males." Thus, when it was discovered that "there are more males among inmates of idiot asylums, . . . this apparently unfavorable fact was at once interpreted as confirmatory evidence of greater male variability; and as such it became immediately favorable to the theory of male superiority. Had it been found," he imagined in the manner of Thompson, "that there were more females among inmates of idiot asylums, how easily it could have been used as evidence of the general inferior quality of female mind." He apparently felt no hesitation in pronouncing in 1916 that "the verdict of present-day science" is an "uncompromisingly negative one: no rational grounds have yet been established that should lead to artificial limitation of woman's activity on the ground of inferior efficiency."[34]

Not all social scientists, even as late as the time of the First World War, were ready to assume Lowie's committed and forthright position. Sociologist Carl Kelsey of the University of Pennsylvania, who, some years before, had opposed racial explanations for behavior, was still ambivalent on the equality of the sexes despite the frequent references in the professional literature to an analytical similarity between the concepts of race and sex. In examining the effects of differences in brain weights in the two sexes, for example, Kelsey concluded that "in the present state of knowledge," social scientists had no reason for assuming "that either sex has any advantage so far as mental ability is concerned." Yet he was skeptical of the research that denied any consequential differences in the behavior of men and women because of menstruation. In general, he thought it "hard to determine whether the differences seen are due to actual differences of constitution or to social and mental traits." He conceded that the daily life of the two sexes, "even in the emotional and intellectual sphere, is so different in modern civilization . . . that we must expect different reactions." He admitted it seemed likely that, as the patterns of life of women and men converged, so would their behavior. And for that reason, he thought, "we must not . . . exaggerate the physical differences," but by the same token, he warned, we must not "make the equally foolish mistake of ignoring them."[35]

Kelsey, in short, contrary to Thomas or Thompson, was not quite ready to write off the differences between the sexes as simply the products of different experiences. All that he was prepared to do was to forget about invidious comparisons: "it is foolish to talk about the inferiority or superiority." Women and men, he concluded, "are different. That is all." He closed with a modified return to the social conditioning argument, a sign of his continued ambivalence. "It is evident that many of the divisions of labor and custom have been based upon artificial or, at least, insignificant reasons."[36]

The ideological roots as well as the radical nature of the social or environmental explanation for sex differences become especially clear when we recognize that not all women social reformers subscribed to it. Women activists interested in social reforms like the suffrage, temperance, and settlement house work—women like Frances Willard and Jane Addams—operated from the assumption that women differed by nature from men. Often denominated "social feminists" by historians, these women emphasized what they considered women's nurturant and caring nature. Society, they were sure, would benefit from those peculiarly female qualities being brought into play in the public sphere. Their approach was not much different from the argument made by feminists like Charlotte Gilman and Lester Frank Ward.

By accepting the traditional idea of women's nature the social feminists thought they would in that way be better able to protect women, particularly working women. By contending that women were different from men and that the special nature of women deserved protection, they could call upon Congress and state legislatures to enact legislation to improve the conditions of women's work and life. A rejection or at least a minimizing of the socialization argument then appearing in the new social science research on sex differences is plainly evident in the famous legal brief in behalf of protective legislation for working women presented to the U.S. Supreme Court in the case of *Muller v. Oregon* in 1908. The women who gathered the social and economic evidence for the brief drew heavily upon the old literature on women's nature, for it emphasized the need to protect working women from threats to the health and character of the nation's mothers. The new environmental explanation or socialization argument for the observed differences between men and women would have undermined the case for such legislation. (Ultimately, of course, the acceptance of the social conditioning argument by both courts and public in the 1970s accomplished exactly what the social feminist reformers of the Progressive era had feared: it eliminated virtually all of the protective legislation that had been enacted in behalf of women who worked outside the home. It is only fair to add that it also eliminated those aspects of the so-called protective legislation that in practice limited women's employment opportunities.)

The women advocates of the new social scientific approach to women's nature and the Progressive era social feminists did agree on one thing. Motherhood, as a state peculiar to women, made women's lives different from men's. For social feminists it justified protective legislation; for the new feminists it limited women's opportunities, a situation that cried out for remedies. W. I. Thomas observed in 1907 that, though women scholars were clearly achieving a high place for themselves in the universities, they are too often "swept away and engulfed by the modern system of marriage," and thus left without a career. Carl Kelsey, too, recognized the conflict when he said woman "should have the right to motherhood or not as she pleases," but then gave voice to the commonly expressed fear that such a choice must have a deleterious effect on society. Few people then saw as feasible—leaving aside the issue of desirability—the combination of motherhood and career, which today is so common.[37]

It was a rare woman scholar who even attempted to work out such a combination. When Helen Thompson married and became a mother, for example, she withdrew from academia, though not from an active role in reform causes and in psychological research under the name of Helen Woolley. The same path was followed by Mary Smith Coolidge, who, upon marriage, resigned her position of sociology professor at Stanford. A notable exception to the pattern was Elsie Clews Parsons, an early student of Franz Boas in anthropology at Columbia. But Parsons was exceptional in other ways, which made her determination to follow a career practical even with motherhood: she was independently wealthy, and her husband was agreeable to her goals. Thanks to both money and husband she was able to pursue field work, publish, *and* bear and rear several children.

Parsons was well aware of the innovative road she was traveling. She made a practice of urging her fellow social scientists to work to permit women to combine career and family. For, as she wrote in 1909, "it is . . . on the fight of the professional woman to get back into the family that the future of the family will depend." The phrase "get back into the family" succinctly captured the self-denying decision the great majority of women scholars had made. At the present time, she acknowledged, the struggle was hard; women did not have the flexible schedules their dual jobs required. "The working schedule of the potential or actual child-bearer," she declared, "must vary from time to time for the sake of both her productive and reproductive capacity." But in both ideological aims and in her actual practice, Parsons was far ahead of her time and most members of her sex. So few women followed Parsons's lead that in her later years Parsons became almost resentful of women for having let men set the constraints within which they were compelled to operate.[38]

No woman social scientist of the time dwelled upon the issue of women's special maternal function more ardently or more lengthily than Teachers College psychologist Leta Hollingworth. Herself married, though without children, Hollingworth painfully recognized the burden that motherhood laid upon women. Both society and individuals, she remarked in 1914 in a professional journal, would profit if women could "find a way to vary from their mode as men do, and yet procreate. Such a course," she bitterly explained, "is at present

hindered by individual prejudice, poverty, and the enactment of legal measures." At a later time, she made a little clearer what she meant by "legal measures." Since society requires children, she contended, "we should expect, therefore, that those in control of society would invent and employ devices for
impelling women to maintain a birthrate sufficient to insure enough increase in
the population to offset the wastage of war and disease."[39] (It is revealing of
attitudes on sex differences at that time in psychology that both of these articles
by a professional psychologist appeared in a sociological, not a psychological
journal.)

As a professional student of sex differences, Hollingworth was keenly aware
of the discipline's old chestnut: the supposed greater variability of males. Might
not maternity help to account for women's apparent lack of variability or proportion of great minds? she asked. "Surely we should consider *first* the established, obvious, inescapable, physical fact that women bear and rear children,"
she responded. For it is a fact "that this has always meant and still means that
nearly 100 per cent of women's energy is expended in the *"performance and
supervision of domestic and allied tasks, a field where eminence is impossible,"*
she wryly emphasized.[40]

Hollingworth's acknowledgment of the hindrances imposed by motherhood
left her, as someone interested in the expansion of woman's role in society, in
an ambivalent position. On the one hand, she could not deny, as she confessed
in 1916, the social necessity of motherhood—she placed it on a level with service in the armed forces. Such recognition also meant that the public and too
many social scientists reasoned from that necessity, and from biology in general,
to an assertion of the existence of "the 'maternal instinct,' which," she regretted
to say, "is popularly supposed to characterize all women equally, and to furnish
them with an all-consuming desire for parenthood regardless of the personal
pain, sacrifice, and disadvantages involved." She quoted statements from leading psychologists such as William McDougall and Joseph Jastrow, in which
maternity was identified as a female instinct. Inasmuch as no "verifiable data"
exist to warrant belief in such an instinct, she urged her fellow social scientists
"to guard against accepting as a fact of human nature a doctrine which we
might expect to find in use as a means of social control."

Hollingworth readily identified examples of such control, specifically in the
laws against abortion, contraception, infanticide, and desertion of children. She
interpreted the many warnings from social scientists against single-child families as subtle encouragement of child-bearing by women. Even works of art with
their frequent depiction of madonnas, she complained, fostered motherhood.
On the other hand, she remarked, few social scientists took much cognizance
of the dangers of childbirth. Her more positive expectation was that the time

would surely come when women would awaken to the way they were being manipulated to bear as many children as possible. It was natural for women to want to have some children, she admitted, but if society wants a surplus, it will have to pay for them, she maintained. She was sure that there were women who would be willing to bear eight to ten children if they were adequately paid for the effort.[41]

Despite Hollingworth's recognition that maternity set women apart from men and placed an undeniable burden on women in their competition with men, she was reluctant to acknowledge any fundamental differences between the sexes. This attitude was reflected in the high standard she set for comparing behavior of males and females. In reviewing the recent literature on the subject in 1916 for a psychological journal, she complained that none of the authors followed proper comparative methods. Instead, she complained, "they proceed to describe all differences between the two groups as sex differences." Logically, she contended, any differences found by that method "should be treated as group differences, unless the author is able to show that the group of males differs more from the group of females than from other groups of males similarly selected."[42]

When she reviewed the professional literature three years later her comment was similarly jaundiced about the validity of research into sex differences. "Nothing consistent" emerges from the comparisons of the sexes in regard to mental traits, she wrote; "in this respect it resembles the work of other years." For that reason, among others, she recommended that "perhaps the logical conclusion to be reached . . . is that the custom of perpetuating this review is no longer profitable, and may as well be abandoned." Yet only the year before, Joseph Jastrow, who had been fighting the battle of sex differences for over twenty years, published an article in which he qualitatively and quantitatively differentiated the "feminine mind" from the masculine.[43] Hollingworth may have given up on tests as a way of settling the issue, but obviously other psychologists had not. Indeed, as we shall see in a later chapter, the contest within psychology over the value and validity of mental testing still lay ahead—but in regard to race, not sex.

Having discovered bias and social control, Hollingworth made them her primary targets in her criticism of the contemporary research into sex differences. So far as she was concerned, there was no evidence of innate differences, only bias. Sometimes that deeply held conviction led her into fresh research rather than the utterance of yet another warning against the dangers and prevalence of bias against women. One example was the way she came to conduct a study of the effects of menstruation. While reviewing the raw data collected by her husband, who was an experimental psychologist, she, being a woman, noticed something that her husband had overlooked. His research was on the effects of caffeine on women's mental and motor abilities. Because most psychologists of the time thought menstruation strongly affected those faculties of women, he asked his women subjects to note the onset of their menstrual periods. Hollingworth noticed, though her husband did not, that the abilities of the women subjects had not been affected by menstruation, as her husband had originally feared. From that observation Hollingworth went on to complete a doctoral dissertation and book on *Functional Periodicity* (1914).

On the other hand, her enduring suspicion that the professional atmosphere was pervaded with bias in favor of males could cause her to issue warnings that can only be characterized as extreme. One example appears in a long letter she wrote in 1922 to Lewis Terman, the Stanford psychologist and designer of intelligence tests. Apparently he had inquired of her about the best way to select subjects for a study he planned on exceptional children. His intention was to ask teachers and parents to identify the subjects. Hollingworth's response was to warn him against sex bias. She wrote of the tendency of parents to discriminate against girls, as when they denied permission for a girl to skip a grade on the ground that she was too delicate to handle the acceleration of learning. No boy would be denied such advancement, she reported. She also alleged that parents feared to make "blue-stockings" out of girls. She said she knew of girls with "relatively stupid brothers" whose parents had not allowed the girls to move ahead of the boys because the brothers "would feel badly. I have never heard of the reverse of this situation," she wrote Terman.

Teachers were no better than anyone else in escaping sex bias, Hollingworth warned Terman. At one time she had asked teachers in a New York institutional home for children to identify those who seemed to be feeble-minded. Many more males than females were so identified, she found, yet when she tested girls who were perceived as feeble-minded, Hollingworth discovered them "as feebleminded as are the males who *are* suggested. Might not the same kind of factors be at work at the opposite end of the scale?" she asked Terman. "I believe that teachers (and people in general) tend to think loosely of 'the most desirable' or 'the best' when asked to designate the most intelligent," she declared. The result, unfortunately, is the "selection of girls who are personally attractive and pretty, at the expense of girls who are highly intelligent, but not so blest with desirable 'feminine' qualities." The best method of selection, she advised Terman, was objective tests.[44]

It does not take a highly active imagination to recognize that Hollingworth's own sex shaped her determination to refute the idea of innate sex differences. Indeed, at times, she almost revealed it in her own words. The influence of that motivation appears in sharp relief when her work on mentally deficient and intellectually advanced children is examined. In that research, to which she turned during the 1920s, social environment or conditioning as a possible explanation for mental achievement or lack of it is categorically denied. An important reason for the inconsistency was her almost total faith in mental testing as a sound measure of intelligence. Thus in her book *Gifted Children* (1926) she concluded that lack of opportunity could explain the lower achievement of women, since tests had demonstrated that women were equal in intelligence to men. "We must assume," therefore, she informed her readers, "that there are powerful determinants of eminence besides intellect," by which she meant social discrimination against women.[45]

When Hollingworth came to poverty or class as possible causes of lower achievement, she displayed neither patience nor understanding. We know, she announced, that the great proportion of the world's wealth is held by a few people, a good sign as she saw it, that "wealth is in positive correlation with intelligence." And if we become more specific and examine the distribution of wealth in the United States, she continued, that too demonstrated that "economic reward is correlated positively with intellect." Or we might look at students in private schools, where attendance costs money, to demonstrate the same correlation since the pupils at such schools have IQs above 100. Then, "at the opposite extreme of income," she noted, "we have paupers; and here we have actual test knowledge of intelligence. Paupers are very stupid as a group, including a few persons of better than average mental capacity. A lengthy bibliography of scientific studies exists," she assured her readers, "to establish this fact beyond a doubt."

Failure to grasp these facts, she complained, has "led to much fallacious thought and propaganda." Some people have even gone so far as to urge that "all children should be obliged to attend high school, because statistics prove that high school graduates have larger incomes than do those who have not been graduated from high school." The direction of the causality, she emphasized, is just the reverse: "It is not the high school education which wins the reward, but the person who is able to win both the education and the reward."[46] Her standard of proper comparative method here fell conspicuously short of the standard she demanded earlier of anyone seeking to ascertain the causes for differences between the sexes.

In Hollingworth's view, the correlation between wealth and intelligence in a competitive economy would be naturally high. As she saw the situation in the United States, however, that high degree of correlation was unfortunately reduced by the actions of labor unions, and legislatures or administrative bodies, which set salaries and otherwise interfered with market forces, thus tending "to equalize the stupid and the intelligent." The correlation also declined when age, sex, and race were introduced, but she thought that was understandable in the case of sex because "a given woman of the same intellectual caliber as a given man is not of the same economic value as the latter, because masculinity is in itself an asset of superior worth." Regrettably, she informed her readers, thinkers of a utopian cast of mind just do not understand "the existing distribution of biological endowment. . . . The immemorial division of mankind into 'lower', 'middle', and 'upper' classes, economically speaking, rests on a biological foundation which guarantees the stubborn permanence with which it persists in spite of all efforts to abolish it by artifice."[47]

As the foregoing suggests, Hollingworth was not only a devoted believer in the decisions of the market, she was also a rather doctrinaire eugenicist. "Modern biology has shown that human beings cannot improve the qualities of their species, nor permanently reduce its miseries by education, philanthropy, surgery, or legislation," she confidentially announced. For such efforts "are palliative merely and leave a worse condition for the next generation to face." An act of philanthropy to a thousand paupers of this generation may offer them some relief, she admitted, but it "bequeaths at least two thousand paupers to be relieved by generations immediately following, for it has enabled a thousand organisms of pauper quality to live and breed." Her compassion for the lot of woman stood in stark contrast to her feelings for the poor. Eugenics, she wrote, has revealed what needs to be done to improve society, but unfortunately our ability to accomplish that improvement is severely limited. "Those whom it is thought highly eugenic to eliminate through lack of offspring are the very ones who most often cannot grasp the message, or, grasping it, are indisposed to comply with its conditions."[48]

Leta Hollingworth died prematurely, at the age of fifty-three, yet the body of work she left to her profession was substantial. A large part of it was well within the hereditarian tradition, as her quoted remarks make clear, an outlook that, in part at least, may explain the high praise Lewis Terman, another hereditarian psychologist, bestowed upon her. "Comparable productivity by a man," he wrote in his introduction to her husband's biography of her, "would probably have been rewarded by election to the presidency of the American Psychological Association or even to membership in the National Academy of Sciences." That failure to recognize her achievements, Terman ruefully acknowledged, was "primarily a reflection on the voting habits of male psychologists."[49] To the very end, in short, Leta Hollingworth was plagued by that male bias of which she was so well aware and had fought so persistently.

Although in the 1920s Hollingworth turned her attention to other psychological and social problems, her work on sex differences remained standard and continued to influence researchers. Indeed, by that decade, the new intelligent tests had pretty well settled the issue of comparable intelligence of males and females. As Lewis Terman himself remarked in 1922, mental tests "have at last vindicated woman's claim to intellectual equality with man." In fact, "among psychologists," he asserted, "the issue is as dead as the ancient feud as to the shape of the earth." And even Edward Thorndike, still a committed proponent of innate differences between the sexes, agreed that intelligence was not among them.[50]

The variability of specific traits between the sexes, once a major bone of contention among psychologists, also declined in the wake of Hollingworth's attack upon variability. Terman is 1922 still thought there were specific traits to investigate as between the two sexes, but even he conceded in the end that lack of opportunity might play a role in accounting for differences. A husband and wife pair of psychologists reported about the same time that they had tried to investigate variability, but came to the conclusion that they had "largely lost interest in the problem." A young educational psychologist, seeking in 1924 to summarize the state of studies on women's nature, repudiated the whole concept of variability, citing Hollingworth's work as conclusive. Echoing Hollingworth's argument from maternity, he related woman's apparent lack of diversity in intelligence to her reproductive role; she had little or no opportunity for the variety of activities in which men participated. One reviewer of the psychological literature in 1927 cited Hollingworth's work as central in causing the abandonment of the concept of variability in study of sex differences. The same reviewer also pronounced Hollingworth's work on the effects of menstruation as standard and uncontroverted.[51]

Hollingworth's emphasis upon the burdens imposed by marriage and motherhood upon women continued to shape the thinking of social scientists, particularly those who were women. Ruth Reed, a Columbia psychologist, for example, in 1923 raised doubts about Edward Thorndike's assertion that there was a maternal instinct. In an investigation of pregnant women, Reed reported that 65 out of 150 women were negative about the prospect of becoming mothers. The evidence is limited, she conceded, but what there is of it,"does not lead us to agree with Thorndike that all females and from early childhood to death have an original interest in human babies." She welcomed what she perceived as a "growing tendency among educated women to break away from the sentimental considerations associated with the bearing of a child." She particularly recommended to her professional colleagues Leta Hollingworth's article of some years before on the hardship of child-bearing.[52]

Again and again in the social science literature of the 1920s, Hollingworth's insistence upon social as opposed to biological explanations for the different behavior of the sexes was restated and reinforced. One such reinforcement was the sociological study revealingly entitled *Social Change with Respect to Culture and Original Nature,* published in 1922 by William F. Ogburn. Like so many other environmentally oriented social scientists of those years Ogburn was a professor at the University of Chicago. He left no doubt where he stood on the matter of sex differences. To see women as shaped in their behavior by their reproductive or sexual activities, he pronounced "a bit mystical. It seems more plausible," he thought, "to seek the explanation in the differences in daily activities of men and women." If women seem to be more personal than men, he added, recognizing a difference often remarked upon, that "difference is either wholly due to culture or else is greatly accentuated by culture."[53]

Educational psychologist Willystine Goodsell at first phrased his support of the cultural approach more tentatively. May not the frequently identified differences between men and women, he inquired, "be found in large measure in the shape of contrasts in the life experiences of the two sexes?" He then turned more bold, ridiculing the validity of some of the examples advanced by Edward Thorndike to support his position of an innate maternal instinct and a fighting instinct in men. Those assertions, Goodsell remarked, may be firmly held by Thorndike, "but it is difficult to discover by what scientific method the author has arrived at his conclusion." Thorndike had appealed to the long history of such utterances and assertions, but Goodsell scornfully doubted that such pieces of evidence "constitute scientific knowledge." One cannot demonstrate anything by an appeal "to 'common knowledge,'" Goodsell complained, for it has "so conspicuously been demonstrated to be prejudiced and unreliable in matters relating to woman's original nature."[54] It is worth noting, too, as a measure of the professional turn away from belief in innate differences between the sexes, that Goodsell was a young professor of educational psychology at the same institution—Teachers College—where Thorndike was a nationally renowned senior professor.

By the late 1920s and early 1930s it was clear that changes in the society at large as well as the widening acceptance among social scientists of the concept of culture was altering the question of differences between the sexes. The growing number of women entering college classrooms and graduate seminars and then going on into the professions and business made a rigid belief in sharp and fundamental differences in the capabilities of the sexes hard to sustain. Increasingly, especially among social scientists, the issue was transformed into a recognition "that women, like men, are individuals," as one social scientist wrote in 1930, "each to be treated as a separate entity, and not merely as a member of a class, and each to be given freedom of action and equality of opportunity so that she may find her level in the educational, economic, social and political world."[55] But, as Hollingworth had predicted, that philosophical outlook came up against the old question of marriage and maternity.

Goodsell, for instance, plainly wished to espouse an individualistic approach to the women question, but he was not convinced that maternity should be left entirely open to individual choice. He admitted that a "maternal instinct is being challenged by contemporary psychologists," but as far as he was concerned, "the question is still unsettled." He also had to admit to a certain fear that if women followed their own individualistic choices, marriage and motherhood might well be rejected by large numbers of educated women. Somewhat timidly he suggested that with "a less rigid economic system" women might be able to combine career and family. In an effort to depict that position as realistic he drew attention to the advice and career of Elsie Clews Parson. In the end, though, Goodsell conceded that it would be a long time "before society cheerfully accepts the married professional woman," a prediction, revealingly enough, he did not feel called upon to deplore.[56]

Putting theory into practice might still present problems, but among psychologists there was little objection to the theory. By the end of the twenties, the periodic review of the psychological literature on sex differences, which first Helen Thompson Woolley and then Leta Hollingworth had written for the *Psychological Bulletin,* was being handled by a male. That his conclusions were thoroughly congruent with theirs measured the substantial gains which the cultural or social conditioning interpretation had made over the years. "Few, if any, of the so-called 'sex differences,'" Chauncy Allen concluded in 1927, "are due solely to sex. Individual differences often are greater than differences determined on the basis of sex." The reason for that, he added, is that "the social training of the two sexes is and always has been, different, producing differential selective factors, interests, standards, etc."[57]

A further measure of the triumph of the cultural approach to sex differences, which Thompson, Hollingworth, Thomas, Boas, and others had been advancing for so long, is that Allen's concluding statement in his review of the literature three years later was identical in wording with that which has just been quoted from his 1927 survey![58] The professional literature in the interval, he admitted, had proliferated and expanded in scope, and books denying the cultural interpretation were still being published. But among professional psychologists, the socialization interpretation of differences between women and men was clearly the prevalent view. Biology, everyone agreed, set women apart from men in regard to reproduction, but women's behavior was affected by reproduction only because the social order had not adjusted sufficiently to equalize the differential effects of biology on the sexes.

Consequently, during the 1930s, the attention of social scientists turned away from sex differences as a primary subject of investigation, though the interest never subsided entirely. Between 1933 and 1935, for instance, the *Journal of Applied Psychology* published no more than three articles on sex differences, and even those dealt with nothing more important than identifying sex from handwriting! Among the few scholars who followed the lead of Woolley and Hollingworth was a diminutive yet astonishingly energetic young anthropologist named Margaret Mead. A graduate student of Franz Boas and Ruth Benedict at Columbia, Mead had made her reputation in the 1920s with the publication of her first book, *Coming of Age in Samoa*. Her study had little or nothing to do with sex differences, though it made a pronounced statement in support of the cultural explanation for differences in human behavior.

On the surface, *Coming of Age in Samoa* was simply another anthropological field study of a people largely untouched as yet by civilization. Mead, however, reported her findings in such a way as to dramatize the cultural sources of a social phenomenon of great popular significance. After she had demonstrated that adolescence in Samoa lacked the tempestuous character it exhibited in the America of the 1920s, she asked the inevitable question: Why is adolescence a period of "storm and stress in American adolescents" and "not a specially difficult period" among young Samoans? Mead's answer to that question opened her career as one of the most persistent and persuasive proponents of the cultural approach to behavior. "First, we may say quite simply," she began her answer to her own question, "there must be something in the two civilizations to account for the differences," for the ages, that is, the biology of two groups, were roughly the same. "But," she noted, "the social environment is very different and it is to it that we must look for an explanation."[59]

As one who not only believed in the power of culture but was also a woman, Mead was inevitably drawn to the issue of sex differences, to the resolution of which over the years she dedicated her considerable research skills and rare social imagination. Thus, in 1935 she published *Sex and Temperament in Three Primitive Societies,* in which she sought to rephrase the issue of sex differences by introducing the concept of temperament, an idea she had picked up from Ruth Benedict's *Patterns of Culture,* published the year before. The theoretical argument underlying the concept of temperament does not concern us except that it was clearly intended to be a substitute for sex in explaining differences between human males and females.

Her principal argument was that, after studying three primitive societies in New Guinea, attitudes and behavior that were often linked to one sex or another were actually interchangeable between the sexes. All societies, she observed, have "institutionalized the roles of men and women," but not necessarily in terms of contrast, or dominance and submission. For she had discovered in her field work that men and women in a given society exhibited the same personality traits. Among the Arapesh, she reported, "men as well as women were cooperative, unaggressive, responsive to the needs of others." However, both men and women among the Mundugumor were aggressive, ruthless, and with strong interest in sex, something about which the Arapesh seemed to have cared little. Her point was that in these two societies, different as they were, "any idea that temperamental traits of the order of dominance, bravery, aggressiveness, objectivity, malleability, are inalienably associated with one sex (as opposed to the other) is entirely lacking."

To round out her case by showing that traits often associated in western societies with males could also be found in females, she introduced the Tchambuli, a people whose women were dominant, impersonal, and bossy, while the men were emotionally dependent. The conclusion was inescapable. Her research, she submitted, "suggests that we may say that many, if not all, of the personality traits which we have called masculine or feminine are as lightly linked to sex as are the clothing, the manners, and the form of head-dress that a society at a given period assigns to either sex." In the preface she claimed to have started out "innocent of any suspicion that the temperaments which we regard as native to one sex might instead be mere variations of human temperament." Yet, at the conclusion of her research, she was convinced that "the members of either or both sexes may, with more or less success in the case of different individuals, be educated to approximate" any temperament. This was possible, she announced, because of "the strength of social conditioning. In no other way can we account for the almost complete uniformity" with which the children in each society were raised.[60]

In Mead's thinking, the broader implications of her study were almost breathtaking. "We are forced to conclude," she wrote, "that human nature is almost unbelievably malleable, responding accurately and contrastingly to contrasting cultural conditions." Or, as she was to restate the point more concretely later in *Male and Female,* even behavior patterns related to biological reproduction had been shaped by culture, as shown by the dramatically different attitudes toward sex by the Arapesh and the Mundugumor. "Learned behaviours," she explained, had "replaced the biologically given ones." Her ideal society, as a result, unlike Leta Hollingworth's, was one in which no behavior pattern was identified with sex or even class. "A society is equally unrealistic," she wrote, whether it insists that only men can be brave, or that only individuals of rank can be brave." Her hope was to maximize the opportunities for the expression of individual differences, or temperaments. What she opposed was the linking of any particular temperament with either sex.[61]

At the same time, she did not shy away from the intellectual or philosophical implications of her conception of a human nature "almost unbelievably malleable." Now that we know "that the personalities of the two sexes are socially produced," we also know that "every programme that looks forward toward a planned order of society" from Fascism to Communism is compatible with it. In a sense, of course, that was the cul de sac into which cultural determinism led its proponents. If human nature is entirely a product of culture, there is no standard against which behavior can be judged; anything and everything are congruent with it. Feminist historian Rosalind Rosenberg, who was clearly troubled by Mead's cultural determinism, contrasted it with an earlier feminist outlook, the strength of which, she recalled, "had been its commitment to an ideal of feminine purpose." Mead's cultural interpretation "left feminism without a unifying vision and both men and women in a vicious circle. They grew up learning to behave toward one another in ways that were often bad for both of them; but, lacking a critical perspective on their lives, they found it difficult to question that behavior," Rosenberg pointed out.[62] That absence of purpose or meaning within cultural determinism was one of the reasons some women— and men—in the 1970s would feel it essential to search for the biological in addition to the cultural roots of human nature.

That the impetus behind Mead's emphasis upon culture or social conditioning was her ideological commitment to individual self-realization for women is obvious from the corpus of her writings over thirty years. As late as 1963 she was still publicly emphasizing individualism. Women in the United States over the last twenty-five years, she complained, "have come to rely more on the definition of themselves in terms of sex, and to lay less emphasis upon finding themselves as individuals."

Less public sources tell the same story. Her anthropologist daughter, Catherine, for example, recalls a conversation between her father British anthropologist Gregory Bateson and Mead some years after he and Mead had divorced. Bateson had suggested that, given the physical diversity of peoples in the world, the possibility occurred to him that they differed in cognitive abilities as well. Mead would have nothing to do with the idea. "As long as people tend to move so quickly from concepts of diversity to concepts of superiority," her daughter reported her saying, "and as long as mental variation is treated in terms of such crude and culturally biased aggregate quantity as I.Q., this question cannot and should not be studied." Unlike Boas and Kroeber, Mead was even fearful of admitting a place for genetic endowment in individuals, though privately, according to her daughter, she "always remained convinced that her own uniqueness was partially genetic." Her daughter recalled, too, that Mead was always apprehensive that "any effort to deal with these matters would lead to distortions by those who evoke the old crude dichotomy of nature versus nurture and misuse biological explanations to justify social facts."[63]

The extreme character of Mead's position in *Sex and Temperament* is measured in the objections her fellow social scientists raised against it. Reo Fortune, her husband at the time and the anthropologist with whom she carried out the research into the three New Guinea societies, publicly disputed the accuracy of her findings. Lewis Terman, the prominent Stanford psychologist, who during the thirties had begun to research sex differences, protested her ignoring of his contrary findings on the emotions and attitudes of women and men. Sociologist Jessie Bernard pointed out that Mead's own evidence contradicted her assertion that among the Arapesh there were no differences in outlook or behavior between the sexes. Mead herself took notice of the reaction in her preface to a new issuance of the book in 1950, calling the volume "my most misunderstood book." Yet the only sign of retreat from her extreme position of fifteen years before was the recognition that "the biological bases of development" of human beings set some "limitations, which must be honestly reckoned with." But by the same token, those biological bases can also be seen "as potentialities by no means fully tapped by our human imagination."[64]

Other evidence, however, hints that from almost the beginning of her career she was ambivalent on the matter of sex differences, and her later work reinforces that judgment. As early as 1932, for instance, she noticed how, in the cultures of the world almost universally, "the daily routine of cooking, care of the house, and care of the children, is left to the women." As a consequence, she observed, women can penetrate a culture more quickly than men; it often took male anthropologists months before they understood "the peculiar cultural preoccupations which distinguish one culture from another." Just for that reason, she explained, a "breakdown of culture is almost always of more vital concern to the men than the women." They are left "empty-brained and idle-handed," while the women must "continue to bear and nurture children, to cook the dinner, sweep the house, and wash the clothes," she pointed out. "It is impossible to strip her life of meaning as completely as the life of the man can be stripped."[65]

Even in *Sex and Temperament* there is some faint recognition that differences between the sexes exist, and that they might even have some positive value. "To insist that there are no sex differences in a society" that has always believed in them and depended upon them "may be a form of standardizing of personality," she suggested. Or, as she expressed the same idea a little later in the same book, "the removal of all legal and economic barriers against women's participating in the world on a footing with men, may be in itself ... a move toward the wholesale stamping out of the diversity of attitudes that is such a dearly bought product of civilization."⁶⁶

By the time Mead came to write *Male and Female* in 1949, her recognition of fundamental differences between men and women—their roles in reproduction—became explicit. In the course of a frank discussion of the difference between the male and female sexual orgasm, a difference that some sociobiologists would rediscover a quarter of a century later, Mead directed attention to a difference in the sexuality of men and women. No society can survive without masculine sexual erections and ejaculations, she reminded her readers. "There seems no simple reason for believing that orgasms in females are of comparable importance for conception, in at least the majority of females." From that observation she thought it reasonable to assume that "the human female's capacity for orgasm is to be viewed much more as a potentiality that may or may not be developed by a given culture, or in the specific life-history of an individual, than as an inherent part of her full humanity."⁶⁷

Differences in the sexuality of men and women were only the first of Mead's acknowledgment that from the sexes' different "reproductive strategies"—as sociobiologists would designate them twenty years later—other differences would branch off. For example, she returned to her earlier idea that maternity gave women an assurance that men never obtained from fatherhood. Women gain identity from child-bearing, she contended, while "the male needs to reassert, to reattempt, to redefine maleness." In every human society, she continued, "the male's need for achievement can be recognized. . . . The recurrent problem of civilization is to define the male role" in such a way that a man may reach that sense of assurance and achievement that women acquire in "fulfilling their biological role" of bearing children.⁶⁸

By the 1960s, she had picked up enough of the new information on animal behavior—ethology—which was appearing in Europe and America to discover other differences between men and women, differences that stemmed from their biological structures. At a conference on war, in 1968, she observed that "it is important to take into consideration the possibility that the biological bases of aggression in the two sexes—in human beings as in other mammals—may differ significantly." Drawing on her knowledge of animal behavior, she concluded that "the female characteristically fights only for good or in defense of her young, and then fights to kill." That being so, she warned that to arm women, "as has been done in this century in Israel, the USSR, Indonesia, and Vietnam, may be a suicidal course." It is simply not in the nature of women to make good soldiers.⁶⁹

One might conclude that Mead had thus returned to the position of the early twentieth-century social feminists who had argued that it was woman's divergence from man that justified, if not required, the suffrage and other means whereby woman should enter into public life to improve it through the addition of her unique female character. Mead's position is better read as a harbinger of things to come: the emergence of the sociobiological approach to the ancient question of the nature of the sexes, and the rediscovery by a reinvigorated feminist movement of the human value of differences between women and men.

Before we can pursue that engrossing turnabout, however, we must return to the early years of the twentieth century to follow another assault upon the biological basis of human behavior, namely, the dismantlement of the concepts of feeblemindedness and instinct, and the rejection of eugenics, each of which had once been seen as major achievements of the social sciences in their search for human nature.

6

Decoupling Behavior
from Nature

Brief is the answer to the question as to what is the relationship between social
psychology and instincts. Plainly, there is no relationship.
Psychologist J. R. Kantor, 1923

The future control of the human race and its civilization lies not through selective breeding of the higher social qualities ... but through their transmission
by social contact and control.
Sociologist L. L. Bernard, 1921

The belief that the sexes were at least as different in mind as in body did not
arise from nor depend upon any experimental evidence. Observation of obvious
differences and tradition had been enough. In fact, as the previous chapter
showed, soon after experimental evidence in the form of intelligence testing was
introduced, the assumption of mental differences between the sexes was called
into serious question and essentially abandoned. For other groups, however, the
introduction of the Binet test of intelligence in the first decade of the twentieth
century laid out a quite different future. The tests provided a means of identifying a whole new category of Americans: the feebleminded, that is, those
adults who tested at a level below that of a thirteen-year-old child. And when
it was further found that a large proportion of those "feebleminded" were in
prison, the connection was soon drawn that such low mental states were a
source and a sign of criminal behavior. Henry H. Goddard's 1912 study of the
Kallikaks epitomized as well as documented an unsettling association between
mentality and delinquency or criminality.

The connection, however, was loosened fairly quickly, principally by the publication of new empirical evidence, such as presented in William Healy's *The
Individual Delinquent*, which appeared in 1915. After examining over a thousand cases of delinquent children, Healy could find little empirical support for
a link between mental deficiency and criminality. His contention was that
delinquency was so complex in origin that it should be studied individual by

individual, not through a single standard of intelligence. His assumption, in sum, was that environment or the history of the individual was just as important in accounting for crime as inherited intelligence. By 1926, Healy and his co-worker Augusta Bronner candidly declared it "hazardous to offer any conclusions concerning the possible relationship of heredity to delinquency." Among children especially, they noted, delinquency often resulted from "bad social situations created by socially unfit parents, the effects of which are not those of biological inheritance."[1]

During the early 1920s studies of prison inmates provided additional evidence for doubting a connection between crime and heredity. Pre-eminent in this respect was the work of Clark University psychologist Carl Murchison. His studies demonstrated that the range of intelligence among criminals was about that of the general population; some criminals even turned out to be above the average intelligence of the population. Or, as Murchison pointedly remarked in his book *Criminal Intelligence,* the average score achieved on the Army wartime tests by criminals in a certain prison "was just 75 per cent higher than the average score of the guards." Murchison's personal appreciation of the intelligence of criminals is reflected in his next sentence: "The only reason the guards continued to live was because the architects of that prison had done their job well." His formal conclusion generalized the point; he found "the criminal group . . . superior to the white draft group" in the Army tests of 1917. His explanation for the difference was no less candid. "It is characteristic of high intelligence to resent conservatism, conformity and social suppression," he remarked. "That is one of the possible explanations of the uniformly high intelligence of the criminal group." He had only scorn for those who associated criminality with feeblemindedness. "It is a most stupid fallacy," he wrote in 1926, "to assume that the criminal, per se, must be feeble-minded." After all, he observed, what constitutes criminal behavior changes over time; to discern what is criminal and what is not "takes a high intelligence. . . . To imagine that the criminal in all ages will perceive and elect such behavior, being feeble-minded is sheer nonsense."[2] Criminality, in short, could not be the simple, inherited phenomenon many students of the late nineteenth and early twentieth century had assumed and argued.*

Healy's and Murchison's emphases upon social or environmental explanations may have broken the link between criminality and feeblemindedness, but

---

* Among some physical anthropologists, a connection between biology and criminality was still acceptable in the 1930s and after. See, for example, Earnest Albert Hooton's *Crime and the Man* (1939) and W. W. Howell's favorable review of it in *American Sociological Review* 4 (Aug. 1939), 603–4. Suggestions of its persistence into our own time appear in James Q. Wilson and Richard Herrnstein, *Crime and Human Nature* (New York: Simon and Schuster, 1985). Wilson is a political scientist and Herrnstein is a psychologist; both teach at Harvard.

that was as far as the disconnection extended. The transformation in assumptions or outlook, for example, did not include a repudiation of intelligence testing as such. William Healy, for example, continued to accept the concept of feeblemindedness. Indeed, as noted earlier, during the half-decade 1915 to 1920, a large segment of the public and the psychological profession in particular seemed almost obsessed with the dangers of inherited feeblemindedness. The overturning of that belief in a biological basis of behavior took considerably longer than the decoupling of criminality from heredity; it consumed most of the decade of the 1920s.

The questioning of usefulness of the concept began with the psychologist J. W. Wallace Wallin. As early as 1911 Wallin urged intelligence test administrators to be cautious in making psychological diagnoses from the results of the Binet test. By 1916 Wallin was doing more than urging caution; he was expressing doubts that feeblemindedness, which the tests were said to be measuring, was in fact a true psychological phenomenon. At that time, the mentally defective population had been divided by Goddard, the leading authority on the subject, into three groups: "morons," or those adults with a mental age between eight and twelve; "imbeciles," those with a mental age of three to seven, and, finally, "idiots," with less than three years in mental age. Feebleminded was usually taken as another designation for moron, that is, the highest level of mental deficiency.

The identification of the so-called feeblemineded simply by scores on intelligence tests Wallin saw as dubious if not dangerous. A test result, Wallin pointed out, was no more than a diagnostic aid, like temperature or pulse in medicine; it should never be seen as an "automatic diagnostican, which will enable the examiner to dispense with a thorough clinical examination or to disregard other clinical findings." Nor do test results, he warned, "obviate the need for technical training on the part of the examiner." After all, even trained testers easily made mistakes, as he knew from his own experience and insistence upon using all "the aids that are at hand." He illustrated the point by telling of a superintendent of a state institution who diagnosed a boy as feebleminded on the basis of the Binet tests. Yet the boy, once taken out of the institution, ended up as head of his class in preparatory school and went on to college. "How many non-feeble-minded children committed to state institutions are permanently retarded by the limitations of the institutional routine no one can say," Wallin warned.[3]

Some tests that Wallin himself conducted reduced the whole concept of morons or feebleminded to near-absurdity. He showed, for example, that even those who did not score above a mental age of twelve were nonetheless coping quite well in everyday life. He administered the Binet tests to a number of successful farmers and housewives in Iowa, all of whom failed to reach a "normal" score. Yet, as he wryly observed, some of them had accumulated considerable wealth—as much as $30,000, a considerable sum at that time—raised successful children, and broke no laws. How then, he asked, can we continue to segregate from society such people on the grounds that they will become criminals or will be unable to cope with life?

Wallin's warnings, however, were not immediately heeded, cogent as they may appear in retrospect. In 1917, for instance, he could only lament the rising proportion of feebleminded being uncovered almost daily in the general population. "Feeblemindedness has become the Nemesis of our times," he sadly observed. Three percent of elementary school children are so diagnosed, he pointed out, and at least half of delinquents are singled out as feebleminded, so that the proportion of such people seemed to be approaching 100 percent! He conceded that "feeblemindedness is one cause, but still only one cause of our social difficulties." He urged psychologists to be careful and responsible in making those designations. "The vast majority of delinquents and criminals," he contended in 1920, "who have been classed as feeble-minded during the last decade" on the basis of the Binet tests, "are no more feeble-minded than many millions of our citizens who are law-abiding, respectable and self-supporting."[4]

Wallin never drew a direct connection between the feebleminded person and the environment in which he or she grew up, though he certainly recognized, as his quoted remarks tell us, that the institutional environment provided to the feebleminded by the state was unlikely to improve a child's Binet scores. Sociologists Irene Case and Kate Lewis made just that correlation in 1918 in their article, "Environment as a Factor in Feeblemindedness," in the *American Journal of Sociology.* The two women studied ten Irish families in which all the members were delinquent in one fashion or another. But, as Case and Lewis emphasized, the true source of the children's delinquency was the parents' alcoholism, and the general lack of economic and educational opportunity, rather than heredity. The two scholars specifically attributed the children's "retardation" to the fathers' alcoholism, which wasted money and weakened the family by terrorizing its members. "The children cannot go to school, and they are put down as 'retarded,'" in consequence of which they lost interest in learning despite the need to make a living on their own. "Sexual delinquency under such conditions becomes inevitable," Case and Lewis concluded.[5]*

---

* Unlike other writers on the subject of heredity and environment in these years, Case and Lewis were careful to distinguish among the subtleties involved in the two concepts. For example, they noted that untoward effects upon an infant from a mother's alcholism were strictly neither the result of heredity nor environment. "It is not inheritance, because there is no organic resemblance between parent and offspring, and furthermore because the result is not according to Mendel's law. It is not environment, for the reason that the stimulus (alcohol) did not act upon the organism which showed the defect, but upon one or the other of the germ cells of the parents which gave birth to

By shifting from heredity to environment as the causal link between parents and children, the two scholars seriously undermined, as they pointed out, Goddard's declaration that feeblemindedness was inherited. If the children had been removed from those negative surroundings, they confidently asserted, "there is no particular reason for inferring that the predisposing sexual delinquencies [of the partents] could not have been overcome" by the children's new and improved social circumstances.[6]

They also flatly and specifically disagreed with Charles Davenport, probably the best-known geneticist and active eugenicist in the nation, on the inheritability of feeblemindedness. The statistics they had gathered on the ten families, the two women scholars argued, "surely offer proof that environment is the chief cause in this particular group of families at least." Like Wallin, Case and Lewis were convinced that "lack of training on the part of these people" was the source of their allegedly low intelligence. "More equal opportunities of education would doubtless tend to make them better members of the community and at least self-supporting."[7]

Actually, by that date, Goddard himself was beginning to rethink the whole question of the degree to which feeblemindedness presented a danger or even a problem for the country. "I am willing to say," he wrote in 1918, "that if we educate properly the moron we may very safely neglect this question of eugenics and marriage for a large proportion of them." And in his book on delinquent children, published in 1921, he designated many of the erstwhile feebleminded children as "psychopathic," a term that he introduced to describe what he now saw as an *environmentally* induced disorder of the mind. For, as he pointed out, genetics tells us that a functional disorder could not be inherited. As a result of that new insight, he concluded that "the problem of juvenile delinquency is solvable. . . . The first step is a change of attitude where we regard the delinquent not as a child to be punished but as one to be treated and trained," as with any other environmentally induced disability. He recommended changing physically and socially the institutional environment into which the children were placed. They should be called "schools," not institutions, and the children should be "admitted" rather than committed. And once the children entered these schools they "should remain until they graduate, i.e., until they are ready to take their places in society." In 1927 he saw the "problem of the moron" as one of education. If we had not tried to educate all children in the same way,

the defective offspring. We need, therefore, to distinguish from Mendelian inheritance those cases of defective offspring which result from the action of environmental factors, acting upon the somatoplasm of the parents and indirectly upon their germ cells. For this effect we have in English no name. The Germans call it *Keimschädigung*. We might call the condition gonadic injury or gonadic abiotrophy" (661n.).

he maintained, "there would be very few, if any morons in our institutions for the feebleminded."[8]

Goddard's shift in outlook undoubtedly derived from the accumulation of evidence that those with ages of intelligence between eight and thirteen—that is, the so-called morons—were quite capable of surviving and even thriving in society. Perhaps the most convincing, if backhanded, piece of evidence for the new view was a rather unsettling finding from the intelligence tests administered to the Army during the First World War. It was that half of those tested registered below normal. The clear implication was that something like half of the American people were feebleminded! That finding by itself raised serious doubts about the value of relying upon scores as a measure of social adequacy. How could feeblemindedness be the serious defect that threatened the nation's well-being, as many psychologists had been proclaiming, if half the people of a reasonably well functioning society were in that defective state? The question alone imposed some needed perspective.

Henry Goddard and others of his persuasion may have abandoned by the early 1920s their fear of being overwhelmed by morons, but Goddard himself never abandoned his belief that feeblemindedness or low intelligence was inherited. True, he had expressed his belief in the educability of morons and admitted the mistake of including them among the truly feebleminded. Yet as late as 1942 he defended his work on the Kallikaks, in regard to both method and conclusions. His so-called "recantation" of 1928 was in fact not a repudiation of old beliefs. His oft-quoted remark that "I think I have gone over to the enemy" referred not to the question of the inheritability of intelligence, as several students of the issue have concluded, but to his new view that the moron— but only the moron—was capable of living and working in society.[9]

Nor was Goddard alone in continuing to believe that intelligence was inherited. As the next chapter will show, the issue remained highly debatable within the social sciences throughout the 1920s, particularly among psychologists. Indeed, as far as many psychologists were concerned, the grip of the concept of the inheritability of feeblemindedness would be weakened only when much more empirical research was available. Some of that came forward with the work at the Iowa Child Welfare Research Station in the 1930s. Experiments there demonstrated that the I.Q. of children born to low-scoring, low-income parents could indeed be raised significantly by improving the home environment of the children. Yet as late as 1941, Robert S. Woodworth, the eminent Columbia University psychologist, was not yet prepared to rule out the inheritance of feeblemindedness. Studies of foster children, he wrote in his book *Heredity and Environment,* show that many children do indeed score higher on tests than their biological parents. "By this test of accomplishment some children of feebleminded parents are proved to have average ability," he recognized. But it would be "going too far beyond the present evidence," he cautioned, to infer "that all or even most children of inferior parents are possessed of average heredity."[10]

When Francis Galton, the "father" of the eugenics movement, died in 1911, the Binet test of intelligence was only beginning to be used in America. Its dissemination over the next decade provided proponents of eugenics with a new and presumably objective basis for determining who was a desirable member of society. Although there is no doubt, as we saw earlier, that eugenics was a popular cause during the years before and after the First World War, social scientists were not overly conspicuous among its advocates. Sociologist Edward A. Ross of the University of Wisconsin, it is true, lent his name and reputation in support of certain eugenic pieces of legislation, and economist Simon Patten of the University of Pennsylvania for a while thought the law ought to encourage eugenic marriages. Sociologists Charles Ellwood of the University of Missouri and Frank Hankins of Smith College also looked favorably upon eugenics for varying periods, Hankins longer than Ellwood. But by far the principal scholarly or professional supporters of eugenics were psychologists—largely the testers or measurers of intelligence—and biologists, especially geneticists. Having discovered the power of Mendelian genetics, many geneticists found it only natural in the early years of the century to support "positive" or "negative" eugenics, that is, encouraging the "fit" to breed and preventing the "unfit" from breeding. Most social scientists, on the contrary, lacked any such "scientific" incentive to advocate eugenics, while finding social reasons to oppose it, given their reformist inclinations and their professional practice of dealing with social groups and social problems.

Invidious comparisons among social groups, for example—the very stock-in-trade of eugenics—could not help but challenge, if not threaten, the socially democratic and reformist values of sociologists and anthropologists. The nationally prominent sociologist Lester Frank Ward, for instance, as early as 1907 pointed out that there was no reason to consider the lower classes as any less worthy genetically than the upper classes. Only the former's lack of educational and other opportunities, he insisted, had left them in that position. Ward, with his deeply held belief in the egalitarian distribution of good genes, may have been too optimistic, but other sociologists also focused on the importance of values as against genes. Mendel's peas may have tasted good and have looked different, wittily commented a sociologist in 1914, but the first quality (taste) is a judgment or value, while the other, color, is merely a description, he accurately pointed out. "Good and bad belong to the world of appreciation of values and are subject to entirely different laws," he reminded the eugenicists.

"What if certain people do stand higher on the Binet tests than others?" he asked. "It is yet to be proved that indicates elemental social value." More than intelligence was at issue in all this testing by eugenicists, contended a Dartmouth sociologist in 1920. He feared that "the whole undistinguished mass of the lowly and obscure are . . . under suspicion."[11]

Some sociologists were not content merely to defend the lower classes against the elitist assumptions of eugenics; they denied that the upper classes, who were usually in the forefront of the eugenicists' concern, were really worth worrying about. "If time permitted," Warren Thompson of Miami University told an audience of sociologists in 1923, "I should like to present . . . some facts that lead me to think that the intellectually superior have . . . been so seduced from natural modes of living and have so insulated themselves from the common stream of human thoughts and sentiments that they are not fit spiritual leaders of mankind." Then he turned the Darwinian tables on the eugenicists, contending that nature "shows clearly that she prefers the lower classes who live simply, who reproduce more or less instinctively, who do not think about the future of the race or of civilization, but who are carrying the burden of the future in the rearing of children."[12]

A year later Thompson appeared in print again to expand upon the same class argument, remarking that "the future belongs to the people who raise children." If the educated cannot take the trouble to bear children, then let those who want children have them. We members of the upper class "should do it gracefully, as it ill becomes us to rail at new immigrants and 'the lower orders' who instinctively understand nature's requirements better than we of the so-called upper classes."[13] After all, his argument implied, if the elite cannot compete reproductively against the lower classes then in a purely Darwinian sense they simply were not the superior class.

Other social scientists doubted the validity of eugenics because of its misplaced faith in the assumption that mental illness or mental traits could be inherited. That is the "old faculty psychology," complained a writer in *Social Forces* in 1926. Just because traits run in families proves nothing about inheritance, he maintained; "a far more creditable explanation presents itself," he suggested, by which he meant "the proved matter of social inheritance *outside* the germ plasm."[14] Social conditioning or culture constituted a large part of the social scientists' objections to eugenics simply because in their judgment explanations in biology and in social science were of quite different orders. Biology could not affect culture or history or civilization. For they were the products of the human mind, and with that biology had nothing to do.

No one made that particular point earlier or more forcefully and combatively, as we have seen, than the anthropologist Alfred Kroeber. "As a constructive program for national progress," he wrote in 1917, "eugenics is a confusion of the purposes to breed better men and to give men better ideals; . . . it is nothing less than a biological short cut to a moral end. It contains the inherent impossibility of all short cuts." Perhaps at best it might be considered a form of social hygiene, but at bottom, he concluded, it is "a fallacy; a mirage like the philosopher's stone. . . . There is little to argue about it." He then drew upon his own distinction between the organic and the superorganic in which each has its own explanatory mode. "If social phenomena are only or mainly organic, eugenics is right, and there is nothing more to be said. If the social is something more than the organic, eugenics is an error of unclear thought." His reasons for arriving at that conclusion he had already stated a year earlier when he had pointed out that the "uninformed man on the street" raised the objections that the educated eugenicist seemed always to ignore: "that eugenics is right enough for hogs but not for men." Too many educated people, he feared, have forgotten the basic truth that "while men are animals, animals are not men, and that however much a human being may have the nature of the pig, he nevertheless has one thing that no pig ever had, namely the faculty for civilization and hence for morality."[15]

Not all social scientists were as clear as Kroeber about the theoretical differences between biological and social explanations. Nevertheless, increasing numbers of them were finding biology inadequate in helping them to understand or ameliorate social problems. Many, therefore, came to see biology as no longer relevant to social inquiry. No amount of environmental change could remove genetic defects, one sociologist conceded, but, he quickly added, it must not be forgotten that an unlimited amount of "eugenics would not avail to solve some of the gravest human problems." He was confident that "it is not to eugenics that we shall look for peace on earth and good will to men." Another sociologist put the matter a little differently, but the message was the same: solving social problems was the aim of social science; eugenics offered little help in achieving that goal. Differences in morality, wealth, and intelligence, issues which were becoming central to sociology, were largely beyond the reach of eugenic measures. The situation of delinquents like the Jukes, he asserted, was similar to that of Negroes; in both cases exclusion from society was the essential cause for their intellectual backwardness or delinquency. Heredity, he conceded, may have played a part, "but the sociologist does not need the inheritance of base characteristics to explain their criminality, prostitution, and poverty." Social circumstances or culture will do a much better job of explaining.[16] For a number of social scientists, in short, their opposition to eugenics derived from its inability to address the issues and provide the solutions which they saw as their primary purpose as professionals.

Although several of the social scientists whose views on eugencis have been quoted above may well have gained their conception of social conditioning as an alternative to inheritance from Franz Boas, Boas himself, in rejecting eugenics, went beyond his own cultural argument. In doing so he revealed a side of his cultural outlook not often exposed. Like Kroeber, it is true, Boas distinguished between the approaches of biologist and anthropologist in seeking the sources of social behavior. Boas particularly emphasized how difficult it was to attempt to draw a line between those behavior patterns that stemmed from environmental circumstances and those that emanated from heredity. He repeated the "agnostic" position he had developed in regard to race differences: "I claim that, unless the contrary can be proved, we must assume that all complex activities are socially determined, not hereditary." Unlike Kroeber, however, Boas also repeated his long-held view that, for the individual, as opposed to the social group, "physical and mental characteristics are hereditary," as the eugenicists maintained, and that by proper selection "certain strains might be selected that have admirable qualities, while others might be suppressed that are not so favored."

That recognition, however, raised a more difficult question and one that quickly separated him from the eugenicists. By what standards should the selection be made? Was it not possible, he asked, that traits thought to be desirable today, would be viewed otherwise in the future? "Such a deliberate selection of qualities which would modify the character of nations implies an overestimation of standards that we have reached, which to my mind appears intolerable. Personally," he admitted, "the logical thinker may be most congenial to me, nevertheless, I respect the sacred ideals of the dreamer who lives in a world of musical tones, and whose creative power is to me a marvel that surpasses understanding."[17]

Eugenics, he flatly stated, ran counter to human nature in that it assumed that human beings could regulate their reproductive behavior in support of a set of socially determined goals. "It is exceedingly unlikely," he maintained, "that a rational control of one of the strongest passions of man could ever succeed." From his own knowledge of human customs and habits, he had concluded that "such an ideal is unattainable, and more particularly, that the emotions clustering about procreation belong to those that are most deeply seated in the human soul, and that they are ineradicable." Boas, in short, was coming close to identifying human instincts. Eugenicists assumed, he continued, "that the ideal of human development lies in the complete rationalization of human life," a development that he thought flew in the face of experience. Ordinary citizens, he predicted, were not likely to accept the restraints necessary for a eugenics program, since even in matters of minor importance, evasion of regulations is "of common occurrence." Surely evasions would be "infinitely more common in questions that touch our inner life so deeply" as those that eugenics seeks to regulate. He was convinced that the general public's "'instinctive' repugnance against eugenic legislation is based on this feeling."[18]

In his final objection to eugenics, Boas, the prime advocate of a cultural interpretation of man, skirted very close to accepting a biological basis of human nature. One of the admitted attractions of eugenics, he acknowledged, was its aim of "raising a better race and to do away with increasing suffering by eliminating those who are by heredity destined to suffer and to cause suffering. Particularly attractive, then, was "the humanitarian idea of the conquest of suffering, and the ideal of raising human efficiency to heights never before reached." To that ideal his response was bold and uncompromising, but its premise smacked of biology: "I believe that the human mind and body are so constituted that the attainment of these ends would lead to the destruction of society." The burden of his objection was that for human beings suffering was at once desirable and necessary. "The wish for the elimination of unnecessary suffering," he insisted, "is divided by a narrow margin from the wish for the elimination of all suffering." Such a goal "may be a beautiful ideal," he conceded, but "it is unattainable." The work of human beings will always require suffering and "men must be willing to bear" that suffering. Besides, many of the world's great works of beauty "are the precious fruit of mental agony; and we should be poor, indeed," he was convinced, "if the willingness of man to suffer should disappear." The worst thing of all, he warned, was that if this ideal were cultivated, "then that which was discomfort yesterday will be suffering today, and the elimination of discomforts will lead to an effeminacy that must be disastrous to the race."[19]

To Boas, "effeminacy" was the tendency of the people he saw around him to reduce suffering in the name of efficiency. "We are clearly drifting toward the danger-line," he feared, "where the individual will no longer bear discomfort or pain for the sake of the continuance of the race, and where our emotional life is so strongly repressed by the desire for self-perfection—or by self-indulgence—that the coming generation is sacrificed to the selfishness of the living." In modern society he saw a repetition of that tendency, which "characterized the end of antiquity, when no children were found to take the place of the passing generations." To the extent that the "eugenic ideals of the elimination of suffering and self-development" are fostered, the sooner human beings will drift "towards the destruction of the race," he gloomily predicted.[20] The irony of Boas's objections was that similar apocalyptic fears animated the eugenicists demands for their program. They saw the danger and the inevitable national decline as emanating from the reproductive reluctance of the educated classes, whereas Boas seemed to embrace all classes in his jeremiad.

Despite the telling and steady flow of objections to eugenics emanting from anthropologists and sociologists, the death knell for eugenics was not sounded by them. That job was accomplished by geneticists as they learned that their science was being both misunderstood and misused by public advocates of eugenics. As early as 1923, Edwin Grant Conklin, professor of biology at Princeton University and a social elitist by preference, alerted the public to the limitation within which any biological explanation of human behavior must operate. To predict the physical characteristics of a child from the character of its parents, Conklin assured the readers of *Heredity and Environment in the Development of Man,* is not too difficult. But "where the character is an extremely complex one such as intellectual ability, moral rectitude, judgment and poise"—the very characteristics that eugenics was intended to foster—"it will probably never be possible to predict." Anyone would be a "bold prophet," he maintained, "who would undertake to predict the type of personality which might be expected in the children of a given union. Some very unpromising stocks have brought forth wonderful products, " he reminded his readers.[21]

The well-known biologist from Johns Hopkins, Raymond Pearl, forthrightly denounced eugenics in a 1927 article he published in the popular *American Mercury* magazine. "The literature of eugenics has largely become a mingled mass of ill-grounded and uncritical sociology, economics, anthropology and politics," he caustically announced, and "full of emotional appeals to class and race prejudices, solemnly put forth as science and unfortunately acknowledged as such by the general public." It is worth recalling, as a measure of the sharp alteration in outlook, that as recently as 1908, Pearl had predicted that "the time will come when not only will eugenics form an integral part of the teaching and research work of the great universities, but also will come to be regarded as a legitimate field for the Federal Government."[22]

Pearl's colleague at Johns Hopkins, Herbert S. Jennings, the nationally prominent biologist, was equally disillusioned with eugenics. Even before Pearl had expressed his misgivings, Jennings had publicly deplored the iron determinism of the extreme eugenicists, contending that only outdated genetics espoused such rigidity. Then, in 1930, he publicly condemned the elitist and racist assumptions that always seemed to underpin the writings of supporters of eugenics. "Both racial arrogance and the desire to justify present social systems find a congenial field in eugenic propaganda," he wrote in an article on "Eugenics" in the *Encyclopedia of the Social Sciences.*[23]

Once the geneticists and biologists began to lose respect for eugenics as a movement, those social scientists who had supported eugenics, principally psychologists, also began to have doubts. By 1931, for instance, even Lewis Terman, always a strong believer in both intelligence testing and eugenics, recognized the handwriting on the laboratory wall. In a letter to Charles B. Davenport, the best-known proponent of eugenics among the nation's geneticists, Terman rehearsed the problems facing the movement in which they had a common interest. "As you know," he began, "there are a good many psychologists and anthropologists these days, also sociologists, who are inclined to argue that the intelligence that individuals develop during childhood and adult life is determined largely if not entirely, by his cultural environment and formal training. If that is true," he reasoned, "then eugenics has no place as far as intellectual differences in human beings are concerned. Its principles would apply, if at all, only to physical traits," he reluctantly concluded.[24]

And it was true that by the 1930s the popularity of eugenics declined precipitously, not only among social scientists and biologists but among the public. Unfortunately, however, the recourse to sterilization of allegedly mentally defective persons, which had been one of the few legislative successes of the eugenics movement, was not abandoned. Laws enacted during the first three decades of the century remained on the statute books in 27 states into the 1970s, and on the basis of them hundreds of inmates of mental institutions were sterilzied long after the movement ceased to hold much interest for either the public or the scientists who had once urged the practice upon legislatures and public alike.

Lewis Terman's cautionary words to Charles Davenport can also serve as a measure of the gulf between psychologists, especially those wedded to intelligence testing, and those social scientists who had rejected eugenics earlier in the century. Terman remained a eugenicist throughout the decade. And the principal reason he and many other psychologists persisted in that association was their commitment to the validity of intelligence testing. The criticisms that had been so effective against eugenics touched only peripherally the assumptions on which testing rested. Eugenics, after all, was a movement to control or regulate the distribution of socially desirable traits in society. The methods of regulation could be called into question without denying the value of an instrument for identifying where the good traits might be found. Equally important was the conviction of testers and others that the potentialities of testing and the practical results that had already flowed from it were notable. The objections that had made it increasingly unacceptable as a measure of feeblemindedness or delinquency hardly justified its being doubted or rejected as a serviceable research tool for other purposes. Thus, during the 1920s, testing continued to be a widely used if controversial means of determining racial and ethnic differences.

Before one can understand how and why those uses of testing also came under attack and were ultimately discarded by social scientists, another ideological battle within psychology needs to be examined first. From this controversy the discipline would sever yet another ancient link to biology: the belief in human instincts.

As the work of William James and other early American psychologists made clear, modern psychology began with the assumption that much of human behavior was dependent upon instinct. Darwinian evolution also supported—indeed insisted upon—the idea that the continuity between animals and human beings manifested itself not only in bodily form but in action or behavior as well. In animals human beings could observe their primitive selves.

Given this close linkage between an acceptance of human instincts and the animal origins of human beings, it is somewhat ironic that the first sustained arguments denying instinct in human beings should come from a psychologist who was a confirmed proponent of animal experiments. The psychologist was John B. Watson, the founder of behaviorism. Watson was actually one of the earliest ethologists in American science. Early in his career he spent a number of months in 1900 observing sea birds in their natural habitat on the Dry Tortugas Islands, off the Florida Keys. Upon his return to the United States he increasingly used animals in developing an experimental psychology to replace what he considered the overly speculative or deductive psychology of his time.

Darwin himself and some of his immediate followers, like George Romanes, in their effort to understand human beings in an evolutionary context had drawn heavily upon animal behavior. The practice served also to demonstrate the continuity between animals and human beings. By the time that Watson began his work, however, the reading of human consciousness or motives into animal behavior had been roundly criticized and largely rejected by psychologists and zoologists alike. "Anthropomorphism" was the condemnatory epithet. It is true that Margaret Floy Washburn, the eminent Cornell psychologist, had published the *Animal Mind* as recently as 1908, but she was almost the last of a moribund breed. Biologists like Jacques Loeb and psychologists like C. Lloyd Morgan represented the new order in which only animal behavior was studied; consciousness, motives, awareness were not to be inferred from behavior—they were subjective when "objectivity" was the essence of good science.

Watson followed that lead, paying attention to behavior only, a practice that soon led him to take a similar approach to human behavior. In 1913, in his first public pronouncement on his new psychology of behaviorism, he said his aim was to demonstrate "the necessity for maintaining uniformity in experimental procedure and in the method of stating results in both human and animal work." For in seeking to develop "a unitary scheme of animal response," he recognized "no dividing line between man and brute."[25]

In using that phrase, it is worth noticing, Watson was not following Darwin as closely as it might appear. In fact, despite his commitment to the study of animal behavior, Watson saw little relevance of Darwinian evolution to his behaviorism. It is true that he intended psychology to be a biological science. "With animals I was at home," he wrote in his autobiography. "I felt that in studying them, I was keeping close to biology with my feet on the ground." Along with Darwin he saw human beings as animals, but he focused his attention entirely upon the responses to outside stimuli that animals and human beings had in common. The *continuity* between animals and human beings, which had been Darwin's primary interest, was considered irrelevant by Watson. As psychologist Wesley Raymond Wells warned in 1923, too many psychologists were "losing sight of the biological, evolutionary background of present human behavior." If human beings are indeed animals, as the biologists tell us, Wells continued, "and if we believe in continuity in evolution, we should bring to the study of human behavior . . . Mendelian inheritance, natural selection, and the like, as the study of biology gives us." To Watson, though, the animality of human beings meant simply that behavior revealed all that could be scientifically known; consciousness was subjective and therefore irrelevant. Moreover, by Watson's time, unlike Darwin's, consciousness was perceived to be a peculiarly human phenomenon, a perception that was the real source of Watson's dissatisfaction with establishment psychology.

As early as 1910, Watson told his fellow student of animal behavior, psychologist Robert Yerkes, that he didn't "believe the psychologist is studying consciousness any more than we are. . . . All of our sensory work, memory, attention, etc., are parts of definite modes of behavior." By 1913 he was openly criticizing his fellow psychologists for continuing to deal with what he scornfully called "philosophical" questions of sensation, perception, imagery, and so forth. His own aim, as a teacher of psychology, he announced, was to see that his students were as ignorant of such ideas as were "the students of other branches of science." What was needed was a new psychology, one which made "*behavior*, not *consciousness*, the objective point of attack." Or as he wrote in his first textbook, "psychology, as the behaviorist views it, is a purely objective, experimental branch of natural science which needs introspection as little as do the sciences of chemistry and physics." It followed, therefore, "that the behavior of man and the behavior of animals must be considered on the same plane; as being equally essential to a general understanding of behavior." The psyche he denounced as a *"deus ex machina"* that traditional psychologists felt a need to introduce when they dealt with human behavior. Ultimately Watson became sufficiently convinced of the absence of any need to study consciousness that he concluded human thought was possible only through vocalization in some fashion or from outside stimuli.[26]

The ultimate goal behind Watson's behaviorism, he made clear as early as 1914, was "the prediction and control of behavior." With that, psychology for the first time would be practical, usable by "the educator, the physician, the jurist, and the businessman." Through behaviorism a better social order would be feasible, a goal that many psychologists found appealing as well as professionally novel. The power of that appeal certainly helps to account for Watson's influence on a wide spectrum of his professional colleagues. Behaviorism, though not always Watson's precise brand of it, soon came to dominate American psychology. It remains to this day a peculiarly American strand of the discipline.

Watson's goal of attaining control over human behavior, however, did not lead immediately to an attack upon instincts in human beings. That came only with his shift in experimental subjects from animals to human infants. In a lecture to an audience of educators in 1917, he reported that he had reduced the number of emotions in babies to just three: fear, rage, and what he called joy or love. In a professional field in which dozens of instinctive emotions had been identified over the years, this was a breakthrough that could not fail to appeal to many psychologists. For Watson, his new approach signaled an important step toward the control of emotions and human behavior. For if psychologists believe there are "hundreds of emotions all of which are instinctively grounded," he pointed out, then the likelihood of controlling them is remote. But if, according to his view, all of these dozens of emotions are "due to environmental causes, that is, to habit formation," control suddenly became feasible.[27] The intellectual shift that stood behind these promises was Watson's personal discovery of the conditioned response. His point was that the association of a particular event with the evocation of one of the three basic emotions spawned all of the large number of allegedly instinctive emotions that psychologists had been enumerating, cataloging, and studying for decades.

Despite his denial of the need to study consciousness, and despite his reduction of the number of emotions to three fundamental ones, Watson was not yet ready to deny instincts in human beings. He played down and even denied certain instincts that McDougall and Edward Thorndike had put great store by, such as the maternal instinct in females and the hunting instinct in males. At the same time he insisted "we are not denying . . . that there are some instinctive factors here." Chapter 8 of his 1919 textbook was entitled "Hereditary Modes of Response: Instinct" and it contained a list of human instincts. He specifically singled out sexual attraction, for example, as instinctive, along with a mother's nursing and fondling of a baby. Nevertheless, signs also slipped out that he was moving in the direction of denying outright human instincts. Thus he could not resist concluding his list of human instincts with the observation that "we are inclined to take the point of view here ... that most of these asserted instincts are really consolidations of instinct and habit." By the second edition of his text, the title of Chapter 8 had been revised though the text remained unchanged; it was now "Unlearned Behavior: Instinct," rather than "Hereditary Responses." By the third edition in 1929, the chapter was entitled "Unlearned Behavior," with "Instinct" now enclosed within quotation marks.[28]

In the interim between the first and third editions of his textbook, Watson had published *Behaviorism,* his popular book that marked his complete break with instinctive behavior in human beings. It contained what is perhaps the most famous single passage in the history of American psychology: "Give me a dozen healthy infants, well-formed, and my own specified world to bring them up in and I'll guarantee to take any one at random and train him to become any type of specialist I might select—doctor, lawyer, artist, merchant-chief, and, yes, even beggar-man and thief, regardless of his talents, penchants, tendencies, abilities, vocations, and race of his ancestors." It is true that he followed this assertion of extreme environmentalism with the quasi-qualification that he was "going beyond my facts ... but so have the advocates of the contrary and they have been doing it for many thousands of years."[29]

Actually, Watson's concession went no distance at all in measuring the extent of his exaggeration. For the fact was that he had little or no evidence on which to base his assertion that he could shape a person into anything simply through the application of the principles of conditioned responses. All he was really doing was denying the validity of the evidence or the arguments of those who asserted an instinctive or a biological basis for human behavior. His approach was almost identical with that which Boas and Kroeber took years earlier in anthropology in substituting culture for biology. Unlike the two anthropologists, neither of whom had denied the importance of heredity in shaping individuals, Watson, in his formulation, had extended the application of his environmentalism to the individual as well as to the social group. It comes as no surprise, then, that he entitled a section of his *Behaviorism:* "Concept of Instinct No Longer Needed in Psychology."[30]

The radical environmentalism at which Watson arrived by 1925 was not his principal catalyst to the profession. That was, rather, his early attack on introspection or consciousness in behalf of a truly experimental discipline. Indeed, in 1913, the very year in which Watson published his lectures on behaviorism, James R. Angell, a leading psychologist at the University of Chicago, and later president of Yale, acknowledged, only half humorously, the spread of the oppo-

sition to the traditional view. "Soul," he noted, has recently been identified as
no longer a part of the psychologists' vocabulary. "The term 'consciousness'
appears to be the next victim marked for slaughter," he suggested, "and as one
of the claimants for its fading honors, we meet the term 'behavior.'"[31]

Behaviorism came to stand for many things in the years after Watson made
it a by-word, but its principal effect upon psycholgists was to encourage experimentation and the repudiation of introspection or consciousness. Few psychologists were prepared to follow Watson in rejecting, root and branch, those
ancient concepts of the discipline or to cease to pay attention to what human
beings felt and thought. For as Angell warned, to insist upon behavior as the
sole object of psychological study "would involve trespassing freely on the pre-
serves of biology, physiology, and neurology on the one side and upon those of
the social sciences on the other." In either case, Angell feared, the psychologist
faces the serious threat of finding himself "annexed, appropriated, and in general swallowed up by the owner of the territory which he invades." For if psychology "abandons the stronghold of consciousness as her peculiar institution,"
the discipline would soon learn that it had sacrificed its autonomy; psychology
would "become a mere dependence of biology or some other overlord," Angell
warned.[32]

Watson's rejection of consciousness may have been a source of his professional influence, but the profession as a whole never went so far as to read
consciousness out of the discipline. By the same token, however, his unrelenting
emphasis upon behavior, to be studied as objectively as possible through experiments with animals and human beings, prompted many psychologists to
rethink a concept like instinct, the justification of which had long depended
upon little more than casual observation, deduction, and Darwinian history.

Even before psychologists had begun to follow Watson into a critique of
instinct, social scientists who were not psychologists had made known their
reservations and sometimes serious doubts about its validity for the study of
human behavior. Sociologist Carl Kelsey, of the University of Pennsylvania,
already influenced by Boas and other anthropologists to question the concept
of race, called instinct in 1910 a "recourse of baffled thinking."[33] About the
same time, Luther L. Bernard, also a sociologist, after examining the extant
literature on the allegedly biological bases of crime, was appalled by the flimsy
evidence it contained. He then undertook to write a major sociological treatise
on the inadequacies of instinct, which he published in 1924.

Psychologists' own criticism of the concept of instinct arose from several
sources. The earliest was that reflected in James Angell's fear that, if consciousness were abandoned, psychology would become a mere appendage or worse of
some other discipline. Even before Angell had made his own fears evident,

Charles H. Judd, in his presidential address to the American Psychological Association in 1909, had rejected instinct on just those grounds. The human target of his criticism was William McDougall, then and later the principal proponent among psychologists of human instincts. Judd made no secret of his refusal to accept biology as a basis of human activity. "The fact is human civilization has not been toward instincts and emotions," he told his fellow psychologists, "but away from them." Instinct may have been the root of language, he admitted, but today language "is as intellectualistic a function as can be found in the world." His central objection to McDougall's insistence upon human instincts was McDougall's "attempt to bring human action back to the fundamental formula for all animal behavior." Assuming a frankly anti-Darwinian and environmentalist stance, Judd denied that human behavior is "aimed at maintaining oneself within the environment," as Darwin had said; "it is aimed rather at complete remoulding of the whole environment." And the proper instruments for the reshaping of the social order to Judd, as to Angell later, were "intellectual comparison and deliberation, not emotion."[34]

Knight Dunlap's dissatisfaction with the concept of human instinct also focused on McDougall's irrepressible defense of the idea, but contrary to Judd's, Dunlap's objections were more pointed and technical, a result, very likely, of his being a colleague of John Watson's at Johns Hopkins. Along with Watson, Dunlap drew a distinction between those kinds of activities that were obviously physiologically based, such as an infant's sucking, and those that he saw as "teleologically defined," that is, actions that were intended to achieve a goal, such as McDougall's instincts for "flight," "repulsion," "pugnacity," and "curiosity." The latter group he thought impossible to defend or to employ in research just because they were teleological, which, as he pointed out, was a subjective judgment, not an objective denotation of behavior. The instincts were assumed or deduced by the psychologist, he complained, rather than explained. They were just assigned names derived from the apparent purpose they served; no explanatory theory was offered. Under such circumstances, Dunlap recommended that psychologists "cease talking of 'instinct.'"[35]

J. R. Kantor, a psychologist at Indiana University, elaborated in 1920 upon Dunlap's objections to the teleological nature of instinct theory. He was willing to accept reflexes as a part of human behavior, but so-called "instinctive behavior" he insisted was actually shaped by thought and habit. Indeed, "most of our ordinary behavior is instinctive conduct, but this does not mean in any sense that complex actions such as we perform are the expression of a few inborn impulses." To believe that they are, he scornfully wrote, was to resort to a form of "scholastic simplicity which is genuinely subversive of all understanding of human behavior."[36]

By 1923, Kantor's vehemence had not abated one whit. His answer to the question of the relationship between social psychology and instincts was that "there is no relationship. Social psychology," he maintained, "is essentially a science of post-infantile human activities and since instincts clearly have no connection with such behavior there is no place for them in social psychology." If, as he believed, psychology, both social and individual, is based on data "of actual concrete responses of organism to surrounding objects and conditions . . . what room is left," he asked, "in a scientific psychology for any animistic or teleological process?"

Human beings may be animals, Kantor conceded, but they are also machines, physical objects, and social beings, designations that in themselves explain the variety of disciplines that collect data and study human beings. That also means, he emphasized, that "we have no right to reduce one type of data to the other." Thus he concluded, as had Angell and Judd in psychology and Kroeber in anthropology, that "the particular cultural responses forming the subject matter of social psychology must be handled on their own level without any admixture into them of biological processes." The analysis of human social behavior, in sum, required a different theory from that employed by biologists to study their subject. Therefore, he repeated, "a relationship between social psychology and instincts is non-existent."[37]

For all of the vehemence of Kantor's determination to expunge the concept of instinct form social psychology, he never quite assumed the extreme environmental posture that Watson shaped for himself. Kantor was talking about human social relations; he said nothing about sources of individual behavior. In that sense he followed not Watson but Boas and Kroeber. No social scientist, however, whether psychologist or sociologist, was more intent upon excluding biology from all levels of psychology than Zing Yang Kuo. His hostility to the concept of human instinct was so deeply held and strongly expressed that McDougall once described him as "outWatsoning Watson."[38]

In truth, Kuo's first article opposing the idea of instinct in psychology, published in 1921 when he was still an undergraduate at Berkeley, was sufficiently incisive to evoke a response from the famous McDougall himself. In his article, Kuo denied even a sex instinct in human beings, something which Watson at that time was not yet prepared to do. Sexual intercourse, Kuo contended, was the result of imitation and habit, his point being "that all our sexual appetites are the result of social stimulation. The organism possesses no ready-made reaction to the other sex, any more than it possesses innate ideas."[39] The more Kuo reflected on his objection to biological bases for human behavior, the more radical his environmentalism became.

By 1922 his quiver of anti-instinct arrows included Dunlap's argument that the concept carried no meaning to psychologists because it offered no explanation, only a name for something that could otherwise not be explained. As a consequence, he complained, instinct "is a stumbling block in the way of experimental genetic psychology." Already far in advance of Watson or Dunlap, Kuo was now ready to deny instincts in animals as well as human beings. Not even Luther Bernard, the fanatical opponent of instinct in sociology, Kuo proudly pointed out, had repudiated instinct so thoroughly.

Drawing upon Watson's emphasis upon experimentation, Kuo looked forward to a psychology in which hereditary factors were entirely omitted inasmuch as no laboratory studies or genetic laws had demonstrated that heredity was a basis for behavior. "The Mendelian experiments deal strictly with definite morphological features," he pointed out in 1924, "and it is sheer nonsense to speak of mental traits in terms of Mendelian ratios, when such traits are not reducible to definite physico-morphological facts." Almost in exasperation, he argued that "the time seems to have come when we can no longer tolerate the tyrannic domination of biology in psychology." Unlike the geneticist, he continued, the psychologist takes the organism as a given. "Behavior is always an interaction between the organism and its environment." Given the history of the organism and stimulation, "the psychologist has the task of determining the response. He needs the concept of heredity as much or as little as the concept of god."[40]

All responses of animals and human beings, Kuo insisted, were in some fashion learned, rather than inherited. "Strictly speaking, except the first movement of the fertilized egg," he wrote in 1929, "there is no real unlearned response. Every response is determined partly by the present stimulation and partly by the past history of behavior of the organism." This was indeed "outWatsoning Watson." It was the most radical statement of the environmentalism that was beginning to pervade all of the social sciences by the opening of the 1930s. (It was also a conception of environmentalism that would be difficult to sustain in the face of evidence from the new ethological studies that would begin to appear in the 1970s.) Biologists might rely upon heredity in analyzing an organism, but those who study behavior, such as psychologists, Kuo maintained, should study "the stimuli that cause behavior. . . . Behavior is not a manifestation of hereditary factors, nor can it be expressed in terms of heredity; it is the direct result of environmental stimulation." As he further explained, behavior "is a passive and forced movement mechanically and solely determined by the structural pattern of the organism and the nature of environmental forces."[41]

If one carries Kuo's last statement about the origins of behavior to its logical conclusion, his environmentalism would seem to leave no place for any group behavior. Since, by genetic definition each individual (except for identical twins) is different, behavior would vary, too, from individual to individual even if the environmental influences were the same. But then, Kuo was an experimental not a social psychologist, and perhaps that explains why few other social scientists espoused his radical form of environmentalism (and individualism).

Other social scientists, especially sociologists, may not have been as radical in their expression of environmentalism, but their rejection of instinct came early and often. Some of them, like Luther Bernard, professor of sociology at the University of Missouri, drew upon biological theory itself in 1921 to pronounce instinct in human beings unacceptable. Anticipating Kuo's objection, Bernard pointed out that "one cannot inherit activities," only structures, "the functioning of which," Bernard explained, "determines the action patterns." Or as he put it later in a book: "We do not inherit abstractions, but concrete biological organs and structures." For that reason he thought "to call ideals or social and ethical values . . . such as goodness, criminality, democracy, or conservatism, instinctive or inherited is . . . manifestly unjustifiable."[42]

Sociologist Robert Gault of Northwestern University also drew on the latest biological information in order to justify his doubts about human instincts. Gault's dissatisfaction, however, did not emerge from or follow Bernard's argument from genetics; rather it derived from his acquaintance with recent work in animal behavior. His point was that some scientific studies had demonstrated the widespread presence of learning among animals, suggesting that "there are but few instincts, properly speaking, and that these are less specific than generalized," by which he meant that there are "natural dispositions that determine within wide limits what habits . . . shall develop" when environmental circumstances are favorable. He then cited studies of birds that had not known their proper species songs when raised in isolation, and experiments in which sparrows were "taught" songs by canaries.[43]

Among social scientists of the time, Gault's insight was rare. Yet from today's vantage point it came close to what a modern ethologist might say about instincts in animals and perhaps in human beings as well. A clear recognition of the influence of learning and environment on behavior, but without a denial of innate propensities, either. Interestingly enough, Gault's unusual ethological approach was cited by a Dartmouth sociologist in 1920, in support of his own conclusion that "whatever man's heredity, it *always bears a contingent character*—life and conduct should be talked of in terms of tendency, *never in terms of rigid inevitability*," by which he meant that heredity did not determine actual behavior but only a generalized response to circumstances. These, too, were words that today's students of animal behavior and sociobiology could easily subscribe to.[44]

The more common objection of sociologists and anthropologists to instinct was that it could not account for the great diversity of human social behavior. It was certainly true, Luther Bernard conceded in his 1924 book-length assault upon the concept of instinct, that certain types of human behavior were duplicated around the world, but that universality, he maintained, constituted no proof that the patterns were inherited. Too many of them change over time, thereby calling into question not only their universality but their biological roots as well. Indeed, the very diversity of social patterns among the peoples of the world, anthropologist Robert Lowie remarked, made it clear that neither biology nor psychology could account for human social behavior. "Psychology knows of no instinct that causes a man to avoid his wife's mother. . . . If we wish for an explanation of the phenomenon," Lowie reminded the biologically inclined, "we must look in another direction: we must connect the cultural facts not with psychological facts but with other cultural facts."[45]

To University of Chicago sociologist Ellsworth Faris, too, the concept of human instincts as psychologist McDougall described it fell to the ground when the diversity of human social patterns was recognized. One African tribe hated and feared the birth of twins, he recalled from his African research, while another was extraordinarily fond of them. Viewed separately, the behavior of each tribe might be identified as the expression of an instinct, he admitted, but taken together the two patterns contradict any such conclusion. As a result, Faris concluded in 1921, "we are compelled to assign the phenomenon to nurture and not to nature." There may be reflexes in children, as John Watson had reported, Faris admitted, but that was as far as he would venture in accounting for behavior in biological terms. American boys may pass through certain growth stages, he added, "but no statement can be true of all men everywhere, so long as cultural inheritances differ so profoundly." His recommendation to his fellow social scientists was to confine their research and study to "social values, social attitudes, desires, wishes, and organizations. . . . Nothing but confusion and disappointment," he warned, "will result from regarding instincts as factual data which can be observed, classified and explained." A few years later his denial of instinct came close to Kuo's extreme environmentalism; he reversed the very causal assumptions that had made psychologists like McDougall accept human instincts in the first place: the cross-cultural prevalence of a given human response. "Instincts do not create customs," Faris declared in 1925. "Customs create instincts, for the putative instincts of human beings are always learned and never native."[46]

The behavior pattern that McDougall thought the innateness of which was least subject to doubt—the maternal instinct—was a favorite target among social scientists for specific rejection. It was at most, Luther Bernard contended, echoing Watson and Kuo, a series of responses "to touch, temperature, and odor stimuli from the child," which, in turn, had been evoked by stimuli from the mother. Faris wondered how anyone could believe in a maternal instinct considering the ineptness of new and untrained human mothers.[47]

Sometimes animosity accompanied the intensity that characterized much of the discussion of instinct in those years. For example, in 1927 sociologist Edward Reuter denounced the term "biological sociology," which some sociologists were then using, as a sure sign "that sociologists sometimes combine words without expressing thought."[48]

As historian Hamilton Cravens observed in his *The Triumph of Evolution*, the rapidity with which the objections to instinct spread among social scientists was striking. Even psychologists, the professional colleagues of instinct's principal advocate, William McDougall, were turning away from it by the middle 1920s. Why did an idea, once so uncriticized as to be a fundamental principle of psychology and other social sciences, crumble so fast once it came under attack? Part of the reason, as Cravens saw it, was the arrival of many new workers in the psychological vineyards, men and women attracted to the novel idea of experimentation in psychological research. And the spread of experimentation was indeed a fertile source for the change in ideas if only because there had been almost no experimental evidence to support the concept of human instincts in the first place. Its prime justification had been the old Darwinian assumption that a continuity existed between animal and human actions. Zing Yang Kuo thought that the deductive approach in psychology went back even farther in history. "It is very unfortunate," he wrote to Robert Yerkes in 1922, "that many of the present-day psychological concepts are still inherited from Aristotle. I think experimental work is the only cure for the metaphysical mind in psychology." Kuo considered the prominence of the deductive approach in psychology as his principal reason for "denying instincts." The very concept he viewed as "unexperimental and I therefore consider it a stumbling block in experimental psychology." The drive to experimentation affected even McDougall. Although he had no experimental evidence to support his insistence on human instinct, by 1921 he, too, was calling for more experimentation in his discipline. "The deductive method . . . has been responsible for most of the monstrosities which have long defaced the textbooks of psychology," he complained, "and has been a principal obstacle to the progress of the science."[49]

Many of the young sociologists and psychologists coming to the fore after the First World War were also influenced against the concept of instinct by Watson's emphasis upon environment and Boas's cultural explanation for human social behavior. Sociologist Ellsworth Faris, for example, cited Watson's experimental refutation of William James's contention that fear in general derived from an instinctive fear of strangers. Watson, Faris noted, had presented a variety of animals to infants without evoking any fear in them, though the babies soon learned to fear a particular animal when it was offered simultaneously with a loud sound. Kantor, too, drew on Watson's work with conditioned responses to explain his own doubts about the need for a concept of human instinct. As we have already noticed, a number of sociologists began to doubt instinct from their familiarity with Boas's concept of culture as the primary explanation for differences in human social behavior. Luther Bernard, for example, may well have become acquainted with Boas's ideas from the mouth of the master himself, since he was studying at the University of Chicago when Boas taught there for a year.*[50]

McDougall's public defenses of his theory of instincts in the 1920s suggest another possible explanation for social scientists' rapid and wholesale flight from the concept. It was McDougall's conviction that, if human instincts were successfully removed from psychological theory, as its opponents certainly intended, there would be "a return to the social philosophy of the mid-nineteenth century, hedonistic utilitarianism, with its belief in the absence of all significant differences between individuals and between the races of mankind, and the belief in the limitless perfectibility of mankind by the processes of education alone." Although McDougall's own social values come through loud and clear in his remarks, he had indeed placed his finger upon a significant motive in the minds of many who rejected human instincts, eugenics, and feeblemindedness. Unlike McDougall, they were often much concerned with reforming society, expanding opportunities, and ameliorating suffering. Eugenics or a belief in human instincts, as more than one social scientist remarked, was not going to help much in achieving those goals. Social conditioning and the recognition of culture as an alternative to instinct promised much more in bringing about the social changes they sought.[51]

Almost half a century later, two psychologists, in seeking to explain what McDougall called "the hegemony of behaviorism in American psychology," seconded his contemporary analysis. It all had to do, Lauren G. Wispé and James Thompson thought, with the commitment of Americans in general to individualism, independence, and democracy. Such values assure an American that he can "'shape' his own destiny." On the other hand, "evolutionary theory, which smacks of predetermined factors over which the individual has no control, or

---

*Bernard may have had an additional reason for rejecting instinct. As he wrote with some pride in 1923, "sociology is at last shaking itself free from biological dominance and is developing an objective and a method of its own. Thus it promises to be a science, not merely a poorly organized and presumptuous branch of biology, as some biologists formerly seemed to regard it." Quoted in Cravens, *The Triumph of Evolution* (Philadelphia: Univ. of Pennsylvania Press, 1978), 121.

concepts like 'instincts,' which connote in the popular mind the idea of immutable behavior" hardly fit comfortably into that American outlook.[52]

Influential as the opponents of human instinct undoubtedly were, they did not quite sweep the field clean. Psychologist Edward Tolman, Kuo's own mentor at the University of California, continued to assert that *"instincts cannot be given up in psychology."* Perhaps the simple version of the concept advanced first by James, then elaborated upon by McDougall, and finally demolished by Kuo could not be saved, Tolman conceded, but some kind of drives or biologically based forces in human beings were surely necessary to the discipline's conceptualization. At the end of the decade other psychologists were raising questions about the determined effort, begun by Watson, to rid the discipline of instinct. One commentator, for example, noted in 1927 that three recent textbooks on social psychology, two by psychologists and one by sociologist Luther Bernard, had rejected the doctrine of human instincts. Yet this seemed to psychologist H. G. Wyatt to be not a true picture of what they were doing in fact. Each may have denied instincts by name, he perceptively observed, yet in fact they had merely substituted another form of inborn drive or impulse, a practice that amounted to smuggling instinct in by the back door, he complained.[53]

Another psychologist, Max Schoen, also picked up the nettle of nomenclature, but more forthrightly. He boldly suggested dropping the term entirely and substituting "native behavior," a general term for innate behavior in all animals, but with the proviso that it was least evident in human beings. The great majority of human behavior patterns, according to Schoen, were *"determined* by environment and training." In that way, he optimistically suggested, both Watson and McDougall would be considered right! Or as another aspiring Henry Clay of psychology phrased the issue by the end of the twenties: "As a science, psychology may sagely be presumed to be free from the bias of biological determination on the one hand and from the environmentalism which characterizes much of recent sociology on the other."[54]

Other psychologists followed Schoen's example, suggesting alternative terms in order to escape the increasingly uninteresting yet seemingly endless controversy. How about "maturation"? suggested a psychologist in 1930, to which soon came the fairly predictable response from another psychologist that "maturation" no more escaped the weaknesses of instinct than any of the other alternatives.[55]

By the mid- and late 1930s, discussion of instinct theory almost disappeared from the psychological journals. One modern psychologist has counted the number of articles in which the word "instinct," as opposed to those in which "drive," "motivation," and "reinforcement" appeared in titles cited in *Psychological Abstracts* by four-year periods. Of the 113 items listed between 1927 and

1930, "instinct" appeared in 68 percent; by 1935 the proportion had fallen to less than 40 percent and by 1950 the figure was close to 10 percent.[56]

The figures measured two things at once. The first was the decline in the discussion of instinct among psychologists; the second was the substitution of another kind of innate or biological explanation for behavior. This last was measured in the psychological literature by the rise in the number of references to terms like "motivation," "drive," or "reinforcement." As will appear in a subsequent chapter, among students of animal behavior the question of instinct traveled a similar underground road only to emerge redefined in the 1950s and 1960s in the work of European ethologists Nikolaus Tinbergen and Konrad Lorenz.

Though the concept of instinct came under severe and effective attack early in the 1920s, that was not true of the idea that races and ethnic groups differed in mental abilities. In a sense, because of the new emphasis upon testing, that concept achieved a fresh prominence in the 1920s. As a consequence, the whole decade would be consumed in removing the idea of racial differences from a central position in psychological research.

One of the difficulties under which proponents of instincts labored in the 1920s was the lack of experimental evidence for their position. Virtually all of the experimental results in psychology, as we have seen, cast doubt upon the validity of the concept. And when to that experience in psychology one added the fundamental shift in assumptions about the sources of social behavior which many sociologists and anthropologists were making, a strong case for human instinct was hard to mount. On the other hand, the assumption that people of different races and ethnicity varied in mental ability suffered not at all from lack of experimental validation. Interested investigators were literally surrounded by a myriad of evidence from dozens of Binet or intelligence tests. To ignore it was impossible; to refute it was not easy.

The administration of intelligence tests to the American Army during the First World War had helped to undermine the use of such tests in discriminating between "normal" and "feebleminded," but that same experience had just the reverse effect on the question of mental differences among races. For the very size and diversity of the sample—almost two million persons—made the experiment highly persuasive. More important, the results from comparative mental testing of races and ethnic groups did not threaten the majority of Americans, as the findings in regard to feeblemindedness surely had. For, unlike the case with feeblemindedness, a majority of the population was not being described as inferior in intelligence, only relatively small minorities were. In the Army tests, native white Americans always scored higher on the average than Negroes or recent immigrants from southern and eastern Europe. And, unlike the case with eugenics, geneticists throughout the twenties did not doubt the existence of biological differences in intelligence among human groups. As late as 1930, Herbert S. Jennings, the well-known liberal geneticist who in the mid-twenties had lashed out at eugenics for its misuse of his science, still thought it likely that human races would differ in temperament just as breeds of dogs did. "It might well be anticipated," he wrote in *The Biological Basis of Human Nature,* that the European whites and the African bushmen would differ in mentality as they do in physical characteristics. Certainly there is no antecedent presumption against mental differences between different races; on genetic grounds the presumption is that such differences will be found."[57]

In sum, even though the concept of human instincts had come under serious criticism in the early twenties, almost the precise opposite was true of the idea that human races and ethnic groups differed in mental abilities. Indeed, as the next chapter will seek to show, mental differences between human groups was a source of contention within psychology throughout the decade. Seen in retrospect, however, that controversy also proved to be the last in the long line of efforts, begun some forty years before by Franz Boas, to deny a role to biology or heredity in accounting for human behavior.

7

Decoupling Intelligence
from Race

I believe that the Negro race has tremendous gifts to bring to this country in
the way of artistic development. I think things come by nature to many of
them that we have to acquire, such as an appreciation of art and of music and
of rhythm, which we really have to gain very often through education.

Eleanor Roosevelt, 1934

As the varied uses made of intelligence tests in determining feeblemindedness
and other mental deficiencies suggest, those who used the tests were not necessarily interested in identifying differences among racial or ethnic groups. The
primary goal of ardent proponents of testing such as Robert Yerkes, Leta Hollingworth, and Lewis Terman, as they often said, was to assist organizations
like the Army, schools, or businesses in making closer fits between individual
talents and occupations and roles in society. Almost to a man or woman they
were convinced that the tests revealed much about the abilities of people. They
firmly believed, in short, that they were measuring something called "intelligence," though they also admitted that the concept was difficult if not impossible for professionals to define—as it remains to this day.

The Army tests had identified sharp differences among races and ethnic
groups, but the criticisms and objections to using them in that way did not
begin with invidious comparisons of races and ethnic groups. Rather, some
broader and more socially explosive conclusions from the Army tests provided
the impetus for the first criticisms of comparative mental testing. Among those
findings was the highly provocative conclusion that the average mental age of
the American soldier was less than fourteen years; another was that a clear
correlation between intelligence and occupation had been identified, a finding
that supported the inference that the social structure of the United States was
soundly based upon merit and inborn worth.[1] Those who were in poorly paid
jobs deserved to be; those in highly paid jobs warranted them.

The testers had found, too, a high correlation between test scores and degree
of education, which they interpreted as demonstrating that "native intelligence

167

is one of the most important conditioning factors in continuance in school."[2] Today, an alternative explanation—that education might have an impact on test performance—seems more plausible, but the testers of the 1920s generally rejected outright such an inference. They were so convinced they were in fact measuring innate ability that as late as 1923, Robert Yerkes, the psychologist who had overseen the development and administration of the Army test, rejected any alternative explanation. Such correlations, he recognized, "might be interpreted to mean that intellectual ability is largely the result of education." But that conclusion "is flatly contradicted by results of research for it turns out that the main reason that intelligence status improves with years of schooling is the elimination of the less capable student." And then, to underscore his meaning, Yerkes added that no more than 10 percent of the population, on the basis of the Army tests, "are intellectually capable of meeting the requirements for a bachelor's degree."[3]

The dissemination of such interpretations of the Army tests soon led to public controversy, often bitter. The harsh negative reactions were predictable given the implications of the findings, especially the general conclusion that almost half of Americans ranked below the level that one might reasonably think was normal adult mentality. That particular finding gained widespread publicity from Lothrop Stoddard's eugenicist tract *The Revolt Against Civilization,* published in 1922. Stoddard had concluded from the Army tests, as he baldly put it, that "the average mental age of Americans is only about fourteen." The issue came to a boil when the journalist and intellectual Walter Lippmann published in *The New Republic* magazine a series of six articles exposing and dissecting not only Stoddard's interpretations of the tests but the unexamined assumptions underlying all mental testing. The idea that the mental age of adults could be that of a child, Lippmann exclaimed, was not only incorrect; "it is nonsense." By definition, he pointed out, "the average adult intelligence cannot be less than the average adult intelligence."[4]

Lippmann then turned to the assumptions that undergirded the tests, beginning with the absence of any real definition of what was being tested. The fact that some persons did better than others in solving a series of "puzzles," as he described the test items, could hardly be accepted as a measure of what the general public considered intelligence. "These puzzles may test intelligence or they may not," he contended. "They may test an aspect of intelligence. Nobody knows." Lippmann harbored no doubt that such tests might well be useful in classifying children or potential employees or even soldiers. "As gauges of the classroom," he readily conceded, "the evidence justifies us in thinking that the tests will grade the pupils more accurately than do the traditional school examinations."[5]

His central objection was to the assumption that something called "intelligence" was in fact being measured. He was sure that some skill was being tested, "but whether this is the capacity to pass tests or the capacity to deal with life, which we call intelligence, we do not know," he flatly observed. Indeed, the tests were so general that he found it impossible to know what specific skills were drawn upon in achieving high scores. The tester, Lippmann pointed out, "is testing the complex result of a long and unknown history, and the assumption that his questions and his puzzles can in fifty minutes isolate abstract intelligence is, therefore, vanity." The testers might better spend their time measuring specific skills rather than trying to measure an undefined, amorphous thing called "intelligence." Besides, he warned, the tests for intelligence opened the discipline of psychology to the risk of "quackery in a field where quacks breed like rabbits."[6]*

As Lippmann's snide remark makes clear, his primary objection to the tests was that they were inadequate measures of "the quality of human beings."[7] Concerned as he was with addressing the fundamental question of the validity of the tests, Lippmann ignored the subsidiary issue of differences among economic or racial groups. The Army tests, however, had given much attention to such comparisons, which soon aroused public and professional interest in them.

Among other things the test had shown that blacks performed less well than whites on both the Alpha tests (for literate English speakers) and on the Beta tests (for illiterate or non-English speakers). And when comparative analyses were made of the nationalities of immigrant soldiers, a pattern of correlation between intelligence and nationality emerged. These results reached the public through both popular and scholarly works. Among the latter was Princeton psychologist Carl Brigham's book *A Study of American Intelligence,* which appeared in 1923.

Brigham's book soon aroused in social scientists a host of questions about the Army tests and the interpretations drawn from them. But even before Brigham's book appeared, professional psychologists had begun to question the testers'

---

* Franz Samelson in his thorough study of this era in intelligence testing firmly rejected the common opinion that prominent testers Lewis Terman and Robert Yerkes were racist or even illiberal in their outlook, but he well recognized the ideological limits within which the two psychologists operated. Both, for example, were outraged at what they considered Lippmann's uninformed criticisms. Yet, as a psychologist himself, Samelson was "impressed by the discovery that more than one nonspecialist like Walter Lippmann and others, seems to have had a better grasp of the real issues involved, in spite of misunderstandings of technical details, than the scientists themselves. War may be too important to be left to the generals . . . ," he concluded. Franz Samelson, "Putting Psychology on the Map: Ideology and Intelligence Testing," in Allan R. Buss, ed., *Psychology in Social Context* (New York: Irvington, 1979), 141. See also Samelson, "On the Science and Politics of the IQ," *Social Research* 42 (Autumn 1975), 482–85, for his recognition of Terman's and Yerkes's liberal outlook.

assumption that they were measuring native or innate intelligence as opposed to learning. The earliest questioning was not in connection with racial or ethnic comparisons but with the alleged relation between class and intelligence, that is, that upper-class people were innately more intelligent than members of the working class. In 1917, for instance, two psychologists reported a fairly elaborate mental testing of children in Columbus, Ohio, according to the occupation of their fathers. The investigators found "a striking correlation . . . between intelligence quotient and occupation groups," that is, the children of fathers with upper-level occupations performed better on the Binet-Simon tests than those with fathers in lower-level occupations. As the two psychologists observed, this finding was quite in keeping with previously reported studies. The two researchers, however, closed their report with a caveat.

They recommend a change in the norms that were used in determining the category into which certain children fell in the array of scores. For, as they wrote, if the test results varied according to social groups, as they had concluded, then "it seems only reasonable" that the norms would vary also, that is, the standard scores that marked where feeblemindedness began or genius emerged. For if norms were not adjusted, "we might be obliged to classify . . . the majority of the children of the unskilled labor group . . . feebleminded if judged by the norms for the professional group."[8] Indirectly, they were suggesting that experience or learning might have an effect upon test performance.

Within two years, another pair of psychological researchers, this time at Indiana University, published a study of the relation between class and intelligence. In line with the study of two years before, the new report identified a positive correlation between high scores and high occupational standing. Nevertheless, the two Indiana investigators could not refrain from observing that as many as 6 percent of the lowest social groups—mainly children of laborers—obtained scores in the top 10 percent of the group. At the same time, they noted, with a certain amount of puzzlement, only 41 percent of the children in the top, or professional, group scored in the upper 10 percent. Such discrepancies, they remarked, "would seem to confirm the statement, made frequently, that in the present day industrial world, the level of work is much below the level of ability of the worker. It would certainly suggest," they added, "that many individuals were not where they belonged—were employed in positions calling for much less than their best." This seemed to contradict what certain interpreters of the Army tests had concluded about the essentially meritocratic nature of the American socioeconomic system.

The two psychologists from Indiana University differed from their fellow testers in another way, though they expressed that dissent only in a footnote. Their finding that children of working-class fathers tended to drop out of school earlier than children of professionals they interpreted as showing that children "who are quite as bright as children from more favored homes have much less educational opportunity." In sum, lack of intelligence was not a reason to leave school, as Yerkes and other testers maintained. Although the two Indiana investigators had no quarrel with the assumption that intelligence was inherited, they recognized in the body of their report that social environment might well be significant in accounting for the differences in intelligence that their study had identified between classes. "It may very well be," they suggested, "that subtle changes of attitude, differences in the total fund of information, and other influences immediately resulting from differences in home environment play no small part in such findings." After all, they continued, "such differences are . . . so fundamental as to be of hardly less importance than differences in native intelligence."[9]

The wife of the male member of the research team was sufficiently disturbed by the possibility that environmental influences had been underplayed, that the next year she published a study of her own on the results of tests administered to urban and rural children. The purpose of the research, Luella Pressey wrote, was to evaluate the "frequent assertion of sociologists that the more intelligent individuals in the farming community are constantly moving to the city," a conclusion fully consistent with the prevailing view among psychologists that intelligence was hereditary, and with the traditional research finding that urban children achieved higher scores than rural children. Pressey's test results in fact verified both predicted differences.

Yet her questioning did not end there. She retested the children with an examination that did not require reading. Again, however, the rural children fell below the urban. In view of those results, Pressey felt compelled to conclude that "the differences . . . between country and city children were really differences in intelligence." Still her doubts would not down. Though stating her conclusions in what seemed an unambiguous fashion, she was not yet ready to "press these conclusions. It is evident that all of these tests are pencil and paper tests," she pointed out. "The country children, young as well as older, are more shy with strangers than are city children. The children from well-to-do homes often have nursery games somewhat analogous to the tests." Thus, even in the face of repeated testing, the meaning of the results were still clouded in her mind by theoretical concerns about the effects of the children's different environments. The very novelty of her theoretical concerns, however, apparently prevented her from challenging the conventional interpretation more vigorously and openly.[10]

Thomas Garth, a psychologist at the University of Texas, was not so circumspect when he took up the same question the very next year. He called attention directly, though in general terms, to environmental influences that might account for differences in the mental test scores of social groups. He reported, for example, on a white boy, aged eleven, who achieved a score of seven on a Binet test, that is, a mental age of four or an I.Q. of 36, where 100 was the average. Nothing, Garth reported, could acount for that unusual degree of mental retardation other than that the boy had been raised on a coal barge in New York harbor with almost no contact with schools or other children. All mental tests, Garth reminded his readers, presupposed a certain background or environment and motivation; if that were missing, he explained, then the results would surely reflect it. "For that reason," he insisted, "we cannot draw scientific conclusions as to the mentality of individuals unless those individuals have been exposed to the environment which the test presupposes. We may continually shut our eyes to this fact," he lamented, "but the fact remains."[11]

The great professional debate of the 1920s over the validity of measurements of innate intelligence, however, did not grow out of the study of class differences. Few studies of the relation between class and intelligence appeared in scholarly journals during the 1920s. Class, after all, has never been a very live source of conflict among Americans, but race and ethnicity have a long history of controversy and conflict, and few periods of that history were more turbulent that the years between the First World War and the onset of the Great Depression. Thus, it is not surprising that the relative intelligence of races, not of classes, was the comparison that dominated professional inquiries into the validity of I.Q. testing. On that comparison, the professional literature swelled enormously in the course of the 1920s.

From the very onset of the testing movement the scores achieved by different races and ethnic groups had consistently varied, suggesting to testers comparable variations in mental abilities. William McDougall, the prominent Harvard psychologist, spelled out the standard conception when he wrote in 1921 that "races differ in intellectual stature, just as they differ in physical stature."[12] Yet the very year in which McDougall confidently announced a biological basis for differences in the intelligence of races, Ada H. Arlitt, a professor of psychology at Bryn Mawr College, was voicing serious professional doubts about the whole idea.

Arlitt's doubts appeared in a professional journal under the appropriate title "On the Need for Caution in Establishing Race Norms." Arlitt's study compared the scores of native white, black, and immigrant Italian children with the social standing of their fathers' occupations. As might have been predicted, she found that the native white children, many of whom were from the higher social status group, achieved better scores than Afro-Americans and Italians. Very few of the children of the superior mental group, Arlitt noted, were to be found in the lower social status category; only one child with a low mental score came from the topmost social status group.

Arlitt's contribution to fresh thinking about testing was her showing that class standing—or social environment—affected test scores more than race. In doing so, her research became the first psychological—as opposed to anthropological or sociological—study to question the assumption that races differed in intellectual capacity. That result emerged from her comparison of median test scores. The median score of the native white children, she pointed out, stood 21.5 points above that of the Italians, and 23.1 points above the Afro-Americans' median score. But a comparison between the Inferior Social Status whites and the Afro-Americans, and Italians of the *same* social status revealed that the disparity in median scores dropped to merely 7 points for Italians versus native whites, and to 8.6 points between native whites and blacks. It was true, Arlitt recognized, that native whites still scored above the other two groups, but quite unexpected was the finding that "the difference in median I.Q., which is due to race alone is in this case at most only 8.6 points whereas the difference between children of the same race but of Inferior and Very Superior social status may amount to 33.9 points." Her conclusion was forthright. "Of the two factors"—that is, race and social environment—"status seems to play the more important part." Rephrasing her point, she drove it home: "there is more likeness between children of the same social status but of different race than between children of the same race of different social status." Her advice to fellow social scientists therefore was that "race norms which do not take the social status factors into account are apt to be to that extent invalid."[13] That sentence captured the central issue with which researchers of race differences would wrestle for the remainder of the decade.*

Indeed, that very same year, Thomas Garth, having moved in his research from investigating class differences in intelligence to racial differences, echoed Arlitt's cautionary advice. After recognizing the conflicting evidence regarding racial differences in intelligence, he urged the public and his fellow professionals to acknowledge "that experimental psychology may eventually render the term

---

* An indication of how slowly social scientists recognized the implications of findings such as those published by Arlitt is provided by some research Arlitt herself reported the following year. In this study Arlitt found that black children of ages five and six scored above white children of the same age and social class, but that, as the students grew older, the whites surpassed the blacks. That shift, Arlitt attributed "to a genuine race difference." On the basis of her article of the previous year one would have thought she would have asked how it came about that black children fell below whites as they grew older. Instead, she proceeded to explain why the superior scores of the black children at an earlier age may have been influenced by factors other than intelligence. In short, she fell back to the traditionalist position of assuming that race was controlling even when scores clearly called it into question. Ada Hart Arlitt, "The Relation of Intelligence to Age in Negro Children," *Journal of Applied Psychology* 6 (December 1922), 378–81.

'inferior races' as innocuous as it seems it has the long honored term 'the weaker vessel' as used in connection with sex differences." Thus, in comparing races, he continued, "it would be safer to take for comparison such racial groups as have had as nearly the same educational opportunity as is possible" and, when interpreting the results, to take into consideration whatever discrepancies in opportunity there may be.[14] "We do not know," he declared the following year, "what the negro or the Indian would do if placed in an absolutely white environment from birth until the date of the test." The still tentative character of his views and the prevailing sentiment among psychologists in general are reflected in his next sentence. "This is no contention that the blacks and redskins as groups are necessarily equal to the white, for we still leave the question open. So far," he added, "tendencies appear to show that as groups they are not, where some allowance for environment is made."[15]

For test-oriented psychologists, the Army tests had constituted a breakthrough in mental measurement. Robert Yerkes, for example, wrote a laudatory introduction to Brigham's *Study of American Intelligence* when it came out in 1923. Yet, almost from the outset, other psychologists found Brigham's effort inadequate as well as misleading. Notable among them was Edwin Boring, whose review in *The New Republic* was highly critical. Boring, like Brigham, was a young psychologist but, in contrast with Brigham, destined to be a leader in the field. Although Boring had served as one of the editors of the report on the Army testing program, he thought the evidence was inadequate to sustain the anti-immigrant and racist interpretations that Brigham drew from the "mountain of data" compiled by the Army test study. "That in this case," Boring wrote, "the mountain could bring forth only a timid mouse may be due to the fact that mountains for all their size do not necessarily have leviathans in them." Earlier and privately, Boring had warned Brigham that he ought not to overlook the possibility that many of the soldiers being tested might not be responsive to what Boring referred to as "American hooray methods." Boring rightly feared, as he wrote Robert Yerkes, that the performance of the foreign-born soldiers might not be properly evaluated since no one knew whether their scores resulted from "their intelligence on their lack of assimilation to the American setting." Another psychologist, Adolf Snow of Northwestern University, dissented from Brigham's conclusion that the tests were measuring innate intelligence and that Nordics were superior to other ethnic types. Snow had no trouble in accepting the finding that men of upper income levels scored better than those on lower levels, but to conclude that was "because of a difference in *native* intelligence," he snorted, "is surely unwarranted."[16]

Kimball Young's review in *Science,* the official journal of the American Association for the Advancement of Science, described Brigham's arguments and evidence as outdated. Young referred to "the anthropological innocence of a young psychologist" because he had ignored, among other things, the work of Franz Boas. That omission caused Young "to wonder whether the book is to be considered science or special pleading." Boas himself, even before Brigham's exegesis of the Army test data was published, had denounced in print the assertions of Afro-American inferiority derived from the tests. Those who know the oppression of blacks in the South, Boas contended, discount any assertion of Afro-American inferiority based on the tests. After all, he pointed out, Northern blacks did better in those tests than many Southern whites.[17]

Boas's last point received specific support a few years later from educational psychologist William Bagley. He noted that literate blacks from Illinois achieved scores above those of literate whites in nine Southern states. In fact, Bagley continued, drawing directly upon the Army test data, the median scores of all Northern blacks surpassed those of whites in Mississippi, Kentucky, and Arkansas. (Yerkes had accounted for the same finding by contending it derived from selective migration of blacks from the South.) Anthropologist Alexander Goldenweiser, a former student of Boas, rejected the conclusions from the tests for reasons that were reminiscent of both Lippmann and Boas. Goldenweiser thought the low mental age ascribed to the Negro by the testers was totally unacceptable; "we could not live with them in this country" if that were true. "The Negro would not be much better than a monkey." The tests may measure achievement, he conceded along with Boas, "but they do not in any real sense measure intelligence."[18]

As Kimball Young's criticism of Brigham suggested, the concept of culture which Boas and other anthropologists had worked out was having an effect upon the thinking of some psychologists. Another instance was provided by J. R. Kantor, a psychologist at Indiana University. In an article on racial differences in intelligence he bypassed the whole question by denying that there was such a thing as innate capacity. The analogy often made between a muscle's capacity and that of intelligence he declared false. "Psychological phenomena are in no sense qualities or faculties of an organism, but really concrete activities" in response to a phenomenon or thing. If one person responds differently from another, then that needs to be explained not by innate capacity but by social or cultural, psychological or biological differences. Since many testers had already abandoned the "idea that tests test a native quality called intelligence" in favor of "the idea that intelligence is what the tests test," it would seem to follow, he suggested, that innateness is no longer a viable idea. "Some anthropologists," he stressed, "have in various public prints sufficiently indicated their appreciation of this point."[19]

Insofar as most psychologists were concerned, however, Kantor was more a voice in the wilderness than a shaper of views. Anthropologists like the young Margaret Mead and the established Robert Lowie had indeed published pieces noting the inability of testers to eliminate the effects of divergent environmental influences upon the racial or ethnic groups being compared. Yet even a psychologist as favorably disposed to their point of view as Kimball Young remained unconvinced. At the conclusion of his article-length history of mental testing, which he published in 1924, he concluded that psychologists still "need some fundamental work to test the point of view of Levy-Bruhl, Boas, Thomas and other sociologists that by their very nature mental reactions are socially determined in large measure."[20]

By the end of the decade, criticisms of the Army tests reached their culmination in a repudiation by Brigham himself. In an unusual and courageous recantation, he informed his professional colleagues, among other things, that in comparing groups he had learned that tests "in the vernacular must be used only with individuals having equal opportunities to acquire the vernacular of the test." And so far as the Army tests were concerned, he continued, they had been so badly designed and administered that little could be reliably based upon them. Because of the errors he had made in his own analysis of "the army tests as applied to samples of foreign born in the draft," he admitted, "that study with its entire hypothetical superstructure of racial differences collapses completely." His point, underlined at the end of his mea culpa, was that "comparative studies of various national and racial groups may not be made with existing tests" and that in particular "one of the most pretentious of these comparative racial studies—the writer's own—was without foundation."[21]

Probably one of the reasons Brigham repudiated his earlier work in 1930 is that over the preceding five years professional journals in psychology had been weighed down with studies on intelligence testing in general and on racial differences in particular. One survey of the literature reported that in the eight years prior to 1925, some 73 studies of racial differences had been published, with each year raising the total. When the author came to survey the literature between 1925 and 1930, he found the number had reached 170 for just those five years.[22]

Despite the plethora of studies on the subject—or perhaps because of it—the issue of racial differences in mental ability at the end of the decade seemed no closer than before to a scientific resolution. One despairing psychologist complained in 1928 that many investigations had "virtually arrived at a position of checkmate. We cannot speak with authority on the subject, but it seems almost impossible to devise any technique which will yield valid results." The old question remained: How to equalize environmental influences affecting the two or more racial groups being studied? "Theoretically—and that means practically as well," he observed, "no piece of work has so far been carried out against which we cannot lodge very grave objections." The problem of differential cultural influences, of course, was precisely the problem that had been identified by Arlitt and Garth at the beginning of the decade.[23]

Another psychologist acknowledged the impasse at the end of the decade, spelling out the divisions within the profession over the question. The first group "accepts the fact of race superiority and inferiority," Dale Yoder of the State University of Iowa pointed out. They want to sustain the idea and therefore seek "additional evidence to support the thesis." The second group, Yoder continued, believed that some races might be inferior to others, but that as yet the assumption had not been "adequately demonstrated." The members of that group he thought were principally interested "in balancing arguments for and against the idea. The third," he added, "is a skeptical group, highly critical of the means used to demonstrate race inferiority and of the results so obtained and generally insisting upon racial equality." He concluded, then, "that the consensus of competent scientific thought" was that of the inability to define intelligence or to neutralize "such factors as education, social status and language," prevented any "proof of racial inferiority or superiority" that would meet the traditional standards for scientific acceptance.[24]

By the end of the decade, however, a changed outlook in the profession was apparent. Two psychologists, writing in 1930, conceded that a few members of their discipline still adhered to a belief in mentally inferior and superior races, but serious doubts "have been literally forced upon practically everyone who has examined the vast array of data which has been accumulated during the past decade," they observed. Revealingly they pointed to the similarity between their own conclusions on the subject and those of Franz Boas. With obvious satisfaction they described the alteration in psychologists' attitudes as striking when "one recalls the dogmatic statements which were made regarding racial differences about ten years ago."[25]

Thomas Garth, who had been investigating racial differences ever since the beginning of the decade as well as periodically reviewing the professional literature on the subject, also remarked in 1930 that the existence of such differences "is no nearer being established than it was five years ago," when last he canvassed the professional literature. Yet despite that impasse, he reported that today "many psychologists seem practically ready for another, the hypothesis of racial equality." Nonetheless, the old problems remained: how to obtain a "fair sampling of the races," the effects of different cultural backgrounds, and the devising of a "test and technique fair to the races compared."[26]

That the inconclusiveness of the available evidence continued to plague Garth becomes apparent in his book *Race Psychology,* which appeared the following year. The book summarized the work completed over the preceding decade on racial differences. Garth's own conclusion came close to the agnostic position that the anthropologists and sociologists had arrived at a decade or more earlier. Despite all the research of the last ten years, Garth wearily observed, "we have never with all our searching . . . found indisputable evidence for belief in mental differences which are essentially racial." The possible influence of culture or environment had never been entirely eliminated.

Later in the book, however, he hinted that the uncertainty which had gnawed at him for so long was still not entirely gone. He recognized that differences in social background might legitimately call into question racial explanations for the lower test scores for Afro-Americans, Amerindians, and Mexicans on the one hand, and those of native whites, on the other. But why, he wondered, was there no need to invoke similar environmental explanations in regard to the test scores of Japanese and Chinese? Their social backgrounds surely differed dramatically from those of native whites, yet Japanese and Chinese children almost invariably achieved scores on a level with those of native whites. "Perhaps," he speculated, "it is temperament which makes the [Amerindians, Afro-Americans, and Mexicans] unable to cope with the white man's test." He went no further in exploring the matter of temperamental differences except to add: "it is barely possible they cannot take the white man's seriousness seriously." Garth's continuing uncertainty emerges again, a little later in the book, when he falls back upon authority to support the proposition that "races are mutable and not permanent." He advised his fellow psychologists and the public at large that "we must either adopt this view or give up fellowship with the anthropologist, the geneticist, the biologist generally."[27]

And Garth was right: biologists had begun to raise questions about the validity of biologically rooted differences between races and nationalities. As early as 1924, the well-known Johns Hopkins geneticist Herbert Jennings denounced as a "vicious fallacy" the notion that environment played no role in the evolution of human beings. He decried the attempt to restrict the immigration of certain nationalities because their genetic makeup was deficient, and he rejected the assertion that environment "can bring out nothing whatever but the hereditary characters," a point which he described as "empty and idle; if true, it is merely by definition: anything that the environment brings out *is* hereditary," he pointed out. "But from this we learn nothing whatever as to what a new environment will bring out. . . . What the race will show under the new environment can not be deduced from general biological principles," he emphasized. "Only the study of the race itself and its manner of reaction to diverse environments can give us light on this matter."[28] That more open conception of biology, however, would not influence the thinking of most psychologists until well into the 1930s.

Changes in conceptions of biological or genetic influence were not the only reasons psychologists altered their attitudes. For despite the uncertainty of Thomas Garth, the issue of racial differences was more or less resolved for most members of his profession by the early 1930s. The resolution, to be sure, was untidy and inconclusive, if only because that was the way the professional literature on the subject read. In any case, assertions of racial inferiority in the psychological literature declined noticeably. What, then, had generated those doubts about racial differences and caused the reservations to be defended with such energy and determination over the course of the decade? Was it merely a matter of a search for scientific rigor and truth? In part, certainly. But more than that stood behind those endeavors. One approach to uncovering those motives is through an examination of the research undertaken by one psychologist who was in the forefront of the attack on racial explanations.

Many psychologists expressed reservations about the reality of race differences during the 1920s debate, but none was more tireless or ingenious in creating those doubts than Otto Klineberg of Columbia University's department of psychology. It was at Columbia that Klineberg first encountered Boas's ideas on race. He had gone to Columbia from his native Canada to study nothing more than psychology, as he recalled in later years, but thanks to a course in linguistics with anthropologist Edward Sapir, he became professionally acquainted with Boas. Klineberg remembered that when he arrived at Columbia he still accepted the idea that differences in the mind and character of racial and ethnic groups were attributable to race. His association with Boas and other anthropologists caused him to shed that view forever. Upon publishing in 1935 his book *Race Differences* he appropriately dedicated it to Boas. In the course of five years after coming to Columbia in 1925, when he was only twenty-six, Klineberg published a series of articles on race differences in which he made it his business to do for psychology what his friend and colleague at Columbia had done for anthropology: to rid his discipline of racial explanations for human social differences.

In the course of those five years Klineberg sought to answer or at least to respond to virtually every argument or piece of research advanced in support of the hypothesis of the racial inferiority of certain groups. Contrary to what might be thought today, the case against which he argued was not weak; indeed, as we have seen, it was supported by the leaders of the discipline. Moreover, in virtually every comparative test between blacks and whites, and between Amerindians and whites, Afro-Americans and Amerindians scored lower than whites. Moreover, as Garth observed as late as 1931, some non-native groups, like the Japanese and Chinese, despite wide differences in social environment, performed on the tests at a level with native whites.

Ingenuity reinforced by determination was clearly evident in the numerous and widely varied studies Klineberg initiated in those years. Undoubtedly his most ambitious project was undertaken to counter a major point made by Carl Brigham in his *Study of American Intelligence* (Klineberg began his research before Brigham had recanted). Under the rather dubious influence of Madison Grant's eugenicist tract *The Passing of the Great Race,* Brigham had argued that Nordics were mentally superior to Mediterranean and Alpine peoples. Brigham drew on evidence gathered in the Army test project to establish the point. But in order to use that massive documentation, he had to make the highly questionable assumption that he could determine the proportion of Nordic, Mediterranean, and Alpine heredity embodied in the various national groups he was studying. For not even Brigham believed that all Germans were genetically pure Nordics or that all Irish were pure Alpines. Indeed, few students of the racial types, including the eugenicists, thought that any type was racially pure. Furthermore, all experts recognized that some Nordics lived in France as well as Germany, while some Alpines lived in Germany, and so on.

Klineberg's research project was probably the most elaborate he ever embarked upon. It entailed his traveling to Europe to locate villages in which the ancestry was as pure as he could find, for in that way he could avoid Brigham's error of simply assigning, almost arbitrarily, a certain percentage of Nordic, Alpine, or Mediterranean ancestry to the various nationalities. The hypothesis that Klineberg sought to test was that racial background was more influential than national history or environmental circumstances in shaping intelligence. If Brigham's concept of racial hierarchy, with Nordics at the top, was accurate, then Nordics, no matter whether they lived in France, Germany, or Italy—the three countries in which Klineberg identified communities for study—should achieve higher scores than those achieved by Alpines and Mediterraneans.

Of the ten communities in which he administered tests to about one thousand boys, seven were rural and three were urban. The urban boys consistently attained higher scores than the rural boys, but "the differences among the racial groups," Klineberg reported, "are small and unreliable." Nordics in Germany, for example, achieved high scores, but French Nordics scored quite low. Italian Mediterraneans registered lowest of all, but French Mediterraneans achieved the highest record of the seven rural communities. As Klineberg concluded, his findings "offered no support to the theory of a definite race hierarchy."[29]

In a study closer to home, Klineberg responded to the assertion that blacks performed less well than whites on tests because they naturally worked more slowly. Starting from the assumption that the slower speed of Afro-Americans resulted from their Southern origins, Klineberg tested a group of black and white boys aged eleven to fourteen in New York City. He divided the black boys into three groups, according to the length of time they lived in the city. His research assumption was that if race was at the root of the difference in test scores, the three groups should score about the same. On the other hand, if the slowness of their performance was environmentally induced by their Southern origins, as he suspected, then differences could be expected and they would be graduated according to the length of time each group had lived in New York City. Small differences did show up among the groups, causing Klineberg to conclude that environment "is to some extent at least responsible for increasing the speed of blacks."[30]

Proponents of racial differences, like Robert Yerkes, as we have seen, explained the higher scores of Northern blacks as against those of Southern whites as a consequence of "selective migration. The more energetic, progressive, mentally alert members of the race have moved northward," Yerkes explained in 1923, "to improved education and vocational opportunities for themselves and their children." Klineberg set to work to scotch that argument, too. He studied the school records of over five hundred black children in Nashville, Tennessee, Charleston, S.C., and Birmingham, Alabama, before they moved north. He found no evidence of selective migration, since "the migrants as a group were almost exactly of the average of the whole Negro school population in those three southern states."[31] Of course, he slyly added, the children's parents, who, after all, made the decision to move, may have been exceptional. But if they were above average, Klineberg wanted to know, why were their children only average if intelligence was inherited, as proponents of selective migration maintained? Then, to clinch his argument, Klineberg tested Southern-born black children in New York City, the results of which demonstrated that, the longer the children had lived there, the higher their scores. Environment alone, not heredity or selective migration, accounted for the gains, he concluded.

Not all the assertions or research results of the proponents of racial differences could be countered or responded to by experiment, but that did not prevent Klineberg from raising objections. For example, when a proponent of racial differences offered evidence from a so-called "baby-test," that is, one given to very young children of both races, Klineberg was quick to offer an environmental explanation for the poorer showing of the black infants, even if he had no counter experimental evidence. He pointed to the positive correlation between the superiority of the white infants in height and weight and in mental achievement, contending that such a correlation "may mean that the pre-natal environment of the Negroes were far inferior, and it is quite probable," he suggested, "that the poorer physical condition will be reflected in the behavior of the infant, upon which the mental ratings depend."[32]

When Klineberg sought to account for slower test performances of Amerindian children, he did not devise an experiment, as he had done in seeking to account for the relative slowness of African-American children. Instead, the explanation he offered revealed not only his technical ingenuity but also his commitment to a cultural or environmental explanation for racial differences. The slowness of the responses of Amerindian children, he explained, derived from their culture. They simply felt no reason to work faster, to seek to save time. They often started on journeys without hurry, or returned a month late from school vacations. To them "what difference does a month make, anyway?" Klineberg asked. On the other hand, an "emphasis on speed . . . is one of the striking characteristics of modern American life. . . . Speed seems to have a place in a competitive society" like that of the United States, he contended, "but there is no economic competition among the Yakima Indians," the people he reported on. They "have no need for speed," he maintained.[33]

Klineberg then went on to spell out the implications of his assumption of a cultural explanation. "If speed of performance is a function of environment," as he now contended, "and if it enters as a very important factor into intelligence ratings, it seems obviously unfair to use it as a criterion of excellence in a study of racial or group differences." This would be true, he continued, "even if differences in speed are native, hereditary," since the importance of speed varies according to cultures. "Speed in itself can hardly be regarded as a good, unless it performs a specific function," he maintained. On the Amerindian reservation, speed "has no particular virtue. As criterion of excellence, it belongs in 'white' American life, perhaps, but it is not for that reason applicable to all other communities."[34] Klineberg's implication was that each and every test, even those that measured something as physical as speed, implied a cultural or moral value. Speed was not a universal good; its value varied across cultures. And since the presence or absence of value affected the level of performance of the persons being tested, no test could really determine if differences across cultures were innate.

Klineberg's determination to prove the lack of racial influences on intelligence is evident, too, in the care he took in his own research projects to remove any possible social or psychological impediments to African-Americans' optimum performance on tests. He employed a black graduate of Howard University and a graduate student in anthropology from Columbia to make personal contacts with his African-American subjects and to administer the tests. The black assistant, he gratefully recognized, "made access into the Negro homes possible" and did much "to reduce the feeling of embarrassment which many Negro children have in the presence of a white tester, and to aid in the establishment of that rapport which is an essential to fair testing."[35]

Even the definition of what constituted a comparable urban environment for blacks and whites was not too small an issue to escape Klineberg's scrutiny. Were the communities in which blacks and whites lived truly comparable, he asked in one study? A town of 20,000 people, in which only three thousand blacks lived, he observed, was not the same environment for both races, particularly if the town was Southern, for in such communities the white districts were closed to blacks. Under such circumstances, the social environment in which the two races lived and worked would not be equal, and the differences would vitiate any comparative studies in racial differences.[36]

At another point, he extended the argument to Northern communities as well. For there, too, he explained, blacks lived under conditions much worse than those experienced by whites. "The real test of Negro-White equality as far as intelligence tests are concerned," he finally concluded in 1935, "can be met only by a study in a region in which Negroes suffer no discrimination whatsoever and enjoy exactly the same educational and economic opportunities." Such a place, he conceded, would be difficult to locate, but "there may be an approximation to it in Martinique or Brazil," he thought. He then cited a study of blacks in Jamaica in which the gap between the mental ratings of whites and blacks was narrower than that resulting from comparisons in the United States. "It is safe to say," he concluded, "that as the environment of the Negro approximates more and more closely that of the white, his inferiority tends to disappear."[37] It is worth noting, as a further measure of Klineberg's sensitivity and commitment to racial equality, that he capitalized "Black" and "White," a practice then almost unknown among white American publishers.

As the foregoing suggests, Klineberg was a committed proponent of the concept of culture, which he had acquired from his colleague and teacher Franz Boas at Columbia. In *Race Differences*, in which he summarized the psychological writing on the subject, he carefully explained the meaning and nature of culture to his readers. The term was apparently sufficiently unfamiliar that he felt it necessary to distinguish the anthropological concept from the conventional definition of "a high degree of cultivation or refinement." For him, he added, the term was free of any value judgment; "it applied to that whole 'way of life' which is determined by the social environment."[38] He urged the use of the concept of culture upon his fellow psychologists.

Along with Boas, Kroeber, and other proponents of environmental influences, Klineberg considered culture superior to race as a means of accounting for the great diversity in human social behavior. "Physical anthropologists," he observed, "regard as the same 'race' the Eskimo who did not understand the meaning of war, and the Plains Indians who made war the center of their entire social organization." Culture can account for the difference in outlook, he maintained, but race obviously cannot.[39]

Indeed, Klineberg was so impressed by the diversity of human behavior documented by ethnologists that he was impelled to discount heavily, if not deny outright, some of the favorite explanations of his fellow psychologists for drives such as sex, aggression, and self-preservation. "Culture can produce and maintain profound differences even in those reactions which psychologists have usually regarded as basic to all behavior," he maintained. "To describe a race or a people as innately aggressive or peaceable, sedentary or nomad, promiscuous or puritanical," he insisted, "overlooks the fact that culture may be entirely responsible." He then cited instances from the anthropological literature to illustrate how culture overrode what many psychologists thought were basic or innate sources of human behavior. "There is no reason to believe that the mammary glands, for example, function differently in Tahiti and Australia," he remarked, "yet in one society there may be a great deal of adoption, and in the other, infanticide." Culture determines the "emotional response, as well as the extent to which the response is overtly expressed, and the particular forms which the expression may take," he explained. Death is a matter of sorrow in one society and one of joy in another. Culture "may demand of a people (like the Sioux), the violent demonstration of grief," while requiring "the suppression of any sign of physical suffering. . . . It may make one people 'emotional' and another 'phlegmatic,' altogether apart from their biological constitution."[40]

As we have seen earlier, Boas left some place in his conception of culture for what anthropologists refer to as "universals," or patterns of behavior that are to be found in most, if not all cultures. Klineberg, however, seemed to have no place for them. He specifically disagreed with Swiss child psychologist Jean Piaget's stages of development for all human beings. "Adolescent conflict," Klineberg contended, "is a phenomenon in our society, not of human nature. . . . Not only the nature of the problems of adolescence, but even their very existence as well as the time of their occurrence, will differ according to cultural influences," he believed. Klineberg's cultural relativism carried him close to the point of denying any definition of normal behavior that transcended cultures. "Ethnologists," he reported in 1935, "have recently made clear that even the line between the normal and the abnormal may be culturally determined," citing an article by anthropologist Edward Sapir. "The external form of normal adjustment is very elastic," he remarked, referring to a recent article by another anthropologist, Ruth Benedict, "and needs to be redefined for every culture in turn." Some civilizations have accepted as the "very foundations of their institutional life" types of personality that "seem to us to be clearly abnormal," he concluded, very likely drawing on Benedict's recent book *Patterns of Culture.*[41]

Having come that far in drawing on anthropology to remove any universal definition of normality, Klineberg backed off a little. He denied suggesting that "the concept of abnormality has an exclusively cultural significance," he assured his readers, for there are some human behavioral patterns that transcend culture. The Hindu mystic may stiffen his body much as a catatonic person does, but his behavior is to be judged by a transcendent standard not by a culturally rooted one. "The Hindu is an integrated, and the catatonic a disintegrated personality." But having made that concession, he slipped back to his original point by noting that for some people homosexuality is the behavior of "the pervert in our society" while it is quite acceptable behavior for "the Siberian shaman."[42]

For all of his doubts about universals among the great variety of human social experiences, Klineberg remained in agreement with Boas in seeing the racial sources of human behavior as unproved rather than ruled out entirely. Though there is "no scientific proof of racial differences in mentality," he conceded, "this does not necessarily mean that there are no such differences." Perhaps in the future, with new methods of inquiry, such differences will be discovered, but in the present state of knowledge they are unfounded.[43]

Klineberg agreed with Boas, too, in denying that "the concept of heredity is of no value." Individual differences, he thought, are often best explained by family heredity. Intelligence itself, Klineberg suggested, was clearly influenced by heredity. Though environment may account for differences between blacks and whites, a part of the differences within each group "could be explained only by the superiority or inferiority of individual or family germ plasm. The fact that persons living in almost the same environmental conditions will differ widely from one another in intelligence and the fact that identical twins living in very different environments will yet resemble each other closely, argue strongly," he contended, "in favor of an heredity basis for part of the differences in intelligence between individuals and family lines." Klineberg was not prepared to go as far in the environmentalist direction as his fellow psychologist and behaviorist John Watson. "There are few psychologists or biologists," Klineberg wrote in 1931, "who would agree with the Behaviorists in denying

heredity any importance whatever even in the case of individual differences."
Human beings are plastic, as Boas had said, but that description, Klineberg cautioned, in the end "does not imply that the plasticity is unlimited."[44]

Klineberg's persistence and determination arouse one's curiosity. What impelled him? What motives stood behind his industrious effort to disprove a racial explanation for social behavior? More important, what stood behind the acceptance by fellow social scientists of his evidence and arguments? Why, in short, did culture triumph over race?

8

Why Did Culture Triumph?

Human nature is not a biological concept; human nature is not a corollary of race; rather it is a cultural or civilizational phenomenon.

Ellsworth Faris, 1927

Man has no nature; what he has is history. . . . Man is an "unknown," and he will not be discovered in the laboratories.

José Ortega y Gassett, 1940

The long and somewhat inconclusive struggle of psychologists against racial differences turned out to be the final battle in the war that Boas had begun some forty years earlier. He had won that struggle in anthropology early on; psychologists had been slower to respond, but the determination and persistence of a Klineberg were quite in the same league with the efforts of Boas and Kroeber. As with them, the question returns: Why was the effort so determined, the energy expended almost unlimited? What stood behind that struggle of the 1920s? What, indeed, were the forces, the reasons behind the triumph of culture across the spectrum of the social sciences?

Certainly a general urge to know, combined with a professional and scientifically derived willingness to accept new information and insights were among the forces behind the transformation in outlook that removed race in particular and biology in general from the study of human behavior. But as we have seen in regard to the shift in outlook among anthropologists and sociologists, professional or scientific attitudes were not the full explanation. One needs to look beyond professionalism and standard science; for the change in outlook was too fundamental, to radical to be accounted for on those grounds alone. After all, we are not dealing here with a long-held, well-substantiated theory (that is, race) which new and conclusive evidence had unambiguously disproved and overturned. Rather we see essentially the substitution of one unproved (though strongly held) assumption by another. Or, to make the point a little less baldly: Culture or history could account just as well, if not better, than race and biology for differences in human group behavior.

187

That argument seemed especially compelling to anthropologists and sociologists when it was shown that over time a race might well display quite different levels of intellectual behavior, as in the case of blacks in Africa and in America, or Arabs in the twelfth century and in the twentieth. For psychologists with their attachment to mental testing, the substitution or alternative was a little different. A growing recognition that in making comparative mental tests not all social or historical differences between the races could be eliminated first pushed psychologists to doubt and then to deny the influence of race or biology in accounting for human behavior. Just because no crucial pieces of new evidence or overpowering argument had been brought forward on either side, we need to look beyond professional or scientific attitudes in seeking to account fully for the great transformation: the triumph of culture in American social science. The story is a complex mixture of happenstance, ideology, individual commitment, and broad social influences.

Prominent among the forces operating upon the psychologists, as among the anthropologists earlier, was the ideology of equality, the belief that an acceptance of racial differences denied equality of opportunity. And again, as with the anthropologists, the psychologists and others were moved by a feeling of guilt about the treatment and status accorded blacks and other racial minorities in their America. The unstated assumption in the thinking of those who doubted racial explanations was that, if biological inferiority were to be proved or established, then the groups designated as racially inferior would be denied opportunities that ought to be theirs by right. The predisposition of social scientists—and Americans in general for that matter—to resist biological explanations for behavior was ruefully recognized as early as 1911 by the eugenicist Charles Davenport. He noted even then one of the prominent objections to eugenics was that people did not want to believe in the importance of heredity "on the ground that it is a pessimistic and fatalistic doctrine." That reluctance certainly underlay Otto Klineberg's statement in 1928 that "we have no right to accept the hypothesis of the innate superiority of any one race over any other" until all other explanations have been excluded. Or, as he reflected on the matter many years later, "I felt (and said so early) that the environmental explanation was preferable, whenever justified by the data, because it was more optimistic, holding out the hope of improvement."[1]

No one revealed more candidly than educational psychologist William Bagley the ideological underpinnings of the turn to cultural explanations. He confessed that he did "not deny racial differences in intelligence levels." There is a "fair degree of probability that the Negro race will never produce so large a proportion of highly gifted persons as will the white races." But in the present state of our knowledge, he continued, "invidious distinctions cannot safely be made" between the various types of whites. In any case, he was confident that education could improve levels of intelligence. "Instead of emphasizing the forces that pull men apart," he explained, he wanted to "emphasize the forces that draw men together. Instead of intensifying biological differentiation," he would stimulate "cultural integration." In contrast to hereditarians like Lewis Terman, Bagley did not want to look for ways to identify highly qualified leaders. Nothing would be more inimical to leadership in a democracy, he contended, "than an overweening consciousness of one's superiority to the common rule of humanity." He called his approach, appropriately enough, "rational equalitarian" in its support of "the ideals of humanity and democracy that have been winnowed and refined through the ages." He denominated the "hereditarian's solution of the problem as . . . openly inhumane and anti-democratic."[2]

Most psychologists were not as candid as Bagley in revealing their ideological purposes. But some observers had little difficulty in uncovering the ideological sources of their colleagues' professional views. Dale Yoder of the State University of Iowa, for example, referred in 1928 to the proponents of culture in the profession as espousing an "egalitarian" point of view, which exhibited "the fire of enthusiastic discovery . . . combined with an idealistic zeal aiming at the unseating" of what they obviously considered "scientific inaccuracy and fallacy." Anthropologist Alexander Goldenweiser acknowledged his ideological commitment when he admitted in 1925 that blacks in Africa may not have achieved the level of science and the quality of religion of Europeans, but for him diversity of social achievements did not call into question the unity of human nature. Will we ever accept "the Mongol, the Indian, the Arab, the Negro . . . as equals?" he asked rhetorically. "Who knows," he responded, revealing at the same time his own value judgment; "But who can doubt that we should." One psychologist's ideological outlook was a source of his exasperation with the endless professional discussions of racial differences. It is not possible, he remarked with obvious irritation, "to kill off, sell, or otherwise dispose of the 'inferior' group." They did not create the situation in which they found themselves, he pointed out, but they certainly need help in adjusting to society. "I should personally not care to spend much time on race testing merely to look for general group inferiorities—or superiorities, according to the one-sided point of view taken," he concluded.[3]

That a commitment to the ideological principle of equality of opportunity was the engine driving the efforts to cast doubt on racial explanations emerges clearly in remarks by that old war-horse of race psychology, Thomas Garth. "Any disposition on our part," he wrote in 1931 in his summary of the literature, to withhold from African-Americans and Amerindians, "because we deem them inferior, the right to a free and full development to which they are entitled must be taken as an indication of rationalization on account of race prejudice; and such an attitude is inexcusable in an intelligent populace." That outlook, it is relevant to observe, was not novel for him. Ten years before, when he was only embarking upon his investigations, he had warned against the influence of prejudice in the field of race psychology. He then thought that some studies of racial differences might well "cause us to hasten to conclusions and thus endanger the so-called inferior races with the stigma of being rated low."[4]

Clearly, as far as Garth was concerned, such an inegalitarian approach would be morally wrong. In fact, at that time, 1921, he laid down a general principle for students of racial differences to pursue in their studies, a canon, as he called it, that clearly reflected his own commitment to an egalitarian social outlook. "In no case," his principle read, "may we interpret an action as the outcome of the exercise of an inferior psychical faculty, if it can be interpreted as the outcome of the exercise of one which stands higher on the psychological scale, but is hindered by lack of training."[5]*

By the end of the decade some psychologists were frankly, even bitterly, identifying proponents of racial differences in intelligence as enemies of the American principle of equality of opportunity. For example, psychologists Paul Witty and Harvey Lehman in 1930 described the study of racial differences as "the dogmas of superiority," and referred disdainfully to those psychologists and educators who employed mental tests "to bolster up preconceived notions of *racial differences in innate capacity.*"[6]

A primary source of such strong opposition to inquiries into racial differences was sympathy with the social aspirations of black Americans. For example, Otto Klineberg, like his Columbia teacher and colleague Franz Boas, made it his business to address black audiences to help them counter the arguments of those who described them as inferior. Thus in 1931 in an article in *Opportunity*, the journal of the Urban League, Klineberg attributed the differences in the test scores of whites and blacks to differences in social environment. The longer blacks lived in the North, he assured them, the narrower the gap between the scores of the two races. George Payne, an assistant dean at New York University, denied there would be any solution to the race problem "as long as the white race assumes that the Negro comprises an inferior group and he must

---

*Historically minded psychologists will recognize Garth's "canon" as a paraphrasing of Conwyn Lloyd Morgan's more famous one enunciated in 1890 in regard to the study of animal behavior: "In no case may we interpret an action as the outcome of the exercise of a higher psychical faculty, if it can be interpreted as the outcome of the exercise of one which stands lower in the psychological scale." Quoted in Robert Boakes, *From Darwinism to Behaviorism: Psychology and the Minds of Animals* (Cambridge, Eng: Cambridge Univ. Press, 1984), 40.

remain the ward of the white man. This fact," he pessimistically concluded, "may postpone the solution indefinitely."[7]

A similar sympathy with blacks emerges from the report of psychologists Paul Witty and Martin D. Jenkins, in which they announced their having located a black child in Chicago with an I.Q. of 200. Having made a systematic search of elementary schools for black children whose scores fell in the so-called genius range, they were clearly proud of their success. Although admittedly doubtful of the validity of such tests, they nevertheless were pleased that they had identified the first black child to achieve a score above 180. Lewis Terman, they pointed out, had located only fifteen persons with scores above 180 and Leta S. Hollingworth had identified seventeen, but neither of them, nor anyone else, as far as Witty and Jenkins knew, had published "an account of a Negro child testing at these extraordinary levels." Their sympathy for blacks also came through in their describing the nine-year-old child "to be of pure Negro stock. There is no record of any white ancestors on either the maternal or paternal side," they pointedly noted. They obviously intended to counter an explanation often advanced by proponents of the mental inferiority of blacks, namely, that high test scores were attributable to an admixture of white ancestry.

Witty and Jenkins, along with many other social scientists at the end of the twenties, looked to culture as the alternative explanation to race. That, however, did not mean they denied a biological basis for intelligence. For as anthropologists Boas and Kroeber and psychologist Otto Kleinberg and others had emphasized, biology or heredity was of central importance in accounting for the behavior or performance of *individuals*. This emphasis upon the role of biology or heredity in shaping individual behavior was reflected in Witty's and Jenkins's discussion of the parents of the "little black genius" whom they referred to as "B." Her father, they reported, was an electrical engineer and her mother a teacher. Several ancestors of the family, they added, were also high achievers. "Without doubt," Witty and Jenkins commented, "the family background indicates superior heredity." The conclusion of their report on "B" once again reflects the ideological underpinning of their professional work. "The fact . . . we can find a Negro child whose I.Q. falls in the very highest range indicates that Negro blood is not always the limiting specter so universally proclaimed in discussions of intelligence measured by the Binet technique."[8]

What seems to emerge from this analysis of the controversy over racial testing, and from the way in which the leading social scientists gave allegiance to the cultural alternative in general by the 1930s, is that, rather than the biological or hereditarian point of view being tenaciously defended, it was rather easily and quickly overthrown or supplanted. That overturning was accomplished not because of highly persuasive empirical evidence, unknown before, but rather

by a willingness to substitute a new assumption or alternative explanation, pow-
ered by an ideological commitment to open opportunity for the socially
disadvantaged.

Many historians of the period, it is true, have called attention to the racism
and ethnocentrism of those years, especially in regard to blacks and recent
immigrants from southern and eastern Europe. But among the social scientists
of the period a different, almost contradictory view seems to have prevailed.
Biological or hereditarian explanations for the differences in human group
behavior, or, more precisely, assertions of differences in mental capacity
between groups, were largely on the defensive for a good part of the period.
From the nineteenth century onward, many American social scientists were
predisposed to favor change and progress, social improvement and reform, an
outlook that came to shape their response to explanations for human behavior.
When given a choice between explanations that facilitated or permitted social
change and improvement and those that fixed the status quo or lengthened the
time required to bring about social changes, American social scientists generally
found the former more persuasive and more congenial. The natural tendency
of their world-view was to prefer an environmental or cultural explanation.
Human nature was not divided; the well-recognized diversity among human
groups derived not from race but from different histories and environments.
For, as was pointed out again and again, the same race often exhibited quite
different social behaviors under different circumstances, showing that changing
environment, not unchanging race, was the more likely explanation for human
behavior.

In Boas's argument, as we have seen, as well as in that of others, the burden
of proof was placed upon those who relied upon heredity or biology. One measure of this is that very few efforts were made to shift the balance, to demand
that the burden of proof be borne by those who asserted an environmental
explanation. As Boas characteristically phrased it, if a biological explanation
could not be conclusively proved, then culture must be the causal element.

By the opening of the 1930s the number of reputable social scientists, including psychologists, who espoused the hereditarian point of view in professional
journals or books dwindled noticeably. One last-ditch defender among psychologists, William McDougall, continued to defend his instinct theories throughout the decade, but his professional standing diminished as he did so.
(McDougall's continuing support of the discredited Lamarckian principle of
acquired characteristics undoubtedly contributed as well to his professional
decline.) As noted earlier, Leta Hollingworth, too, retained her faith in eugenics, along with her colleague at Teachers College, Edward Thorndike. Indeed,
at the end of the 1930s Thorndike's optimism about what could be achieved

through eugenics seemed undiminished. "By selective breeding supported by suitable environment," he argued in *Human Nature and the Social Order* in 1940, "we can have a world in which all men will equal the top ten percent of present men. One sure service of the able and good," he advised, "is to beget and rear offspring. One sure service (about the only one) which the inferior and vicious can perform," he concluded, "is to prevent their genes from survival."[9]

At least two prominent sociologists also retained a publicly expressed belief in the power of heredity or biology throughout the 1920s and later, though among sociologists they were rarities. Their writings sometimes revealed the irritation or exasperation their professional isolation provoked. But that, too, offers a measure of the ease with which culture triumphed.

One of the dissenters was the Russian-born Pitirim Sorokin, who specifically discounted in 1928 an environmental or cultural explanation for what he described as the lack of accomplishment of blacks. He contrasted American Negroes with Russians and ancient Greeks. Blacks, he contended, "have not to this time produced a single genius of great caliber" except, perhaps, for a "few heavyweight champions and eminent singers." After all, he noted, Russians suffered under serfdom and Greeks under slavery, "yet these slaves and serfs of the white race, in spite of their environment, yielded a considerable number of geniuses of the first degree, not to mention the eminent people of a small calibre." Other criticisms Sorokin made of the cultural interpretation were no more sophisticated. Among them was the erroneous observation that Boas's study of bodily changes had been "subject to very severe criticism" and was hardly conclusive. Predictably, he concluded that "the factor of heredity plays an important part in determining the traits and behavior of individuals and groups."[10]

Considerably more thoughtful were the doubts of Frank Hankins, a Smith College sociologist who was sufficiently respected professionally to be elected president in 1938 of his national professional organization, the American Sociological Society. Throughout the twenties Hankins stood out as a sociologist who resisted the easy acceptance of the cultural explanation of behavior. He had begun his career, it is relevant to observe, under the influence of Francis Galton, the founder of eugenics, and the English biological statistician Karl Pearson, a beginning that ever after left a strong hereditarian cast upon Hankins's sociology. Despite that background, in 1926 he published *The Racial Basis of Civilization*, which even anthropologist Robert Lowie could recommend "as one of the sanest treatises extant on the vexed subject indicated in the title." Moreover, in the same book Hankins launched a devastating attack on Carl Brigham's study of intelligence, and the invidious comparisons eugenicists had drawn among immigrant groups. Contrary to Brigham, Hankins praised racial mixture, arguing that, from a biological point of view, if nothing else, such mixture increased the likelihood that more geniuses would be produced because of the increased diversity of the genetic material. Thus "well-endowed Italians, Hebrews, Turks, Chinese and Negroes," he advised, "are better materials out of which to forge a nation than average or below average Nordics." He also ruled out any assumption that a given individual from a so-called inferior group or race should be regarded as inferior. "Science, democratic faith, and humanitarian sentiment," he maintained, "join in not condemning a man on account of race, color, or previous condition of servitude."[11]

Hankins could accept Boas's argument in *The Mind of Primitive Man* that races were not arranged hierarchically, but he shunned those "numerous pious wishers filled with humanitarian sentiment" who concluded "that all races of men must be considered equal." To Hankins, the difference between races was the difference in the number or proportion of individuals of superior intelligence they contained. Or as he expressed it: "the frequency of superior individuals born within the group is of the greatest significance for the role of that group in cultural evolution." Both Boas and Kroeber, of course, also recognized individual differences in intelligence, but neither of them agreed with Hankins's interpretation. To them, the range of individual variation within whole races or groups of people was assumed to be about the same, hence differences in achievement among groups were to be explained by their diverse experiences or histories, not by their diverse distributions of geniuses as Hankins contended.

Hankins's approach to racial differences and his strong belief in the inheritability of intelligence caused him to perceive Boas's conception of culture as simple environmentalism, despite Boas's and Kroeber's recognition of the role of heredity in the life of the individual. Therefore it is not surprising that Hankins, too, like Sorokin, took exception to Boas's famous bodily changes study, the results of which he explained as deriving from surreptitious intermixture of types, an explanation that Boas had explicitly countered in his report. Hankins admitted his explanation was "pure speculation, but so also is Boas's unspecified environmental explanation," he added with a touch of irritation.

It is not without significance, either, that Hankins also lacked Boas's faith in the potentialities of black people. Agreeing with Sorokin, Hankins thought that the historical burdens under which blacks labored did not adequately account for their lack of achievement. Other groups had also suffered in the past yet they had managed to do quite well, he noted. Sorokin had compared the experience of blacks with that of Russians and ancient Greeks; for Hankins the comparison was with Jews, who, he remarked with gross exaggeration, have been "more vigorously hated than the negro in many of the United States during the last half century." He considered the natural inferiority of blacks to whites so obvious that he did not think it "needed the argumentation" that he was offering. But it was required, he concluded in the end, because of the influence of the "'Boas School,' which has succeeded in conveying the impression that it believes the races equal in inherited capacities."[12]*

Hankins retained well into the 1930s the eugenic beliefs with which he began, especially in regard to the distribution of intelligence by class. He entertained no doubt that intelligence and social status went hand in hand, not simply because the rich or governing class had superior intelligence by their ancestry, but because the opportunities for education and the openness and competitiveness in American society enabled those with high intelligence, regardless of their social origin, to get to the top. "It would seem a preposterous joke of fate," he remarked a little testily, "if all our effort had availed nothing and the positions of honor and responsibility were occupied by men and women no abler than day laborers." And just because he believed that those who had reached the top were the more intelligent, he lamented their reluctance to breed, to pass on that intelligence to more offspring, especially since those at the lower reaches of society were reproductively more successful. Indeed, in 1926 he was already seeing around him signs of intellectual deterioration which he was convinced stemmed from the reproductive recalcitrance of the upper classes, an example of which was that "Freudianism and psychoanalysis have an increasing vogue."[13]

Hankins's naive faith in the openness of American society in the 1920s and the social Darwinism that followed from it are not reasons for remembering him. Rather, what calls our attention to his ideas was his resistance to what he perceived as his fellow social scientists' excessive emphasis upon cultural explanations. He recognized early on that the cultural explanation as it was then being defined left no place for heredity or evolution in accounting for human social behavior. At times, apparently out of frustration with what was occurring among social scientists, he could be a little excessive himself. Thus he denounced Luther L. Bernard's book-length attack on instinct as "eighteenth century utopianism all over again." Many years later, in looking back on the nature/nurture controversy of the twenties, he identified his "most important contribution to be the concept of organic response as over against the concept of organic plasticity, so widely accepted by sociologists."

Unfortunately, it was just that persistent interest in biological explanations that also predisposed him to overlook what he euphemistically referred to in 1937 as the Nazis' "racial excesses" while praising the attention the Germans were giving to the question of good inheritance and population improvement.

---

* Despite the inclusion of Boas's name, Hankins contended, quite erroneously, that Boas himself did not subscribe to the egalitarian view. In Hankins's eyes Boas's view was simply that blacks were not as inferior as popularly supposed. See Hankins, *Racial Basis of Civilization: A Critique of the Nordic Doctrine* (New York: Alfred A. Knopf, 1926), 323.

Not surprisingly, when sociological textbooks were reviewed in 1940, the author of the survey found that Hankins's text devoted about three times the space to racial matters that the other texts provided.[14]

Hankins and Sorokin were upholding a position that had once been popular, but was no longer. The exasperation that came through in some of Hankins's writings measures both the triumph of culture and the ease with which it was accomplished. A similar exasperation could be sensed, too, in the description one psychologist gave of the outlook of those who, like Hankins, were still holding out for some role for race or biology. "The race superiority enthusiasts appear to be on the defensive," psychologist Dale Yoder noted as early as 1928, "and they seem to be striving furiously to amass more of the usual evidence to support their thesis," often ignoring, he revealingly added, that it "is the quality rather than the quantity of evidence" that will decide the issue. After reviewing the literature on racial differences in 1930, psychologist Paul Witty suggested that the widespread doubts about racial superiority had "been literally forced upon everyone who has examined the vast array of data which has been accumulated during the past decade." Those studies, he added, "lead to conclusions similar to those expressed by a number of sociologists," among whom he included Boas. The change in attitude on the part of psychologists and educators, he contended, "becomes increasingly significant if one recalls the dogmatic statements which were made regarding racial differences about 10 years ago."[15]

Professionally and intellectually, psychologists by the end of the 1920s escaped from the impasse of race differences by emphasizing the interaction of environment and heredity. Neither was decisive, each was dependent upon the other. The attempt to determine which was dominant had proved fruitless. By then the interactionist view was well recognized among natural scientists as geneticist H. S. Jennings's popular but authoritative book *The Biological Basis of Human Nature* made clear in 1930. In Chapter 6, for example, Jennings provided a number of examples from the work of biologists to demonstrate how environmental circumstances affected the expression of genes in animals and plants. And when in the succeeding chapter, he asked what was the relative importance of heredity and environment, his response was that no single answer was valid. "For good results," he advised, "both fit materials and appropriate treatment of these materials are required; good genes and fit conditions for their development."[16]

About the same time, educational psychologist Gladys Schwesinger was arriving at a similar conclusion for social scientists. In her 1933 book *Heredity and Environment* she concluded that, in any organism, "heredity and environment . . . both contribute, and . . . neither alone is sufficient, each being always dependent on the other." It is even possible, she continued, "that the functions of one can at times and in varying degrees be taken over by the other." As a result, she assured her readers, the old question of which was more important "belongs in the scrap basket."[17]

Actually, more was affecting the change among psychologists than the mere accumulation of studies and the impasse that they produced. As in the case of changed views among members of the other social disciplines, how an investigator personally felt about blacks frequently determined where a psychologist came down in the controversy over racial difference. During the twenties, American social scientists of all kinds, not only psychologists, gained an unprecedented opportunity to observe blacks in a fresh and often transforming way. For that was the decade of the so-called "Great Migration," the movement of literally hundreds of thousands of blacks out of the rural South into the cities of the North. Second only to the massive immigration from eastern and southern Europe, the Great Migration (and its expansion during the Second World War) was perhaps the most significant social change among Americans in the first half of the twentieth century. The migration of black families began just before the First World War, accelerating as jobs opened up for blacks in northern manufacturing centers for the first time. As a result of the restrictions imposed on international migration by the outbreak of the European war in 1914, a major source of labor for American industry—white immigration from Europe—was dramatically cut back. (In 1914, 1.2 million Europeans had entered the United States; in 1915, the number fell to 327,000, a drop of 73 percent.)

How, then, did this northward movement of black people affect the thinking of social scientists on the question of racial differences? For one thing, it made the question a national, rather than simply a southern matter. As sociologist Donald Young, writing in the *Annals of the American Academy of Political and Social Science,* observed in 1928, no longer could social scientists fall back upon "Lord Bryce's widely accepted theory that the Negro's 'problem' would solve itself through his automatic segregation in the warmer Gulf states which were supposed to be better adapted to his tropical nature." Furthermore, Young pointed out, the new employment opportunities for African-Americans in the Northern cities widened the experiences of blacks, thereby changing attitudes about their capabilities and potentialities. "Popular ideas about Negro health and the ultimate 'extinction through degeneracy' of the race," Young noted, "have suffered severe shock in the light of improved medical science." A new race consciousness born of the migration, he thought, had enhanced confidence among blacks, one result of which was their movement into the professions, again causing changes in attitudes and ideas among whites as well as blacks.[18]

The entrance of Afro-Americans into the professions, especially education and scholarship, did more than simply provide role models or new perceptions

of blacks. Some black scholars now made contacts with their white counterparts, enlisting their support in fighting race distinctions. W. E. B. DuBois, for
example, wrote both Franz Boas and biologist Jacques Loeb with just that purpose in mind during these years. Black social scientists themselves published
critiques of racial explanations, sometimes in league with whites, sometimes on
their own. Martin Jenkins, for example, who worked with Paul Witty on the
discovery of the black girl with the exceptionally high I.Q., was African-American; later he become president of Morgan State University. Horace Mann
Bond, soon to distinguish himself as a historian, published in 1927 a study of
exceptional blacks in Chicago in order to refute the so-called "mulatto hypothesis," which argued that only blacks with some white ancestry could achieve
high scores on intelligence tests. Howard University psychologist Charles
Thompson in 1928 wondered in print why some Southern whites were not seen
as inferior mentally since their I.Q. scores were lower than those of Northern
blacks.[19]

The issue of the *Annals* in which Donald Young's remarks appeared was
devoted to "The American Negro," the contents of which contained articles by
almost a dozen prominent or soon to be prominent Afro-American scholars,
some of whom were social scientists. Among them were W. E. B. DuBois, the
historian, sociologist, and editor of *Crisis,* the journal of the NAACP; Alaine
Locke, a leader in the Black Renaissance of the period; the author James Wel-
don Johnson; and E. Franklin Frazier, a rising young sociologist trained at the
University of Chicago. This bringing together of black scholars could not help
but foster changes in the attitudes and perceptions of white social scientists
about the capabilities of African-Americans.

New conceptions of Afro-Americans were catalyzed by black Harlem. Har-
lem's Renaissance, which celebrated, among other things, a vibrant and hitherto
ignored cultural tradition frankly drawn from African roots was transforming.
Reflective of the new prominence of blacks in American culture and its effect
upon social scientists was the reference by the prominent sociologist Ellsworth
Faris in 1923 to the writings of the Negro author Claude McKay. Faris thought
McCay's poem "If We Must Die" expressed the defiant feelings of "millions of
Negroes [who] have read and memorized" it. Many people, Faris continued,
have assumed that blacks "have an exaggerated instinct of submission. But there
is no submission here." African-Americans, Faris concluded, were now on the
march to a better life.[20]

Two psychologists writing in 1929 saw more than that: "There is apparently
developing in New York under the more severe struggle for existence," they
speculated, "a highly selected negro population which represents the best genes
in the race—whether pure or mixed with white and Indian blood." They were

careful to say, however, that there was no reason to expect any correlation between intellectual quality and "degrees of white characteristics." The new opportunities provided by New York to the newly arriving blacks, the two psychologists thought, might well "yield a negro of high intelligence, even surpassing, it may be, the general level of whites, who are less handicapped by social discriminations."[21] Blacks in Chicago projected a different but no less potent message. Thanks to the Great Migration, black Chicagoans in 1928 elected the first African-American member of Congress from a district in a northern state.

One Southern white sociologist thought that the movement of Afro-Americans out of his region catalyzed even those blacks who remained behind. The revival of the Ku Klux Klan in the early twenties, he thought, signaled a new self-awareness of Southern blacks as they saw a Northern alternative. At the present time, Guy Johnson warned, only a minority has fled the South, but the new self-consciousness can only spread with time. It is not clear how this new image of Afro-Americans by the 1920s affected the thinking of southern white sociologists like Guy Johnson and Howard Odum, but it cannot have been negligible. After all, widespread expressions of discontent among Afro-Americans led naturally to questions about the source of their complaints, something that the old Southern social order had rarely been compelled to confront.[22]

Southern black scholars, too, obtained a voice during the 1920s, providing an additional impetus to change in outlook among American social scientists and in the atmosphere in which they worked. The first issue of *Social Forces*, a scholarly journal founded and edited by Howard Odum at Chapel Hill, contained a piece by E. Franklin Frazier. In a subsequent issue Abram Harris, an economist at Howard University, vigorously rejected the argument that the African heredity of blacks accounted for the inability of black and white workers to stand together; there was ample history, he wrote, to explain quite adequately blacks' suspicion of white working-class overtures and promises. One of the earliest criticisms in a professional journal of the Army tests was advanced in 1922 by black educational psychologist Martha McLear of Howard University.[23] During the 1920s descriptions and analyses of the social sources of the alleged backwardness or deficiencies of Afro-Americans found ready forums in new black publications like *The Crisis* of the NAACP and *Opportunity* of the Urban League and in the pages of national liberal magazines like *The Nation* and *The New Republic*.*

---

*One indication of the complications and ironies in the debate over intelligence testing in the 1920s and the 1930s was that blacks, who were often among the leading critics of mental testing, nonetheless were also well represented among those who used such tests themselves in black schools and colleges to measure achievement of black students or to identify differences in social back-

Blacks were not the only disavantaged group among social scientists to testify against race as a category of social analysis. A psychologist of Chinese descent, S. L. Wang, protested in 1925 the use of standardized tests of intelligence among minorities, since most of them lacked the familiarity of natives with English. A sign of the way in which criticisms of race moved through the social sciences is provided by Wang's observation that he had been a student of S. L. Pressey, one of the earliest doubters, as noted earlier, of the validity of testing without careful attention to the effects of diverse environmental circumstances.[24]

Another psychologist of Chinese descent mounted a much more powerful criticism of tests for racial differences in 1929. Hsiao Hung Hsiao followed Wang in raising questions about the testing of Japanese and Chinese because of their lack of familiarity with English. But then Hsiao listed a number of cultural differences that had to be evened out before valid conclusions as to comparative intelligence could be accepted. Among them he included "social status ... , variation in age and grade range of the groups compared, [and] the failure to adapt materials to racial differences in ways of thinking."[25]

Undoubtedly the most influential objections to racial comparisons came from European immigrants, if only because their objections were by far the most numerous. Scholars of Jewish descent who had long been held at a distance or excluded entirely from American colleges and universities were now coming to the fore. The beginnings of the pattern go back as far as Boas himself, but by the second decade of the twentieth century, immigrants or children of immigrants were increasingly prominent in American education in general and in social science in particular.

Criticisms of recent immigrant groups like those in Madison Grant's popular *Passing of the Great Race* and in Carl Brigham's scholarly *Study of American Intelligence* sparked pointed and often devastating reactions from scholars of immigrant background. In 1924, for example, Maurice Hexter of the Federated Jewish Charities of Boston and Abraham Myerson, professor of neurology at

---

ground. Horace Mann Bond, for example, while dean of Dillard University in 1937, warned against indiscriminate attacks on I.Q. tests since they had proved useful in measuring the great variation in the achievement of individual black students. Historian William Bonds Thomas has written at length and with some bitterness on that historical situation, culminating in the painful observation that some of the results of those tests were used in the 1950s by "Southern attorneys attempting to thwart the 1954 desegregation efforts in *Brown v. Board of Education*." William B. Thomas, "Black Intellectuals, Intelligence Testing in the 1930s, and the Sociology of Knowledge," *Teachers College Record* 85 (Spring 1984), 496. On this subject see also his two other articles: "Guidance and Testing: An Illusion of Reform in Southern Black Schools and Colleges," in Ronald K. Goodenow and Arthur O. White, eds. *Education and the Rise of the New South* (Boston: G. K. Hall, 1981), and "Black Intellectuals' Critique of Early Mental Testing: A Little-Known Saga of the 1920's," *American Journal of Education* 90 (May 1982), 258–92.

Harvard, faulted Brigham's book for failing to take into consideration differences in culture between certain immigrant groups and native Americans. Another recent Jewish immigrant, and an established psychological researcher, Gustave Feingold, brought the same point to the attention of the readers of the *Journal of Experimental Psychology*.[26] A number of the prominent social scientists who took part in the crusade against the use of tests in support of racial differences were immigrants or children of immigrants: Boas, Kroeber, Klineberg, Goldenweiser, Sapir, and a more recent student of Boas, Melville Herskovits.

Franz Samelson, a modern student of the subject, argues with some cogency that during the 1920s the central concern of social scientists shifted, in large part because of the change in the social origins of many of them. In analyzing the causes for the shift in outlook, Samelson minimizes the role of new information or of conclusive findings. "Empirical data certainly did not settle the issue," he observed, "one way or the other," any more than the controversy over the heritability of intelligence among respected psychologists in the late 1970s settled that issue. "The vast majority of psychologists," he suggested, "have never taken a close look at all the data, but accepted, by and large, received opinion supported by appropriate pieces of evidence then and now." Although no one has documented the ethnic background thesis for the change—or failure to change—he remarked, it is certainly "arguable that a change in the pattern of ethnic backgrounds among psychologists contributed significantly to the shift." Hamilton Cravens in his study of natural and social scientists in this period has similarly pointed to ethnic and social differences as being a source of the shift in outlook.

Accompanying the change in ethnicity, according to Samelson, was a shift in intellectual perceptions. Prior to 1920, he contends, "the deterioration of American intelligence and genetic potential had been called the problem of overriding importance." By the end of the decade, however, sociologists and other social scientists were seeing race relations as occupying that central position. And one reason they shifted was that by the 1920s social scientists and to an increasing degree the public at large came to recognize that the Americanization movement of the early twenties—that is, the effort to submerge differences completely or to assimilate the immigrant—was not working. Instead, cultural pluralism, or acceptance of differences among Americans, was the only viable principle for such a racially and ethnically diverse population.[27]

Samelson's insight into the causes of the shift fits well with the social sources of the political changes going on simultaneously in the late twenties. The nomination of Alfred E. Smith for president in 1928, political historians now recognize, marked the emergence of a new politics of ethnicity. Smith was widely perceived as a representative of immigrants, largely because of his Catholicism, although his parents had actually been born in the United States. Even though Smith was defeated, the outpouring of immigrant support for him at the polls became the social basis for the sociopolitical coalition that brought Franklin Roosevelt into the White House four years later and kept him there for the rest of his life. That coalition of voters depended in large part upon the movement into the political mainstream of the new city-based European immigrants and their children, and the blacks of the Great Migration in Northern cities. It marked, in short, the coming into political power of a new ethnic and racial pluralism, creating in the process a cultural as well as a political atmosphere within which a recourse to biology or heredity in accounting for human differences would be increasingly difficult to condone.

Changes in the ethnic makeup of the social science community and in the political and social atmosphere constitute a large part of the explanation for the shift in outlook on race, but not the full explanation. The onset of the Great Depression in the early 1930s surely influenced thinking about social causality as well. Samelson, for example, believes it pushed psychologists toward a more leftist political and social outlook, an ideology that had traditionally played down race and biology while emphasizing the social sources of human differences. Whatever may have brought the change about, a 1939 survey of psychologists certainly documented the shift. The author of the survey remarked that though at one time it had been held that tests "measured biological differences" among various groups, "practically no one now believes this." One historian has contended that the stock market crash and the ensuing Great Depression made it difficult for ordinary citizens and social scientists alike to see a clear correlation between economic status and intelligence, since any person could, and many worthies did, go under in the debacle. In such circumstances it was not difficult to look to the social environment as the cause of poverty rather than to innate deficiencies within a group or individual. It became plausible to believe now, as it may not have been in more prosperous times, that the standard complaints about the status of African-Americans were better explained by reference to the times and circumstances, rather than to the effects of race and biology.[28]

Along with the Depression, too, came the fateful news from Germany that the Nazis were putting into practice in an increasingly horrible way eugenic ideas about race purity and population improvement. Some American eugenicists welcomed the early efforts of the Germans to sterilize and otherwise control people thought to be eugenically deficient. Paul Popenoe, for example, a popularizer of eugenics, defended the Nazi program to geneticist L. C. Dunn as late as January 1934, contending that the program was not racially motivated.

The aim, he insisted, was to eliminate all "undesireable elements among the Aryans, whatever these are, than to hit any of the non-Aryan groups," he assured Dunn.* That year, however, over 56,000 persons were sterilized in Germany for mental and other defects; by the middle of 1936, the total had reached 150,000. As late as 1936, after a lengthy visit to Germany, one scholar defended in a leading American sociological journal the German efforts to improve "the biological and racial qualities of the German people. . . . These measures," she contended, "are not arbitrary experiments." From her own observations in Germany she was "convinced that the [sterilization] law is administered in entire fairness and with all considerations for the individual . . . and for his family."[29]

Soon thereafter, however, even last-ditch defenders of eugenics began to recognize the enormity of the Nazi program as it escalated into the murderous horror of the Holocaust. When asked after the war why he thought eugenics died so quickly in the United States, a wiser Paul Popenoe said that "the major factor . . . was undoubtedly Hitlerism." Geneticists and social scientists alike went public in opposition to the Nazi racial assaults; Franz Boas, not surprisingly, was in the vanguard of the movement to mobilize social scientists in opposition to the Nazis. One measure of Boas's activities and of the reaction of American social scientists to the Nazi racist theories and actions was the passage, by a unanimous vote, of a resolution denouncing racism at the 1938 meeting of the American Anthropological Association. The resolution recognized races as physical entities, but went on to say that "psychological and cultural" meanings, "if they exist, have not been ascertained by science." Anthropology, the resolution continued, provided no scientific grounds for discrimination against "any people on the ground of racial inferiority, religious affiliation, or linguistic heritage." As one leading psychologist observed in 1940, the "passing of such a resolution by a scientific body is unusual." It was an early but not the last indication of the impact Nazi practices had on American scholarly thinking about race and biology in human affairs. That impact can hardly be overestimated in explaining why during the 1930s and 1940s concepts and terms like "heredity," "biological influences," and "instinct" dropped below the horizon in social science.[30]

---

*Dunn was less credulous than Popenoe. After a visit to Germany, he wrote a colleague in 1935 that he had observed "some of the consequences of reversing the order as between program and discovery." The German government, he concluded, was using medical and genetical information to enhance its power over the nation and not to improve the population. The German example, he concluded, "should certainly strengthen the resolve with which we generally have in the United States to keep all agencies which contribute to such questions as free as possible from commitment to fixed programs." Quoted in Kenneth M. Ludmerer, *Genetics and American Society: A Historical Appraisal* (Baltimore: Johns Hopkins Univ. Press, 1972), 128.

The deep impression the Nazi racial policies and practices made upon the thought of the American scholarly community, once their full enormity became known, is especially well illustrated by the response of geneticists. Although they were natural scientists, the geneticists, according to historian of science William Provine, seem to have been as influenced by social and political considerations in their thinking about race as were social scientists. In the late 1930s, Provine noted, most geneticists assumed what he called an "agnostic" position on the question of cross-breeding among human races. They neither recommended it nor condemned it from a genetic point of view. In 1946, however, two leading geneticists, Leslie C. Dunn and Theodosius Dobzhansky, forthrightly contended that "the widespread belief that human race hybrids are inferior to both of their parents and somehow constitutionally unbalanced must be counted among the superstitions." Yet, as Provine observed, "the scientific evidence on race crossing had not changed significantly between 1938 and 1946. There simply was not a decisive study on race crossing during that time."

Provine also provides a striking example of the way in which the knowledge of the Nazi horrors affected social scientists' thinking about race and biology. The example was the handling of the UNESCO statement on race in 1950 and after. The first committee appointed by UNESCO to draw up the statement was headed by the well-known American cultural anthropologist Ashley Montagu. The statement his committee drew up in 1950, however, was not issued officially by UNESCO because, among other things, it seemed to put biology on the side of human brotherhood. The unissued 1950 statement said, for instance, that "biological studies lend support to the ethic of universal brotherhood, for man is born with drives toward co-operation, and unless these drives are satisfied, men and nations alike fall ill." That was a position to which few geneticists or physical anthropologists could subcribe.

A second statement on race was then drawn up, this time by geneticists and physical anthropologists. It took a less extreme position, but still minimized the role of race in accounting for human behavior. It found "no reliable evidence that disadvantageous effects are produced" by race crossing, and that "available scientific evidence provides no basis for believing that the groups of mankind differ in their innate capacity for intellectual and emotional development." The impression left by these negatively worded statements, Provine noted, was "that races were alike in hereditary mental traits."[31]

That statement was sent to 106 prominent physical anthropologists and geneticists for their reactions; 80 of whom responded. Twenty-three accepted the published form; 26 agreed with the general argument, but disagreed with particulars. The remainder disagreed substantially, principally to the provision implying equality of mental traits among races. No one seriously objected to the statement on race mixing. "The entire reversal" on race crossing, Provine concluded, "occurred in the light of little new compelling data for students of actual race crosses." When Provine asked ten prominent geneticists for the reasons for the reversal, "not one believed that new evidence on race crossing was the primary reason." Provine himself concluded that "the most important [reason] was the revulsion of educated people in the United States and England to Nazi race doctrines and their use in justifying the extermination of Jews. Few geneticists," he maintained, wanted to agree with the Nazis "that biology showed race crossing was harmful. Instead, having witnessed the horrible toll, geneticists naturally wanted to argue that biology showed race crossing was at worst harmless." In short, as Provine phrased it, "geneticists did revise their biology to fit their feelings of revulsion."[32]

One measurable consequence of the triumph of culture over biology among social scientists in the 1930s and 1940s was the precipitous decline in the number of articles in professional journals discussing heredity and racial and sex differences. By 1935–40, for example, the *Journal of Applied Psychology* carried no articles on those subjects though in the 1920s they had been almost a staple of its tables of contents. The *American Sociological Review,* founded in 1936 and thus reflecting the latest interests of the profession, offered in its first volume ten articles on some aspect of culture, but nothing on race, heredity, or sex. When a sociologist in 1940 surveyed some twenty textbooks in his field, all of which had been published in the 1930s, he found that race was discussed in all of them in varying degrees, but he concluded that "sociologists do seem to be approaching an agreement that the innate intellectual and temperamental differences between the races are small, insignificant, and doubtful."[33]

Broader measures of the wholesale adoption in the 1930s of the concept of culture or environment as the primary shaper of human behavior were the relevant articles appearing (or not appearing) in the *Encyclopedia of the Social Sciences,* first published in 1932. "Feeblemindedness," once a frequently mentioned term among psychologists, did not appear at all in the fourteen volumes. The authors who were selected to write the articles on controversial concepts of the 1920s reflect the new predominance of cultural explanations among social scientists. The article on "Instinct," for example, was written by sociologist Luther L. Bernard, who had written a book in 1924 categorically rejecting the concept. Franz Boas was the author of the article on "Race," and the young Otto Klineberg, not Lewis Terman, the elder statesman of the subject, contributed the piece on "Mental Testing." The article on "Heredity" was not written by a social scientist, but its content mirrored the decline in interest in biological influences nonetheless. "There is as yet no conclusive evidence of genetic differences between races," wrote geneticist Alexander Weinstein, "because no study has eliminated the differences due to environment." Weinstein's next sentence revealed that the ideological implications behind the cultural interpretation had influenced some geneticists even before the Nazis came into power. "Most claims of genetic mental superiority are inspired," he suggested, "not so much by scientific data as by a desire on the part of some classes, nations or races to justify their subjection of others."

Individual expressions of the acceptance of a cultural interpretation of human behavior were often all-encompassing as well as unambiguous. One sociologist, for example, stressed the role of environment in shaping heredity. Even at birth, George Lundberg wrote in 1931, heredity was not the sole influence operating on an individual. For behind each person at the moment of conception, he contended, were "just as truly the selective and conditional influences of environment as at any subsequent time." Indeed, the very combination of genes in any given individual, Lundberg insisted, "is as truly the result of responses to environmental conditions as his subsequent selection of what college to attend."[34]

No work spread the word of culture's triumph more broadly or effectively than Ruth Benedict's *Patterns of Culture,* published in 1934. As first a student and then a colleague of Franz Boas at Columbia, Benedict brought before both the lay public and her fellow social scientists a powerful and positive assertion of the value of culture in thinking about human social organization and behavior. She made no secret of her ideological or moral purpose. At the outset she observed that "modern existence has thrown many civilizations into close contact, and at the moment," she regretted to report, "the overwhelming response to this situation is nationalism and racial snobbery." Consequently, she warned her readers, there has never been a time "when civilization stood more in need of individuals who are culture-conscious, who can see objectively the socially conditioned behaviour of other peoples without fear and recrimination." Indeed, in her mind a clear recognition of the "cultural basis of race prejudice is a desperate need in present Western civilization."

Her message was the power of culture and the weakness or irrelevance of biology in shaping human societies. Biology may shape the societies of insects, she told her readers, but it has nothing to do with human social systems. In primitive man "not one item of his tribal social organization, of his langauge, of his local religion, is carried in his germ cell," she pointed out. As a result, any one or all of those traits can easily be picked up by members of any race. Blacks in Harlem, she pointed out, use the last three units of the turnover in the stock market to make their bets, though the stock market has played no part in their lives in the south nor in Harlem. The blacks' reliance on numbers game, she stressed, "was a variation on the white pattern, though hardly a great departure. And most Harlem traits keep still closer to the forms that are current in white groups," she added, to clinch her case for ease of cultural transmission. Even customs that seem to be derived from biological forces, like puberty rites, she continued, are not determined by biology. It is "not biological puberty, but what adulthood means in that culture [which] conditions the puberty ceremony." Even universal patterns of behavior, those that occur in virtually every culture, like animism or exogamy, she contended, should not be viewed as "biologically determined." Rather they are probably "exceedingly old human inventions." In truth, such universals, she thought, "may be as socially conditioned as any local custom."

Yet for all her emphasis upon culture and its separation from biological influences, Benedict stayed with her mentor Boas in seeing heredity as molding individuals through family lines. Beyond the family, however, heredity as a determinant of human behavior she pronounced to be nothing more than "mythology." Bodily shape or race, she repeated, "is separable from culture, and can for our purposes be laid to one side except in certain points where for some special reason it becomes relevant." It is significant that she identified no such points.[35]

By recognizing a role for heredity in the making of an individual, Benedict stopped short of assuming a totally cultural determinist position. In fact, her very concept of "cultural patterns" argued that societies differed according to the dominant personality traits they followed, a concept, as noted earlier, Margaret Mead borrowed from her. "The vast proportion of all individuals who are born into any society," Benedict contended, "always, and whatever the idiosyncrasies of its institutions, assume . . . the behavior dictated by that society." That happens, she quickly pointed out, not because the dominant personality trait contains "an ultimate and universal sanity" about it. Rather, it happens because "most people are shaped to the form of their culture because of the malleability of their original endowment." In short, "the great mass of individuals take quite readily the form that is presented to them." However, she added, not all individuals have equal endowment or degree of plasticity. Hence some individuals do not conform, and when they do not they are frequently thought to be insane or pathological, for the tendency of the dominant culture is to define itself as absolute. In another culture with a different dominant pattern, however, the deviant's "pathology" might well be admired or cherished.[36] In sum, Benedict's cultural relativism, in theory at least, could accommodate behavior patterns that departed from the dominant culture.

By the 1940s, however, the concept of culture as the shaper of human behavior had matured sufficiently that anthropologist Leslie White could find little or no place for an individual who deviated from the dominant culture. For him culture was all-controlling. The word "heredity" may have appeared with only a few references in the index to Benedict's book, but in the index to White's 1949 book *Science of Culture,* the term did not appear at all. The individual, on the other hand, found a favored place in White's book simply because White was determined to show the subordination of the individual to culture. He criticized by name his fellow anthropologists Alexander Goldenweiser, Edward Sapir, and Clark Wissler for assuming that individuals played a role in shaping civilization. Their assertion might appear to be self-evident, he conceded, but in fact it was but "an expression of the primitive and pre-scientific philosophy of anthropomorphism." Individual differences are no more help in explaining cultural differences, he insisted, than are racial differences. The individual has no part in shaping culture; on the contrary, "it is the individual who is explained in terms of his culture, not the other way round."[37]

To White, the whole concept of the individual was altered by what he christened his "culturalogical interpretation." "Instead of regarding the individual as a First Cause, as a prime mover, as the initiator and determinant of the culture process," he insisted, "we now see him as a component part, and a tiny and relatively insignificant part at that, of a vast, socio-cultural system that embraces innumerable individuals at any one time and extends back into their remote past as well. . . . For purposes of scientific interpretation," he concluded, "the culture process *may be regarded* as a thing *sui generis;* culture is explainable in terms of culture." The individual is merely a surrogate, never a principal.[38]

At first sight, White's culturological interpretation seems little more than an expansion upon Kroeber's "Superorganic," for that, too, had stressed the separation of civilization or history from the influence of the individual. Kroeber, however, retained a place for biology in shaping the individual, and he specifically denied that culture smoothed out individual differences or overrode heredity in the family line. White, however, left no place for the individual. In 1963, for example, he specifically criticized Boas himself for having thought that an individual could be out of phase with his culture in any way. (He might also have leveled the same charge against Benedict, of course.) Boas, White wrote, "obviously had no understanding of the origin and substance of his ideals. He believed that they originated within himself, not only independently of his culture but in opposition to it." Boas could not have been more wrong, White maintained. No one could grow up "outside a cultural tradition," for in doing so "he would have had no more ideals than a gorilla."[39] Whatever contradictions there may have been between Boas's ideals and his culture, they arose, according to White, not from within Boas but from the contradictions within the culture itself. For cultures can be divided against themselves, White explained.

White, unlike Benedict, was a cultural determinist. Culture was the true and virtually the sole shaper of human beings and their social order. In White's view there was no longer any point in talking about human nature. "The fallacy or illusion here is, of course," he wrote in 1949, making a play on words, "that what one takes for 'human nature' is not *natural* at all but cultural. The tendencies, emphases, and content that one sees in the overt behavior of human beings are often not due to innate biological determination—though such determinations do of course exist—but to the stimulation of external cultural elements. Much of what is commonly called 'human nature' is merely *culture* thrown against a screen of nerves, glands, sense organs, muscles, etc." Five years later, the well-known anthropologist Ashley Montagu echoed White's cultural determinism when he wrote that man "has no instincts, because everything he is and has become he has learned, acquired, from his culture, from the man-made part of the environment, from other human beings."[40]*

At the end of the 1950s, George Stocking, Jr., made a survey of some forty social psychological studies concerning minorities. He reported that in "not a single instance" did a writer suggest "that the personality characteristics of Negroes might be in any part the result of innate racial tendencies." As Stocking remarked, the authors obviously did not think the question of race was worth raising.[41] The acceptance of the culture concept was equally thorough among sociologists. Indeed, a president of the discipline's major organization, the American Sociological Society, in 1961 carried the idea to such length as virtually to eliminate biology as a limitation on human potentiality. The president was Robert Faris, the son, appropriately enough, of Ellsworth Faris, who forty years before had welcomed the repudiation of instinct and race as modes of sociological analysis. The son, in his presidential address, picked up where the father left off. There was a time, Robert Faris recalled, when "ability was generally held to be fixed in biological inheritance and improvable, if at all, only by a glacially slow and impractical eugenics program." But today that outlook has been supplanted by the view that society generates the level of ability. "We no longer heed the doctrinaire testers who pronounce specific individual limits for potentialities in mechanical ability, language ability, artistic ability, and mathematical ability," he reminded his fellow sociologists. "Their ceilings have been discovered to be penetrable. Slow readers are being retrained," he reported.

---

*Five years later, Ashley Montagu qualified that cultural determinist statement, but only slightly. "With the exception of the instinctoid reactions in infants to sudden withdrawals of support and to sudden loud noises, the human being is entirely instinctless," he wrote in the second edition of *Man and Aggression* (New York: Oxford Univ. Press, 1973), 10.

"Barriers in many fields of knowledge are falling before the new optimism which is that anybody can learn anything." As a result, he concluded, "we have turned away from the concept of human ability as something fixed in the physiological structure, to that of a flexible and versatile mechanism subject to great improvement." Limits to performance, he admitted, "may eventually be shown to exist, but it seems certain that these are seldom if ever reached in any person, and in most of the population the levels of performance actually reached have virtually no relation to innate capacities."

Faris's belief in the power of culture and its hopeful promise for the future appropriately culminated in metaphor. "In the present opera on the nature and destiny of man's genius, we have heard only the opening bars of the overture, but the music suggests that the production will someday be a success, and that the amount of effort we put into it will make a difference in the time required." Biology and genetics had failed to offer much hope. "But now," he was pleased to report, "we perceive that an important part of the relevant causation of abilities is essentially sociological in nature, and that control is most likely to come through penetration of this aspect of the subject."[42] Progress, in short, was almost unlimited, if only sociologists and other social scientists would devote their minds to the task.

Even as Faris was spelling out the unbounded opportunities culture had opened up for human beings, other social scientists were beginning to express reservations about the breadth and depth of the potentialities of the concept. Could it explain or account for everything that constituted a human being; was its dominance among social science theories justified? Was the disjunction between human beings and other living organisms as total as the culture concept implied? Had culture extinguished all the influences of biological evolution, which still, after all, defined human beings as animals? Thinking and talking animals, to be sure, but still animals. Where, in all of the emphasis upon culture and its creation of a unique human animal, did Darwinian evolution fit? Was it not contradictory to speak of a belief in the Darwinian explanation for evolution and yet to see little or no continuity in behavior between ourselves and the rest of the animal world? Could it be true, as Wallace had told Darwin, that human beings through their achievement of culture had succeeded in carving an unbridgeable chasm, in erecting an impenetrable wall between our animal ancestors and ourselves?

By the 1950s and 1960s new developments in biology caused some social scientists to begin to think about such questions. The result would be a revival of interest in biology as the other component in human nature. For some questions there would be a return to earlier answers; for others the responses would be fresh and imaginative, sometimes unsettling and even threatening. At once familiar and foreboding was the return of the towering figure of Charles Darwin, the man who had started the whole thing in the first place.

III

Remembering Darwin

9

Biology Redivivus

> The premise which cannot be stressed too often is that what the hereditary determines are not fixed characters or traits but developmental processes. . . . It is lack of understanding of this basic fact that is . . . responsible for the unwillingness, often amounting to an aversion, of many social scientists . . . to admit the importance of genetic variables in human affairs.
>
> *Geneticist T. Dobzhansky, addressing a meeting of anthropologists in 1963*

A convenient measure of the triumph of the concept of culture among social scientists may be found in a single book published in 1944 and in the work of a single man who came into prominence in the 1950s. The book is Gunnar Myrdal's *An American Dilemma;* the man is the famous Harvard psychologist B. F. Skinner.

Myrdal's book was a two-volume study of the place of blacks in American society, a project financed by the Rockefeller Foundation and written by a leading Swedish social scientist. Drawing upon the whole range of the social sciences for its conclusions, the book accepted without doubts or qualifications the concept of culture in accounting for the deplorable economic, social, and political position of black Americans. Racial explanations found no place in Myrdal's more than 1400-page compendium of facts and interpretations; *An American Dilemma* quickly became the standard work on race relations in the United States. The reformist impulse that stood behind the work was candidly acknowledged by Myrdal when he described the attacks which American social scientists had made upon racial explanations for human behavior. "Social research has thus become militantly critical," he was pleased to note. "It goes from discovery to discovery in various areas of life. . . . By inventing and applying ingenious specialized research methods, . . . this research becomes truly revolutionary in the spirit of the cherished American tradition." As he wrote at the end of his book: "We have today in social science a greater trust in the improvability of man and society than we have ever had since the Enlightenment."[1]

In a true sense, the book can stand as the epitome of the transformation through which the social sciences had passed since Franz Boas, more than half a century earlier, first began his explication of the concept of culture as the essential explanation for differences among human groups. As sociologist Howard Odum wrote in 1951, nothing had changed more in his field than attitudes toward race and ethnic groups. To early sociologists, he recalled, "race was an elemental and relatively unchangeable heritage," but today "these assumptions no longer predominate. . . . Rather, race is interpreted as a complex of societal conditioning in which culture is considered to be a more dominant factor than biology."[2]

The rise to prominence in American psychology of Burrhus F. Skinner of Harvard University in the 1950s was another, though rather different symbol of the removal of biology from social scientists' explanations for human action. Skinner, it is true, became well known, inside and outside his profession, for his behavioral experiments with pigeons. His use of animals was similar to John Watson's: the pigeons were trained through conditioned responses to perform certain acts, thereby demonstrating the behaviorist principle that the source of behavior, human as well as animal, came from outside the organism, not from its biological history. In his experiments, Skinner assumed, as Watson had, a biological organism, but its actions were reactive, not original. He never doubted that biology imposed limits on what an animal could perform, but, within that behavioral repertoire, an animal's activity was shaped by the demands and examples of the environmental situation in which it existed. His famous experiments with pigeons were intended to demonstrate just that. The so-called Skinner Box, which he contrived for the rearing of human infants, including his own daughter, was conceived on the same principle. By completely controlling the environment in which an infant was reared, its behavior could be shaped as the experimenter wished.

Yet at the very time that these and other signs of a culturalist triumph at the expense of evolutionary biology were making their point, some social scientists were having doubts and not a little regret at the way things had worked out. Their questions and reservations would not echo with equal force in all of the social sciences until the 1960s, but by then there could be no mistaking the intention of a growing number of social scientists to take a hard second look at earlier decisions to extirpate biology and heredity from explanations for human behavior.

Psychologists made the first move. Their field, more than any other, conceived of itself as a "science." Furthermore, psychology had been the discipline in which the battle against racial differences had been the last to be won. And so it came as no surprise that biology "returned" earliest there. To be completely accurate, biology had never really left the discipline. Behaviorism may have finished off "instinct" as a primary principle, but the idea that intelligence was biologically rooted had never been abandoned. The debate in the 1920s over racial differences in intelligence had largely removed the issue of group intelligence from professional discussions, but the question of the sources of differences in individual intelligence easily survived the onslaughts upon biology during the 1920s and 1930s. The principal reason for the survival was the willingness of psychologists to pay heed to geneticists, whose work by the 1930s was increasingly sophisticated as well as responsible and respected, now that the leaders in the field had severed their connection with eugenics.

The extent to which a biological approach survived within psychology can best be seen in Robert Woodworth's report *Heredity and Environment,* which he published in 1941 under the auspices of the Committee on Social Adjustment of the prestigious Social Science Research Council. Woodworth was a leading psychologist, then teaching at Columbia University along with Franz Boas. Indeed, in 1910, Woodworth had been one of the earliest psychologists to oppose the use of race in explaining an individual's capability. He was, in short, hardly an old-line hereditarian.

On the opening page of his 1941 report, Woodworth noted the division within the social sciences over his subject. Biologists, he observed, tended to stress genetics, while sociologists emphasized the role of environment. "Psychologists are more divided in their interests," he observed, and so among them "the controversy between hereditarians and environmentalists is most acute." The substantive part of his report was a summary of recent research on twins and foster children. The purpose of the research had been to ascertain the extent to which intelligence was biologically based. A significant finding in several studies had been that children of parents of a low socioeconomic level performed much better on intelligence tests than anticipated. Woodworth's comment was indicative of his belief in the important influence of heredity. "Instead of saying that these children have made good in spite of poor heredity," he pointed out, "we must conclude that their heredity was good or fair in spite of the low status and unsatisfactory behavior of their own parents." Clearly, in Woodworth's mind, intelligence was largely dependent upon heredity, even when that was not the most obvious conclusion to be drawn.

The same conviction was apparent when he conceded that even children of feebleminded parents—as opposed to those who were merely of a low socioeconomic status—sometimes performed well. But, contrary to what some environmentalists might conclude from such evidence, Woodworth quickly added that it would be "a scandalous exaggeration of the known facts [to] assure a gifted young couple that they could do as much for the next generation by adopting any 'normal' infant as by having a child of their own." With convictions like that about the importance of heredity, it is not surprising that Woodworth advised social researchers to cooperate with geneticists. "Without such collaboration," he warned, "the social investigator will be likely to work under false assumptions and to miss important leads."[3]

Woodworth's advice was taken to heart soon after the end of the Second World War when a biological laboratory called a conference of psychologists in September 1946 on the subject of "Genetics and Social Behavior." The innovative character of the symposium was reflected in the small number of people invited and the informality of the presentations. The occasion, significantly enough, was the inauguration of a new program at the Jackson Biological Laboratory at Bar Harbor, Maine, on "psychobiological and sociobiological" studies. (That was surely one of the earliest references to "sociobiology" in the professional literature.) Although the impetus for the meeting came from biologists, the participating psychologists had clearly been waiting for the opportunity, even if many of their fellow psychologists were not. For, as one of the participants remarked later in the published report, "it will come as a surprise to many psychologists that all five of the distinguished contributors to this Symposium [all of whom were psychologists] have emphasized the role of heredity in the determination of behavior.* Not one," he pointedly noted, "has insisted that environmental factors are of primary significance in shaping the psychological characteristics of the individual." All the speakers, he added, acknowledged the importance of environment, but "the interesting thing is that all of the speakers place their stress on heredity. . . . Twenty-five years ago," he noted accurately enough, "the situation was quite different." He then proceeded to call the roll of the great opponents of instinct of the 1920s: Watson, Kuo, Kantor, Bernard, Dunlap, and Faris, who, in his opinion, had reacted much too violently against William McDougall's concept of instincts.[4] The attack upon the behaviorists and enemies of the concept of instinct had begun.

Within less than a year another conference on heredity and environment, organized this time by experimental psychologists, was held at Princeton University. Again the critics of human instincts were declared to be extremists and the role of environment described as much more limited than many thought. Indeed, Leonard Carmichael told the group "there is nothing that I know of" in recent studies of early human behavior that would "deny the possibility . . . that human nature is nine-tenths inborn." In fields like "social psychology, sociology, and cultural anthropology," he admitted, "this hereditarian point of view has not been fashionable." But he predicted that in the years ahead those fields would be wise to investigate "inborn individual and racial *differences* of

---

*Among the well-known psychologists present were: Karl Lashley, Frank Beach, and Calvin Stone, each of whom gave papers, and Gardner Murphy, T. C. Schneirla, and Robert Yerkes.

behavior." And if they do, "a knowledge of heredity and the science of genetics and of what may be called the embryology of behavior" will surely prove essential. Environment, Carmichael conceded, could not be ignored, but it must not be "asked to try to carry the whole task of explaining all of human nature."[5]

If psychologists were in the forefront of the return to biology, anthropologists were close behind. And that was true even though, as anthropologist Clyde Kluckhohn told a meeting of psychologists in 1948, his discipline had provided psychologists with "the concept of culture." Now, he said, "we ask from you in return a formulation of the human nature which is the raw stuff that all cultures act upon." His reference was to biology, for he was "firmly convinced that every kind of psychologist needs a substantial training in biology, and the same is true for the anthropologist." The common interest of the two disciplines in bridging "the gap between the organic and the socio-cultural," Kluckhohn was convinced, accounted for their "toughmindedness and for the realization of the full complexity of behavioral phenomena. . . . Some social psychologists and some cultural anthropologists have tended too much toward a complete environmentalism," he conceded, but he now thought them to be in the minority. Most anthropologists, he erroneously informed his audience, have "steadily insisted upon the relevance of genes, biological structure and process maturation and other organic facts."

Kluckhohn was doing more than suggesting a rapprochement with psychobiology; in emphasizing human nature he was really restructuring the emphasis within the two disciplines. "Both psychology and anthropology could well begin to lay as much stress upon the similarities in all human beings as they have . . . upon the differences," he advised his audience. An anthropologist, looking at the present state of psychology, he regretted to say, "must express some fear lest 'human nature' go the way 'mind' has gone."[6] His point was well taken, for the concept of culture had indeed been "invented," in effect, to account for the differences among human groups. The new biology, the biology of Darwinian evolutionary theory, and genetics, however, dealt with the whole species, not with the races or subgroups within a species.

Neither Kluckhohn nor the psychologists who were drawing upon biology in accounting for human behavior at that time paid any significant attention to evolutionary theory or the animal roots of human beings, to which Charles Darwin had attributed much explanatory importance. Margaret Mead, however, for all of her earlier emphasis upon cultural explanations, by the end of the 1940s was beginning to direct attention to man's animal past. For example, when she came to discuss the behavior of males in her 1949 book *Male and Female,* she drew upon animal behavior in identifying what was peculiarly human about the behavior of males cross-culturally. The title of her chapter on human fathers captured her point: "Human Fatherhood Is a Social Invention." Her argument, of course, was that among the great apes—human beings' closest animal relatives—fathers played no role except as generative beings. They did not stay around once that role had been fulfilled. Her unstated assumption was that because human beings were descended from animals, behavior patterns were expected to be biologically transmitted. When one was not, as in the case of "fatherhood," then it could be assumed to be culturally derived. Similarly, as we saw in an earlier chapter, when she came to discuss the psychology of sex—in particular, the female orgasm—she again drew upon animal behavior in defining what it meant to be human, that is, to be a culture-creating being.

A review of ethologist W. C. Allee's *Cooperation Among Animals with Human Implications,* which appeared in the *American Anthropologist* in 1952, offers another measure of anthropologists' generally negative reactions to comparisons between human and animal behavior. Anthropologists, the reviewer advised, should "not cling to notions of man's essential otherness as to feel antagonistic to the view that man's societies have something in common with those of other animals." The reviewer thought that Allee's presentation of the latest work in ethology might well be useful to anthropologists. To do so, however, would require, as Kluckhohn had said earlier, that anthropologists place a new emphasis upon human nature, as opposed to human differences.

Even someone like Alfred Kroeber, who, forty years earlier, had insisted upon resolutely separating history and biology, by the 1950s was calling for the study of human nature, or what distinguished humanity. Looking back from 1955, he drew a thumbnail sketch of the history of the rise of the concept of culture, in which he had played a central role. The earliest anthropologists, those of the 1895–1915 period, he recalled, had simply assumed a single human nature standing behind the many cultures, one consequence of which was their search for human "universals." That belief, however, had long been superseded, he noted, by telling criticism and the introduction of "cultural relativism." Today, and quite properly, he thought, there was a renewed interest in what it meant to be a human being.

One way of analyzing the question, he suggested, was through a comparison of animal and human behavior. Still loyal to his earlier position, Kroeber could not recommend experimentation since that would not be historical. Instead, he advocated process, or what he called "the relation of cultural history to organic history," a practice he saw embodied in the recent work of European animal behaviorists like Konrad Lorenz and Nikolaus Tinbergen. Studies of sub-human behavior patterns that are "similar to, or anticipatory of human cultural behavior patterns" he thought might well be helpful. Investigations of pecking orders, or dominance behavior, among animals would similarly be relevant in seeking to identify the boundaries of human nature. The social behavior of birds and other animals he described as crying out for study for the same reason. Invoking the example of Darwin's writing of the *Origin of Species* after twenty years of patient accumulation of data, he urged anthropologists to do the same with human and animal history. The rising interest in human nature, he added, makes it "clear that we cannot permanently ignore the basic genetic part of our psychology."[7] Having once created a concept of culture that was to be free of any connection with or reliance upon biology, Kroeber was now ready to recognize the biological roots of human nature.*

Among the few works by anthropologists that Kroeber thought worthy of mention as following a sound approach to the study of human nature was David Bidney's *Theoretical Anthropology,* published in 1953. Interestingly enough, Bidney's book was a rather devastating critique of Kroeber's concept of the "Superorganic," in which the irrelevance of biology to cultural analysis had been insisted upon. It was that insistence upon the lack of any connection between the two that Bidney singled out as the central flaw in the Superorganic position. To substitute "a superorganic fatalism," he remarked, was just as objectionable as the "organic fatalism" which the proponents of culture had sought to eliminate. The idea that culture stood above human beings, yet molded them, "while developing according to natural laws of its own," Bidney categorically pronounced "the prime example of the culturalistic fallacy. It ignores the question of the human origin of culture and regards cultural phenomena as if they were autonomous, efficient agents of themselves." For him, human nature was "logically and genetically prior to culture.... In other words," Bidney continued, "the determinate nature of man is manifested functionally through culture, but is not reducible to culture." Ortega y Gassett's dictum that "'man has no nature; he has history,'" told only a part of the story for Bidney. "Adequate self-knowledge requires a comprehension of both nature and history," he contended. As for his own conception of man, Bidney defined it as "Omnis cultura ex Natura," a radical revision of Robert Lowie's old rallying cry for the proponents of culture: "Omnis cultura ex cultura."[8]

In connection with Kroeber's revised outlook, it is pertinent that a student of his, Earl Count, went further than any anthropologist in the 1950s in seeking to establish fundamental links between biology and the social sciences. In an article that would be frequently cited in subsequent years as a landmark on the

---

*Kroeber's own revival of interest in the biological sources of human behavior apparently goes back several years earlier. Earl W. Count, a student of Kroeber's at Berkeley, wrote that his pioneering article "The Biological Basis of Human Sociality," "was undertaken in 1951 under the kindly prodding of Professor Kroeber. Its existence is herewith gratefully attributed to him." Count, *American Anthropologist* 60 (1958), 1049n.

subject, entitled "The Biological Basis of Human Sociality," Count carefully placed human beings within the spectrum of animal evolution. He also introduced the term "biogram," by which he meant an organism's "way of living," a combination of its behavior and its physical form. Boldly, he announced that anthropology's mission was to "account for the emergence of a culturized biogram out of a prehuman, nonculturized biogram." A part of that accounting, he continued, would require a recognition of the similarity between animal and human behavior. "It is," he believed, "already possible to present a very reasonable case for the supposition that man, with all his capacity for self-conditioning, has not escaped from an innate vertebrate biogram; the evidence," he suggested, "comes from ethology, psychology, psychoanalysis, and neurology."[9]

If the revival of biology required anthropologists to qualify the concept of culture, among psychologists the revival demanded a re-evaluation of the once reviled concept of instinct. As seen already, experimental psychologists had begun the re-evaluation soon after the close of the Second World War when a small band of them had reopened the question. Frank Beach, then a Yale psychologist who had been present at the early meetings, confidently announced in the 1950s that his point of view had become that of almost the majority in the profession. "There are militant opponents of the instinct doctrine among present psychologists," he admitted in 1955, but "it is undoubtedly correct to say that the concept of instincts as complex, unlearned patterns of behavior is generally accepted in clinical, social and experimental psychology." The publication in 1952 of a comprehensive review of no less than 250 items on the "Social Behavior of Vertebrates" in a major psychological journal suggested yet another avenue along which biology was being brought back into psychology.[10]

For many ethologists, animal social behavior usually meant the activities of animals in their natural or near natural habitat. That was not the kind of social behavior study that psychologist Harry Harlow of the University of Wisconsin conducted. Nevertheless, his work constituted an important step in the rehabilitation of the concept of instinct. Or, put more accurately, a step toward a recognition that animals exhibited an innate behavior that could not be explained by the behaviorist's conditioned response. At the same time, Harlow's work was yet a further example of the growing interest in the study of social behavior among animals. Harlow's findings were reported in his presidential address to the American Psychological Association in 1958. The title of the address, "The Nature of Love," revealed, to the point of excess, his acceptance of an intimate connection between animal and human behavior.

Harlow's series of experiments with infant rhesus monkeys has since become renowned. By constructing artificial "mothers" for the infants, Harlow was able to show that something more was involved in an infant monkey's behavior than simple conditioned response. For when the monkeys were given a choice between two artificial, or "surrogate," mothers, they chose the warm, terry-cloth figure rather than the cold, wire-constructed figure even though it "offered" milk. Despite the intended "reinforcement" from the milk, the babies clung to the milkless terry-cloth figure. Conditioned response theory, of course, predicted that the reward of food would override alternatives. Instead, contrary to expectations, some other—presumably internal—influence was operating. Could it be an element in "monkey nature," an innate preference, a source of behavior that was immune or at least independent of environmental influences like rewards? After Harlow's experiments with surrogate mothers, the stern behaviorism of John Watson had some explaining to do. As will be evident a little later, contemporary European ethologists like Nikolaus Tinbergen and Konrad Lorenz were already at work accounting for what Harlow referred to as a "surprise."

The flurry of mail to the *American Psychologist* that the publication of Harlow's address produced was revealing. Those who were eager to see their discipline become more biologically oriented lavishly praised Harlow's efforts. Those of the older persuasion ridiculed the idea that there could be any tenable similarity between a monkey "mother"—real or contrived—and a human mother since the latter was "defined by the culture altogether." How could one talk—even in a title—about the "nature of love" from an experiment with monkeys? Rather than biology having anything to do with the matter, one defender of culture remarked, the fact is that "the social definition dominates and transforms biological reflexes."[11]

Despite such resistance, a re-evaluation of the concept of instinct continued to animate psychologists, especially when new and unexpected evidence appeared. In 1961, for example, a pair of psychologists who trained animals, Keller and Marian Breland, reported on some strange behavior among their animals, actions that strikingly reinforced the conclusions that Harlow had arrived at with his monkeys. The Brelands recognized that the unusual behavior they had observed in their animals cast doubt on the validity of a basic behaviorist principle. Their perplexity was enhanced by their recollection that only ten years before they had publicly expressed their confident belief that with proper methods and patience they could train virtually any animal to perform any behavior or trick. Now, in 1961, they felt compelled to admit that they found themselves fighting "a running battle with the seditious notion of instinct." Over the years they had trained some six thousand individual animals from thirty-eight different species, but for some time they had begun to notice that a large proportion of them experienced "breakdowns of conditional operant behavior," that is, the animals were ceasing to follow their training after having been apparently well conditioned. The Brelands described several examples in which an animal reverted, as the Brelands interpreted its behavior, to the food-seeking activities of its species. More needed to be known about an animal, they concluded; its "instinctive patterns, evolutionary history, and ecological niche" were essential information if it were to be understood. "The notion of instinct," they confessed, "has now become one of our basic concepts. . . . When behaviorism tossed out instinct . . . some of its power of prediction and control were lost with it."[12]

By the end of the 1960s the return of instinct to psychology was being confidently asserted. "Instinct was sometimes treated as a crude superstition" in the discipline, one psychologist wrote in 1968, and to such an extent that "evidence in favor of it was attacked or suppressed." Even careful experimental studies, he recalled, "could go virually unnoticed. The situation is now different," he reported. "One begins to see McDougall's name again and some mention of his ideas with a tinge of respect." In short, he concluded, "instinct seems to be coming back into favor."[13]

Signs that a return to biology in psychology might involve more than a reevaluation of instinct also appeared around this time. One was the commissioning by the program committee of the American Psychological Association of a paper on heredity, in order, as the author observed, to "introduce or reintroduce the general experimental psychologist to heredity." He was advised by the committee, significantly enough, to conceive of his presentation "as instructional in nature." Another indication of the trend to biology was the observation by psychologist Randall Eaton of Purdue University that the new field of ethology had much to offer to the old sub-field of comparative psychology. "Biological theory" also, he reported, "is finding a role in the explanations of human behavior," a consequence of which, he predicted, is that "American psychology is becoming more than a science of learning."[14]

By the late 1960s and early 1970s even some sociologists and political scientists, otherwise far removed from biology, were beginning to see connections between their discipline and the biological sciences. The earliest sign among sociologists was Bruce Eckland's article "Genetics and Sociology," which appeared in the discipline's oldest and leading journal, *American Sociological Review,* in 1967. The subtitle, "A Reconsideration," neatly reflected the novelty of his recommendation. Eckland fully recognized the steepness of the hill he proposed to climb. "This paper," he alerted his readers, "is directed particularly to those who believe that the ties between sociology and genetics either 'have been' or 'should be' buried." For those less rigidly committed, he described the article as an illustration of how "genetic principles might find their way into the sociologists' repertoire."[15]

Eckland's pioneering appeal to his fellow sociologists was followed a few years later by a book written in behalf of Eckland's cause by a sociologist and a psychologist. Whereas Eckland had directed attention to genetics, the two authors pointed to the relevance of ethology to social science inquiry. They saw animal behavior as a means for enriching and deepening their study of human nature, much along the lines that Margaret Mead had pioneered twenty years before. "We will look for ordered progression of social characteristics across closely related primate species," they reported. And if they appear, "then it seems reasonable to consider them human species characteristics rather than effects of cultural diffusion." Another sociologist, John T. Doby, in his presidential address to the Southern Sociological Society in 1970, was even more specific in directing his colleagues' attention to ethology, calling attention to the work of Konrad Lorenz, which he thought would throw fresh light on why human beings behave as they do. There is an evolutionary relation, he insisted, between animals and human beings; we have a "genetic structure blueprint" just as all species do. Human beings, it is true, have a quite different structure from other species, but it still is "a mixture of mechanisms, some biological, some cultural . . . and some not."[16]

The first political scientist to direct the atttention of his colleagues to the place of biology was Albert Somit of the State University of New York at Buffalo. His article, "Toward a More Biologically-Oriented Political Science: Ethology and Psychopharmacology," which appeared in 1968, made evident that one source of Somit's interest was the new ethological work; another was some fresh findings on the way drugs affected human awareness and consciousness. Two years later, Thomas Thorson broadened the connection between biology and politics by suggesting that Darwinian evolutionary theory would be valuable in arriving at a theory of political and social change. Human affairs, he believed, were much more likely to be correctly understood from the perspective of evolutionary theory than from physics, which had long been the model, although a hardly appropriate one, for political science.

Some years later, political scientist Roger Masters spelled out the differences quite specifically. "Like biology—and unlike classical physics—the social sciences study *populations* of organisms that change over time. Like biology—and unlike classical physics," Masters continued, "time is an essentially irreversible variable of decisive importance in most of the phenomena analyzed by political scientists. Like biology and unlike classical physics—the perfectly controlled experiment is difficult if not impossible in political science. Like biology—and unlike classical physics—some form of teleological or functional reasoning seems inherent in political life." Finally, he concluded by noting that "political science studies complex systems (human societies) which are self-replicating organizations of information," as biology also does, and physics does not.

A combination of biology and politics was sufficiently engaging to political scientists that, when in 1975 an international conference was convened on the subject in Paris, a number of American political scientists participated. Before the seventies had come to an end, even some economists, as will be seen a little later, had begun to see a relevance between sociobiology and their own social science.[17]

For political scientists and economists the opening to biology was not a return to an abandoned concept like instinct, as occurred in psychology, or a revival of a discredited theory as happened in anthropology. Instead, it was the recognition of a relevance barely thought to exist in the younger days of the disciplines. At the same time, there was not any particular catalytic event precipitating the new sense of relevance of biological information and concepts to social science. After 1975, to be sure, arguments for the relevance of biological theory and data to the social sciences became commonplace, even insistent. For in that year Edward Wilson published *Sociobiology: The New Synthesis*. The book was an extensive and learned compendium of the latest information on animal social behavior, but its final chapter laid out in strongly assertive, even imperialistic terms the relevance of sociobiological theory for a true understanding of human sociality. We shall return to the impact of Wilson's book on the social sciences; the point here is simply that long before Wilson's book appeared, the return to biology in the social sciences was well under way.

Wilson's book, to be sure, was highly influential in bringing the new ethology and evolutionary theory to the attention of the public and many social scientists for the first time. But it was not the precipitating cause for the accelerating interest in the relevance of biology to the social sciences. That, as we have seen, had been going on for at least two decades before Wilson's provocative book appeared. The question, then, is why did social scientists, beginning in the late 1940s, become increasingly interested in the relation between biology and their particular social science? What had happened to bring them to reconsider their traditional emphases upon social, cultural, or environmental influences in explaining human actions? In describing that mounting interest, some of the reasons behind it already have been implied. The question, though, is of sufficient importance to be examined more specifically and directly from the remarks of those who experienced the change in outlook.

At the outset, it is worth confronting an explanation that is often made, namely, that a conservative political climate has fostered the revival of biological ideas and concepts in the social sciences and elsewhere. "The current theories about the biological basis of social structures," wrote feminist Marian

Lowe in 1982, "are of use to those who want to preserve and strengthen the dominant political and economic interests. One result," she complained, "is that a great deal of media attention is given to biological theories that offer naturalistic explanations for the distribution of wealth and power in this society."[18] Such arguments, however, overlook the simple fact that social scientists began to be interested in bringing biology back into the human sciences as early as the 1950s and then through the 1960s, when the political atmosphere can hardly be described as conservative. The explanation is further called into question when it is recognized, as will be noted with some specificity in a subsequent chapter, that many of the proponents of a recognition of the role of biology in human behavior were and are personally liberal, rather than conservative, in political outlook.

New information from the biological sciences was clearly an important impetus behind the reawakening of interest in biology among social scientists. As the Brelands found out from their animal training, and Harry Harlow from his work with monkey "mothers," animals could no longer be seen as devoid of innate behavior as Watsonian psychology taught. Moreover, like the Brelands, many social scientists were themselves sufficiently informed about the new European ethology to make connections between it and their own work. Both psychologists and anthropologists, as early as the 1950s, were increasingly aware that the old subject of comparative psychology, which had long included the study of animal behavior, was now undergoing profound revision.

In 1951 Nikolaus Tinbergen, a Dutch-born ethologist working at Oxford University, published his revisionist *Study of Instinct*. That work left no doubt that much of animal behavior was instinctive, even if the book said nothing about human behavior. More important, the book offered little support to the overly simple and narrow conception of instinct associated with William McDougall. The other great European ethologist whose work was read as early as the 1950s by a number of American social scientists was Konrad Lorenz, the Austrian founder of ethology. Lorenz had been publishing articles on instinctive behavior in animals since the late 1930s and had collaborated with Tinbergen in some of the latter's earliest ethological studies of the Greylag goose. By 1973, their work in animal behavior, along with that of Karl von Frisch, the discoverer of the so-called language of the honey bee, had gained such wide recognition that the three were awarded the Nobel Prize, even though there was no specified award for ethology. (Their award was technically in medicine.) Social anthropologists and animal psychologists, as one might anticipate, were among the earliest social scientists to see the relevance of the ethology of Tinbergen and Lorenz; their work, however, did not catch the eye of political scientists and sociologists until the late 1960s.

When Tinbergen's *Study of Instinct* was reviewed in the *Psychological Bulletin* in 1952, the reviewer recognized that "a book with 'instinct' in the title" would arouse doubts; yet, significantly, he pronounced it a means "whereby the American student might become better acquainted with the increasing number of European studies of animal behavior and with the somewhat different approach to common problems" than that of American comparative psychologists. To reassure the dubious, he stressed the rigor of the Europeans' methods as contrasted with those of the nineteenth-century amateur animal behaviorists, and remarked on the better controls they employed in their experiments with animals as against those McDougall used.[19]

Konrad Lorenz's work was frequently referred to, though not always favorably, by psychologists attending a conference on instinct held at Princeton in 1952. T. C. Schneirla, a leading animal psychologist, was one of the unimpressed, but he conceded that Lorenz's argument that animals are innately programmed to learn certain things and not others "appeal[s] to American psychologists." By 1971, University of Washington psychologist Robert Lockard was convinced of the merits of Lorenz's finding. "The old concept of an animal as having some degree of intelligence and thus able to learn nearly anything in accord with its endowment," he observed, "is giving way to the view that natural selection has probably produced rather specific learning mechanisms that correspond to ecological demands."[20]

Frank Beach, psychologist at Yale and later Berkeley, had long been interested in animal behavior, so it was not surprising that the new findings of the European ethologists almost immediately appealed to him; repeatedly he urged their work upon his colleagues. Indeed, as early as 1950, Beach had made scholarly fun of his fellow psychologists for concentrating their experiments on the ubiquitous Norway rat instead of looking at behavior in other species, and generally at social behavior in animals. He reminded them that animal behavior was not simply a matter of heredity or environment, but a combination of both. Nonetheless, "psychologists have come to realize that interspecific differences in behavior are traceable to associated differences in the species genotype," by which he meant that behavior, like bodily shape, was inherited.[21]

For some sociologists, the impetus to take a fresh look at the role of biology derived from Lorenz's popular book *On Aggression,* which appeared in 1966. Sociologists generally resisted or rejected outright Lorenz's contention that certain animals and human beings shared an innate tendency to fight. Yet even a sociologist like Donald W. Ball, who found the idea of innate aggression unacceptable, admitted that he was attracted by the broader implications in Lorenz's insistence upon a connection between aggression in animals and human beings. "The definitive answer to the continuities question," he concluded, "is not at hand," referring to the transmission of behavior from animal ancestors to human being. "It is still open to conjecture, speculation, opinion, and persuasion." Emory University sociologist John Doby, on the other hand, in his presidential address to his fellow professionals, frequently cited Lorenz's work to illustrate the insights that ethology would bring to the understanding of human behavior.[22]

As it turned out, Lorenz's conception of aggression was found by other ethologists to be weakly supported in both theory and observation and soon disappeared from the ethologists' research agenda.* Nevertheless, Lorenz's provocative theory awakened some social scientists to the possible relevance of animal behavior to their study of human actions. Indeed, the whole question of the relation between animal and human behavior gained a considerable amount of public attention in the 1960s from the publication of a number of popular books on the subject, a few of which even reached the best-seller lists. Among them was *African Genesis* by erstwhile dramatist Robert Ardrey in 1961, and his sequel *Territorial Imperative*, which he published in 1966. Anthropologist Lionel Tiger's *Men in Groups* appeared in 1969, to be followed by *Imperial Animal* in 1971, written in collaboration with his fellow academic anthropologist Robin Fox. Finally, a well-known British ethologist, Desmond Morris, reached the best-seller list with his *Naked Ape: A Zoologist's Study of the Human Animal* in 1967. So far as one can tell, none of these books exerted much influence on American social scientists. Yet together they undoubtedly helped to create a climate of thinking—a receptivity to facts and ideas of biology in connection with human behavior—that further encouraged some scholars to look afresh, and perhaps even with a little more professional confidence, at the relation between animal and human behavior.

Certainly that was true of some political scientists, who were attracted to biology through the works of Ardrey and Tiger, as well as Lorenz's on aggression and Tinbergen's on ethology in general. As political scientist Albert Somit remarked in 1968 as these popular books were coming out, "the most decisive factor was the transformation of ethology from a relatively obscure to an almost indecently glamorous specialization." The work of Tinbergen and Lorenz, he thought, offered a powerful alternative to seeing political and social behavior as simply "the product of learning and social conditioning." Fred H. Willhoite, Jr., a young political scientist at Coe College, found Lorenz's and other ethologists' conception of man as a cultural, social, and political animal especially provocative because it abandoned the idea that human beings were simply

*Indeed, by the 1980s the trend was in the opposite direction with an emphasis upon the cooperative or even "peacemaking" tendencies of the larger primates. See, for example, Frans de Waal's delightful *Peacemaking Among Primates* (Cambridge, Mass.: Harvard Univ. Press, 1989).

"blank slates" on which the environment wrote. For that reason alone he rejected anthropologist Ashley Montagu's conception that a human being was "nothing but the form in which his particular culture molds his plasticity."[23]

Recent relevant work of ethologists was one reason why anthropologist Earl Count looked favorably upon biological influences on human behavior patterns. Ethological studies showing similarities in behavior of apes and human beings he considered so convincing that he believed they threw the "burden of accounting scientifically" for the common behavior patterns on those who denied any "fixed action patterns" to human behavior.[24]

Anthropologist Victor Turner recalled that his recognition of a place for biology in human activities began also with his acquaintance with new work in ethology. It was at a conference in 1965 on "ritualization of behavior in animals and man," he recalls, that he became aware of a shift in his thinking. The greats in ethology such as Konrad Lorenz, R. A. Hinde, and W. H. Thorpe were there, along with their counterparts in the social sciences and humanities: Erik Eriksen, Meyer Fortes, Maurice Bowra, and E. H. Gombrich. Anthropologist Sir Edmund Leach spoke for the non-ethologists when he proclaimed that "'ritual, in the anthropologist's sense, is in no way whatsoever a genetic endowment of the species.'" As Turner remembered, "I took up no public position at that time, since I was secretly, even guiltily impressed by the ethologists' definition of 'ritualization' which seemed to strike cords in relation to human ritual."[25]

Ethology was only one aspect of biology that aroused social scientists' interest in the usefulness of biological ideas. Indeed, since the Second World War, biology itself had reached a new stage in its development, a situation that laid the theoretical basis for the transformation of ethology and evolutionary theory in general. Among biologists the change in theory has come to be called the "Modern Synthesis," by which is meant the bringing together of Darwinian evolution and its theory of natural selection with the science of genetics. Not until roughly the time of the Second World War had that unification been completed. And until it was accomplished, principally through the work of Sewell Wright in the United States and R. A. Fisher in the United Kingdom, natural selection and evolution stood apart from genetics and other branches of biology. The new genetically grounded theory of evolution through natural selection was given popular explication in 1942 as well as its name with the publication of Julian Huxley's book *Evolution: The Modern Synthesis*.

The modern synthesis carried obvious implications for physical anthropology's study of the origins and evolution of human beings. Indeed, as early as 1950, those implications brought about a conference at Cold Spring Harbor, New York, on the "Origin and Evolution of Man." For some anthropologists it looked as if the disparate sub-fields within their discipline had a chance of being united through a study of "the relationship between cultural factors and genetic factors in the evolution of human differences and adaptability," as one anthropologist described the effort years later. As a result of that conference, she remarked, anthropologists now draw upon "population genetics, biological ecology, developmental biology, medicine, nutrition, physiology, and biochemistry." In the future she thought recent findings in neurology would be especially relevant to anthropologists, for she saw neurology "as one of the fundamental links between biological and cultural anthropology."[26]

One physical anthropologist, William C. Boyd, took pride in the gains made in his field as a result of new knowledge of inherited blood types, a sub-field of genetics. Where once anthropologists had looked to cephalic indices to identify races, they now had a more stable and precise measure in the frequencies of blood types. Thus, Boyd pointed out, anthropologists have established that gypsies are not related to East Indians, that Lapps are Europeans, not Mongoloids, and that American black people derive almost a third of their ancestry from whites.[27]

Advances made in understanding the mechanisms and influences of heredity were among the principal reasons why sociologist Bruce Eckland saw the need for his discipline to rethink the place of heredity. Recent genetical studies of identical twins reared together and apart, he wrote in 1967, had convinced him that intelligence was largely inherited and that sociologists interested in the study of social mobility, among other things, needed to integrate those findings into their research projects.

Social psychologist Gardner Lindzey was similarly impressed by the "enormous advances—theoretical, instrumental, empirical" in genetics over the last fifty or sixty years. He saw their gains outshining those of any other behavioral science. Social scientists in the past, he complained, had placed too much emphasis upon the "plasticity of behavior, social amelioration, and the overriding importance of environmental variation as a determinant of behavior." Today, those social scientists drawing on genetics in their work exude an "air of excitement, novelty, and challenge" that was "in marked contrast to some sectors of the social science world." For their part, Lindzey remarked, geneticists well recognize now the role of environment in shaping behavior; they know that the genotype affects behavior "only in interaction with environmental determinants."

Even racial differences, Lindzey thought, could now be studied without the ideological dangers or political fireworks that had accompanied such work in the past. Indeed, he was so confident of the new scientific genetics that he thought if earlier investigators of race differences "had possessed more knowledge of genetics . . . , there would have been less emphasis on the study of racial differences in general intelligence," which, he was convinced, was a subject too entangled with culture and language to be profitably studied. Two years later, Lindzey's confidence in the usefulness of genetics to psychologists was fully endorsed by Jerry Hirsch in a "tutorial" to his fellow psychologists. "The battle to overcome ignorance and the behavioralistic opposition to according heredity its proper place in the behavioral sciences," he announced "has been won effectively and decisively."[28]

New knowledge about heredity was not the source of political scientist Albert Somit's interest in biology; rather it was new studies in psychopharmacology. And though he had been intrigued by the recent reports from animal behavioralists, he was not yet convinced that human ethologists had established the biological sources of the behavior they described. However, new evidence in psychopharmacology on the effects of drugs, he thought "confirms . . . the ethological contention that there is a direct link between biology and behavior in man as well as in other forms of life."[29]

New knowledge, valuable as it surely was in encouraging social scientists to take a fresh look at the biological roots of human behavior, was not in itself the primary impetus behind the new openness toward biology. New information, after all, appeals most powerfully to those for whom the present state of knowledge in a given field is perceived as being unable to lead to fresh outlooks or theories or to answer novel questions. Dissatisfaction with the reigning theories or approaches were heard in all of the social sciences at the very time that biology was opening up novel avenues of inquiry and formulating new insights. The two developments—inadequate knowlege and theory in the social sciences and new and exciting answers and theories in the biological sciences—converged in the 1950s and 1960s. The product was a renewed interest in the biological roots of human action.

As implied already, among psychologists the inadequacy of strict behaviorism was undoubtedly the most prominent reason for a renewed interest in biology, particularly as it related to the concept of instinct. As early as 1947, psychologist Leonard Carmichael asserted that "it now seems almost certain that . . . the pendulum of [behavioristic] theory has swung too far." Watson's famous rejection of innate sources of behavior is no longer tenable, Carmichael asserted. Psychologist Jerry Hirsch, too, found Watson's behavioralism inadequate because it had no place for genes in accounting for human behavior. "Except for [William] James," one psychologist complained, "American psychology has *not* concerned itself with the *phylogenetically* adaptive role of psychological processes. Evolutionary theory," he regretted, "has played an increasingly minor role in psychological theory and method." Both behaviorists and vitalists, he complained, "overlooked the inherited determinants of behavior." Skinner-

ism, he continued, has unfortunately perpetuated the environmental views of
Watson by portraying "the organism as *quasi-tabula rasa.*"[30]

Among political scientists, too, behaviorism was the inadequate theory that
caused some members of the guild to look to biology for new ideas and theories.* Their behavioralism had nothing to do with the existence or absence of
instincts, to be sure, but the implicit theory of human nature held by American
political scientists was Watson's behaviorism. For as one biologically oriented
political scientist described the situation in his discipline in 1972, most seemed
"to believe that one's genes do not begin to develop until after birth—and then
only as the result of the impact of environmental forces upon them." Such a
tabula rasa approach to human nature, commented John C. Wahlke in his presidential address to the American Political Science Association in 1979, was more
than outmoded, it was "what earlier political philosophers would call a flawed
conception of human nature and modern biobehavioral scientists would call an
inadequate and erroneous model of the functioning individual human
organism."[31]

Political scientist Roger Masters was brought to an appreciation of ethology
by his recognition of its relevance to his work on the political theory of Jean-
Jacques Rousseau. Rousseau differed from John Locke, Masters pointed out, not
only in his conception of the social contract from which all government presumably sprang but in his seeing continuity between animals and men, thereby
departing from Locke's assertion of the environmental origin of human action.
In his *Second Discourse on the Origins of Inequality,* Rousseau deliberately
examined animal behavior in an effort to find the sources of human nature.**
His study of Rousseau, Masters recalled, compelled him to "cross the bridge
between [C. P.] Snow's two cultures," that is, to look into the science of biology. Masters then embarked upon the study of biology, ethology, and evolutionary anthropology. From that inquiry emerged his discovery "that social
behavior is a natural characteristic of human beings. . . . Society is a widespread
phenomenon in the animal world—and hence that explanations of human soci-

---

* David Easton's presidential address before the American Political Scientist Association in 1969
was a clear sign of the dissatisfaction of many political scientists with the behavioral approach to
political science, but the alternative Easton suggested had nothing to do with biology. His advice
was to turn from "description, explanation, and verification" to a new political science of commitment and values. See his "The New Revolution in Political Science," *American Political Science
Review* 63 (Dec. 1969), 1051–53.

** A modern reader cannot help but be struck by Rousseau's anticipation of Darwinian evolutionary ideas. Man came "from the hands of nature," he wrote in one place in the Second Discourse,
"an animal less strong than some, less agile than others, but all things considered, the most advantageously organized of all." Like other animals, he continued, man was shaped by nature, which
"treats them precisely as the law of Sparta treated the children of citizens: it renders strong and
robust those who are well constituted and makes all the others perish." Jean-Jacques Rousseau, *First
and Second Discourses,* Roger D. Masters, ed. (New York: St. Martin's, 1964), 105.

ety as something totally conventional and man-made is inconsistent with biology."[32]

Dissatisfaction with a primitive behavorism or environmentalism was also Glendon Schubert's gateway to biology. It was that dissatisfaction which had long caused him to prefer Freud to Marx as a guide to the springs of human action. The advantage of biological theory was that it rejected "the presumption that our political behavior as a species began 2500 years ago in Athens." For in Schubert's judgment, "the roots of political behavior go back not thousands, but millions of years." Political man, Schubert quipped, did not spring "from the forehead of Socrates." Acknowledging man's animal roots, Albert Somit added, rendered political theory more realistic by providing a basis for the irrationality that was at once a legacy from that past and an active ingredient of human behavior today.[33]

"For half a century," remarked political scientist Steven Peterson, "most social scientists have taught and conducted their research on the assumption that human behavior is almost totally learned and that our genetic make-up contributes little, if anything to our behavior." The new ethology, he correctly observed, "flies directly in the face of that tradition." And it was just that dissatisfaction with the traditional approach that moved sociologist Gerhard Lenski in a new direction in his textbook *Human Societies,* published in 1970. "The most important contribution to primate ethology," he remarked, quoting from the work of Harvard primatologist Irven DeVore, "is the finding that 'all species of monkey and apes live in social groups.'" The finding, Lenski remarked, "tends to support the hypothesis that the societal mode of living is something man inherited from his primate ancestors, not something he invented" for himself.[34]

The inadequacies of his own discipline on the question of culture and biology became apparent to anthropologist Victor Turner only gradually and reluctantly. "The present essay is for me one of the most difficult I have ever attempted," he confessed in 1983. This was so, he continued, because he was questioning "some of the axioms anthropologists of my generation—and several subsequent generations—were taught to follow. These axioms," he reported, "express the belief that all human behavior is the result of social conditioning." And he well recognized that "a very great deal of it is, but gradually it has been borne home to me that there are inherent resistances to conditioning," by which he meant that human beings are not empty slates upon which anything can be written by outside influences. "One of these distinctive human features," he suggested, "may be a propensity to ritualization of our behaviors, from smiling and maternal responsiveness onwards."[35]

Sociologist Allan Mazur and psychologist Leon Robertson asked, Is it "in the interest of social scientists to maintain an 'environmental approach'? If they are to have credibility in the long run," they answered, "we think not." Present sociological theories of human behavior, they believed, fell short of meeting the requisite standards. "The credible scientist builds theories which are accurate predictors of the phenomenon in question," they observed. "To ignore a possible major contributing factor may result in a theory that is not predictive. To ignore biological contributions to behavior," they concluded, "is foolish." Sociologist Bruce Eckland experienced similar unhappiness with a totally environmental explanation for human behavior. Has the anthropologists' view of the equality of all races gone too far? he asked. "Is the plasticity of man so unlimited? Is the genetic basis of man so uniform and, therefore, inconsequential?" he asked.[36]

Anthropologist Jerome Barkow's interest in the possible relevance of the new biology stemmed from his belief that earlier efforts by some anthropologists to identify biological sources of motivation were not biological but mechanical. The genuinely biologically oriented question to ask, he recommended, is why the organism "behaves one way and not in another, at any given moment, rather than where its 'energy,' its 'push' or 'pull' comes from," all of which had been concerns that had occupied the attention of the earlier theorists. The emphasis of an earlier anthropology on a search for "the distinctive characteristics of each social group," Barkow thought, was also rendered obsolete by the "Darwinian perspective," which sought out "the transcultural behavioral components and processes which all human groups share and which distinguishes them from the social groups of other species."[37]

No social scientist found the underlying assumptions of his own discipline more wanting than the prominent Yale anthropologist George P. Murdock. Both cultural and social anthropology got off on the wrong track, he contended in 1972 in a rather bitter article entitled "Anthropology's Mythology." The mythology referred to was nothing less than the underlying assumptions of the discipline. He described his own repudiation of that mythology as "total," principally because early in its history, anthropology had accepted Herbert Spencer's idea of the superorganic, that is, "the social aggregate as the preferable unit for study" instead of the individual. After years of accepting the idea of culture, Murdock now concluded that the concept was seriously flawed, that it was the individual who made culture, not the other way around; the individual was the anthropologist's proper object of study. An emphasis upon the individual, he now thought, "bears a close family resemblance to that generally accepted by biological scientists to account for the development and differentiation of the living organism." Nor is it likely, he remarked, as he continued his attack, "that this parallel is accidental, for man, despite what we are wont to call his culture and his society, is nevertheless fundamentally a biological organism." Nor is it simply accidental that biologists have arrived at their present consensus. For in the past they have been testing and discarding a similar "series of illusory concepts, such as those of vitalism, which seem to me strictly comparable to what I have called the myths of anthropological theory, and which a lifetime of professional trial-and-error has finally forced me to reject."[38]

Even before Murdock had announced his dissatisfaction with the inadequacies of anthropological theory, Eliot Chapple, another member of the discipline, had signified his own complaints in *Culture and Biological Man,* which he had published as early as 1970. As he later explained, "the study of what are presumably innate behavior patterns has proved to be an important counterbalance to the long-standing belief that the newborn animal was a blank sheet upon which the learning process wrote out its own formula." Behavioral genetics, he argued, "is now regarded as fundamental in understanding man as well as the lower animals." Stanford anthropologist George Spindler moved more cautiously to a recognition of biology. "We are a long way from anything resembling a coherent integration of biosociocultural evolution," he wrote in 1978. "But if we are to escape the double bind of our cultural overdeterminism," he was convinced, "we are going to have to go beyond culture and even ecology, to biochemistry, to physiology and neurology, to genetics—to biology in the broadest sense of the term."

Spindler fully recognized that "at this moment biogenic explanations are not politically popular," if only because many social scientists associated an assertion of a biological component in human behavior with a revival of social Darwinism or, what was worse, racism. But since, as he explained in Darwinian terms, biology and culture are both adaptive "and the two interact in any human adaptive process, there is no ultimate reason for a political rejection of the implications of a biogenic view."[39] Some social scientists clearly resented the limits that such fears seemed to place upon their inquiries or their freedom to disseminate their ideas. Sociologists were especially prone to such fears, remarked Bruce Eckland in 1967. A sociologist is usually politically liberal, or he or she "at least believes in a free, equalitarian society," the values of which seem threatened by biological explanations of human behavior. Or as political scientist Elliott White remarked in 1972, "the idea of 'limits' on anything is anathema" to many social scientists. The counter assumption "that we are all equal in ability—would be were conditions equal—must be one of the most unempirical 'givens' in all of empirical science," he thought. Fears of and objections to racism, Albert Somit contended, had delivered the "coup de grace" to earlier efforts to introduce biological thinking into political science and continued to be an obstacle in the minds of many social scientists.[40]

As late as 1982 anthropologist C. Loring Brace still saw fear of a return to the old "race concept" as an obstacle to a recognition in his field of "the biological nature of the humanity that is the object of study in our discipline." Once that fear can be finally exorcised, he anticipated, anthropologists will at last be in a position to apply "to *Homo sapiens* the systematic perspective of Darwinian biology. It may seem curiously late in the history of science," he admitted, "to be making a beginning of such an effort—a full century after Darwin's death—but the opportunity has yet to be exploited in systematic fashion." Sociologist Allan Mazur was somewhat more positive about the issue. He admitted that "a quarter-century ago biology was expunged from sociology, Nazi racism having knocked out whatever remained after the blows of cultural relativism and operant conditioning." But by the early 1980s he thought "biological explanation is popular again," though most of the research, he regretted to say, is being done outside sociology.[41]

Undoubtedly the most frequently expressed philosophical outlook moving social scientists beyond an environmental or cultural explanation for human behavior was an appreciation of the Darwinian emphasis upon the continuity between animals and human beings. As psychologist Robert Lockard rightly pointed out, "the hypothesis of mental continuity" was the essential thesis of Darwin's *Descent of Man.* "Humans are animals, perhaps very special animals," he admitted, "but animals nonetheless. Just as other fauna on this planet are products of organic evolution and are now understood in the framework, humans evolved and deserve understanding in this context." Political scientist Fred Willhoite was no less convinced that "an ethological approach represents an attempt to take seriously man's animality, to view him as a particular kind of animal within the evolutionary order of nature." To believe otherwise, added anthropologist Jerome Barkow, is to conceive of human beings as "the mysterious results of some saltatory process." The truth is, Barkow continued, that, like all animals, "we have been 'programmed' by evolution to form social bonds, acquire language and culture, etc., and so to become human. Natural selection has generated in us," as it has in other animals, "a finite range of capacities and constraints, requirements and potentials."[42]

A need to recognize the continuity between animals and human beings also stimulated sociologists to look favorably upon a revival of biological explanation in their discipline. If we want to know why people marry, explained Joseph Lopreato, we need to look beyond the immediate circumstances; we need to recognize that the competition for mates is a phenomenon throughout the animal world. "There is no evidence," he contended, "that human beings do not partake of that competition," too. Similarly, communication and systems of hierarchy and dominance among primates, sociologist Donald Ball argued, are eminently pertinent in understanding human social behavior because they "forcefully assert man's continuity with nonman." Ball, in fact, was prepared to move considerably beyond the position of most proponents of the continuity between men and animals. He noted, for example, "that culture is not limited to humans alone. The question," he presciently observed, "is no longer whether culture among infrahumans exists, but in which animals, how much, and in what form (where, to what extent) and vis-à-vis what areas of social behavior or organization?" This was asserting Darwinian continuity with a vengeance, but not without a good deal of ethological support.[43]

Whatever particular reasons or sense of the inadequacies of their own disciplines may have brought social scientists to seek a return to biology, their recurrent view was that evolutionary theory offered new and challenging insights. As political scientist Thomas Wiegele wrote in 1979, in returning to biology "we are not dealing with a momentary aberration or a fad that will wither away in due time. The twentieth-century impact of the life sciences on the understanding of social behavior across all species has been awesome," he contended. "The necessity to incorporate biological considerations into our understanding of social man will not disappear." Therefore he urged that at least some departments of political science "launch pilot programs to train students to work at the disciplinary juncture of the life and social sciences." Robert Lockard thought that if psychologists would "cross their experimental sophistication with evolutionary biology, a vigorous and fertile hybrid would result." The interest over the last decade "in the biological causes or bases of human behavior," anthropologist Frank Livingstone observed in 1980, "has led to increased communication among biological and social scientists . . . and should lead to a more rigorous definition of the issues and surely some advances in our understanding of human behavior." Contrary to what some anthropologists may think, Livingston continued, the present discussion of the relation between biology and social science differs significantly from the nature/nurture controversy of the 1920s. The present debate does not approach the influence of heredity and culture as a question of one or the other; the primary concern, rather, is the degree or proportion of influence of each. All participants in the discussion fully acknowledge that interaction is fundamental. To anthropologist J. Hartung the advantage of looking "long and hard at Darwin's paradigm" was that it would be "useful to know what we are up against" when we seek to improve the social order.[44]

Sociologist Gerhard Lenski thought the new ethology offered a model where "the scattered and unorganized findings of the *social* sciences may be brought together within a single theoretical framework. . . . A rapproachement between the biological and social sciences," he thought, "was long overdue." After all, he reminded his fellow professionals, "human societies *are* part of the biotic world, and by denying or minimizing this fact we impoverish both theory and research." Two psychologists, in identifying the trends within the field of behavioral genetics in 1981, seconded Lenski's point. They predicted that "behavior genetics will play a major part in the slowly growing rapproachement between sociology and the biological sciences." In the near future, they anticipated, "social, behavioral, and biological scientists" will work together on projects.[45]

For all their emphasis upon the value of a return to biology, social scientists were not dogmatic about the way in which biological concepts should be integrated into the study of human behavior. Political scientists do not have to decide whether patterns of human action are innate or cultural, John Wahlke assured them in his presidential address to the American Political Science Association in 1979. All that was needed was a recognition of "the inseparable independence of both, and to distinguish those cases where people are behaving in ways characteristic of all human organisms acting in similar circumstances" as against those that are merely individual. No need to dispute about learning versus inheritance, nature or nurture, genetics or culture, he said. Simply seek out possible universals among human beings that might have some analogues with those identified by animal ethologists. Even then, though, the analogues would be no more than working hypotheses, not conclusions. "Ethology, he assured his colleagues, "is a source of *questions*," not answers, to unasked questions.[46]

Wahlke's fellow political scientist Albert Somit was not at all sure that ethology would offer much in the short run to political understanding, but he thought it might stimulate fresh ideas about political behavior. Nevertheless, ethology was forcing political scientists to consider "(or reconsider) the possibility that *some* political phenomena are due to *some* measure of our genetic programming." If that rethinking could be widely disseminated it would constitute a gain comparable to that achieved by Marx, who "compelled non-Marxists to look to material and economic factors," for it will have forced "upon us the same 'open-mindedness' with regard to biological factors."[47]

Other political scientists were equally careful to assure their colleagues that an acknowledgment of the role of biology in human actions did not commit them to denying all other influences. As products of an evolutionary history, Fred Willhoite told them, "we are genetically 'programmed' to learn and persist in certain kinds of behaviors much more readily than is the case with other possible behaviors." But at the same time, he continued, "this does not necessarily mean that a particular behavior is *inevitable,* rather that heredity significantly affects the *probability* of its development." We cannot ignore the possibility that biology affects human behavior, another political scientist pointed out, but in recognizing that influence "we need not engage in a blind acceptance of a biological determinism that pays no heed to human purposive activities."[48]

The pride political scientists justly took in drawing upon a rich variety of causal factors in their work made many of them especially fearful that a return to biology would be reductionist. In the opinion of Thomas Wiegele, the fear was quite unwarranted inasmuch as the new biopolitics would do no more than expand the possible factors to be taken into consideration. It would not eliminate any traditional ones. On the contrary, he suggested "the reductionist label could be applied more appropriately to present-day 'traditional' political science, which too often reduces the explanation of political behavior to exclusively rationalist considerations." Indeed, to Roger Masters, a primary appeal of ethological theory for political scientists was that it provided a bridge to biology "that is *not* determined, *not* reductionist, and *not* inconsistent with existing traditions for analyzing human life." Modern science, he continued, has largely abandoned the nineteenth-century view of a deterministic world. Instead it has taken a probablistic approach, something especially "necessary in the study of animal populations, especially if they reproduce sexually; as population genetics has shown, the presence of genes in a population is hardly an all-or-nothing factor, since selection operates on the entire gene pool as well as on component organisms." There need be no fear of determinism, he continued, since "a biological approach to human behavior can be presented in a probablistic and nondeterministic form." Besides, added Mazur and Robertson, "the issue of reductionism is a matter of empirical research. Whether or not principles regarding cultural, political, or other social phenomena can be reduced to psychological or biological principles is not a question of values but a hypothesis to be investigated." Social science, they concluded, would no more be dismantled by biology, than physics has dispensed with chemistry.[49]

Early on, psychologist Gardner Lindzey tried to meet the fear of social scientists that bringing back biology into their thinking would threaten the independence of their disciplines or reduce the variety of explanatory modes open to them. "The methological sophistication of behavioral scientists," he assured them, "has increased to a point where they no longer need fear that their entire area of concern may be engulfed by" the biological sciences. "The limits of reductionism are now sufficiently well understood so that such effort (whether involving biological, psychological, or social reduction) no longer warrants serious thought." Consequently, he concluded, a social scientist has achieved sufficient "intellectual freedom . . . to recognize that there may be potential gains from placing the concepts, problems and methods of his discipline beside comparable elements within the biological sciences."[50]

Finally, in seeking to uncover the reasons why many social scientists were attracted to biological ideas, the power and success of the biological sciences cannot be ignored. As noted already, these were the years in which genetics had not only shaken off the dubious connection with eugenics but was also making giant strides in understanding the complex mechanism of heredity. During those same years the inner mystery of heredity was unraveled with the discovery of the double helix of DNA, evolutionary theory and ethology came into their own. Biological science's newly exhibited certainty held enormous appeal for social scientists who had come to doubt the direction research was taking in their own fields, or the value of their own disciplines' theoretical structures. In short, many social scientists were undoubtedly encouraged, even inspired, to think about a return to biology by their hope to obtain a comparable certainty, a similar "scientific" lawfulness, in their own fields of inquiry.

Although such an appeal seems a reasonable interpretation, not many social scientists referred directly to it. It can, however, be inferred from the manner in which some of them expressed the appeal that biological ideas held for them. To political scientist John Wahlke, for example, the unchanging character of "law" clearly had a comforting ring. In his presidential address he said that "the people whose political behavior political scientists study are, after all, no more exempt from the laws of behavioral dynamics than from the laws of gravity." A similar outlook underpinned another political scientist's interest in biological concepts. In the very first sentence of his article on Darwinism and sociocultural evolution, John Langton's admiration for a "scientific" approach is evident. "The principal task of this paper," he confessed, "is to demonstrate that the social sciences have the capacity to make the evolution of sociocultural systems as comprehensible as Darwin made the evolution of species." To political scientist Peter Corning, the social sciences, with their "environmental (cultural) determinism and a functionalism without a theory," no longer had a right to call themselves "sciences." They had "become disconnected from the only scientifically acceptable explanation of the origin, nature, and 'purpose' of human life."[51]

Sociologist Walter Wallace made no secret of his wish to have his discipline considered a science. His book, *Principles of Scientific Sociology*, which he published in 1983, argued that "sociology is, in its actual practice as well as in its abstract design, one of the natural sciences—that is, much more akin to biology, chemistry, and physics than to philosophy, poetry, or religion." Included in his showing "the scientific nature of the discipline" were the findings of sociobiology, because "the scientific analysis of social phenomena follows exactly the same general principles whether the objects of that analysis are human or non-human organisms." Sociologist Joseph Lopreato made his preference for a scientific outlook evident, too, when he suggested that the term "biocultural science" was superior to "human sociobiology." Among George Murdock's reasons for finding traditional anthropology hard to accept was what he considered the deplorable lack of agreement among its leading figures, a lack that he thought was absent from the natural sciences. "It is inconceivable," he remarked in 1972, "that four men of comparable standing in any established field of science . . . could differ so radically from one another on basic theoretical issues. One can only conclude from this that what [Meyer] Fortes, [Leslie] White, [Alfred] Kroeber and I have been producing is not scientific theory in any real sense, but something much closer to the unverifiable dogmas of differing religious sects."[52]

One of the natural consequences of the growing conviction among social scientists of the relevance of biology to their disciplines was the production of numerous scholarly papers, from which much of the foregoing has been taken. Another was the organizing of conferences and the founding of journals devoted to discussing the ways in which biology, ethology, and genetics could enrich or deepen social scientists' study of human social behavior. Comparative psychologists held a symposium on the threat of sociobiology to their sub-field at the meetings of the American Psychological Association in Toronto in 1978; the Southern Political Science Association organized the first conference on bio-politics in 1967. When the journal *Ethology and Sociobiology* was founded in 1979, its psychologist editor justified the establishment of yet another scientific publication on the ground that the articles on the subject were then scattered among "perhaps twenty different journals" in disparate fields, an inconvenience that the new journal was expected to mitigate. "Whatever the eventual verdict on the relevance or irrelevance of evolutionary theory to the understanding of human behavior," he acknowledged, "a powerful theory exists and many students of behavior are trying to see what the theory can do."[53]

The *Journal of Social and Biological Structures* and *Politics and the Life Sciences* are two other recently established vehicles for the dissemination and exchange of research on sociobiological or ethological interpretations of human behavior. A few universities have even established institutions or programs to advance the integration of biology and social science. Northern Illinois University, for example, supports a Center for Biopolitical Research as part of its political science department, while the department of political science at the State University of New York at Stony Brook has a Laboratory for Behavioral Research, pursuing similar interests. In sum, the stimulus from ethology and evolutionary theory was already becoming institutionalized by the opening of the 1980s.

In the course of the late 1970s, as we have seen, the whole subject of sociobiology gained new visibility through the publication of Edward Wilson's seven-hundred-page book on ethological research, *Sociobiology: The New Synthesis*. It was his last chapter that brought the subject to public attention and notoriety. The chapter bore the provocative title "Man: From Sociobiology to Sociology"; its substance was the relevance of animal ethology to the study of human behavior.

For social scientists who were serious about the inclusion of biology in their field, the provocative aspect of Wilson's book were the preceding twenty-six chapters in which he admirably summarized and analyzed the contributions and insights of the rapidly growing field of ethology and population genetics. The 2500 items in his bibliography amply testified to the wealth of research that had been published in the preceding twenty or thirty years. The items themselves made clear that a whole new dimension had recently been added to the study of animal behavior. The founders of modern ethology—Tinbergen, Lorenz, von Frisch, and others—had worked closely and sensitively enough with their animal subjects, but their primary work was largely experimental; they devised and tested hypotheses about animal behavior. The more recent ethologists, on the other hand, were more likely to be workers in the animals' native habitat, patiently and silently observing, reporting, and analyzing the natural behavior of the animals. In the course of their research they learned what no experimental ethologist could: how wild animals related to one another over time and to the environment in which they lived. With their work a whole new dimension was added to the understanding of social behavior among animals. To a large degree it was the work of this new breed of ethologists, men and women like Jane Goodall, Irven DeVore, George Schaller, Hans Kruuk, Diane Fossey, Shirley Strum, Cynthia Moss, T. J. Maple, Sarah Blaffer Hrdy, and literally hundreds of others who supplied the information on the social behavior of animals which riveted the attention and reshaped the thinking of many social scientists in the course of the 1970s and 1980s.

Having examined how and why a renewed interest in biology developed among social scientists in the years after the Second World War, it now is necessary to examine some of the concrete ways in which social scientists made use of the new insights from ethology and sociobiology. What, in short, had been learned from animals about human social behavior?

The next four chapters are intended to provide an answer. The first is a case study of a problem in human behavior that has bewildered social scientists for more than a century, but for which sociobiology now promises a solution. The following three chapters will provide a broader survey of concrete ways in which social scientists in a variety of disciplines have drawn upon Darwinian evolution and sociobiology to better understand the nature of human beings, or what it means to be human, as Margaret Mead once phrased the matter.

10

The Case of the Origin
of the Incest Taboo

The prohibition of incest is the process by which nature transcends itself. . . .
It brings about and constitutes in itself the advent of a new order.

Claude Levi-Strauss, 1949

Since the professionalization of anthropology in the nineteenth century, probably the most fascinating, not to say relentlessly obsessive concern of students
of human behavior has been the prohibition against marriage or sexual relations
between certain close relatives. The so-called incest taboo must be expressed in
those general terms because it is not a simple phenomenon, despite the single
identifying term. Phrased in that broad way, however, the taboo is one of the
few undisputed "universals" of human behavior. It appears in one way or
another in all known societies, historically and culturally. It needs to be said,
however, that, of particular prohibitions, only that against the mating of a son
with his mother obtains in virtually all cultures. Marriages or matings between
other pairs of relatives may be prohibited in one society, while in another that
same relation may be permitted or even encouraged. In no human society, incidentally, is the incest taboo confined to the members of the nuclear family.

Early on in the study of different cultures, anthropologists sought to account
for the emergence of this prohibition, the apparent universality of which flew
in the face of the enormous diversity of cultural forms which the discovery of
each new culture drove home with ever increasing insistence. The taboo was
intimately connected with kinship patterns in these societies. In fact, its existence was understandable only when one understood who was related to whom.
It soon became apparent that the taboo lay at the heart of what anthropologists
called exogamy, that is, the practice of seeking mates outside of the immediate
family or social group.

Lewis Henry Morgan, sometimes called the father of American anthropology, noticed the phenomenon in the course of his mid-nineteenth-century study
of North American Indian tribes. He accounted for the prohibition of incest by

reference to a basic principle of animal breeding, namely the deleterious consequences that seemed to follow from the mating of animals too closely related. Morgan's suggestion of a biological explanation was rejected by most anthropologists of the time, despite the powerful impetus given to biological explanations by Darwin's contemporaneous accounting for human evolution, and which so profoundly influenced social scientists in dealing with other aspects of human behavior. Much of the objection stemmed from the recognition that in some cultures the prohibition extended to persons who were only remotely related genetically or sometimes not at all.

The explanation that came to dominate anthropology during the late nineteenth century and well into the twentieth was that advanced in the late 1880s by the English anthropologist Edward B. Tylor. His explanation for "marrying out," as he called the consequence of the taboo, was that the prohibition served to broaden or extend human connections beyond the family. Gaining alliances with others through intermarriage was a source of protection and a means of survival in a world that was very dangerous for proto-human beings.

The principal challenge at the time to Tylor's explanation for marrying out was advanced by Edward Westermarck, a self-taught Finnish anthropologist, in his *History of Human Marriage,* published in 1891. A confirmed follower of Darwin's conception of human evolution, Westermarck resisted the idea that such a profound and widespread phenomenon as exogamy could result from an apparently rational decision by primitive human beings in order to gain allies or make connections with others. He offered what he considered a more plausible and basically Darwinian explanation, one that linked human and animal behavior as Morgan's explanation had implicitly done. "If we want to find out the origin of marriage," Westermarck wrote at the very beginning of his book, "we have to look outside human beings themselves. The path that will lead to the truth," he contended, was "open to him alone who regards organic nature as one continued chain, the last and most perfect link of which is man." It is no more acceptable, he insisted, to "stop within the limits of our species when trying to find the root of our psychical and social life, than we can understand the physical condition of the human race without taking into consideration that of the lower animals." He was therefore almost contemptuous of those anthropologists, of whom Tylor was the best known, who postulated a rational or conscious act on the part of early man to marry out, by making rules or establishing customs against certain kinds of behavior. "Law may forbid a son to marry his mother, a brother his sister," he wrote, "but it could not prevent him from desiring such a union if the desire were natural. . . . The home is kept pure from incestuous defilement," he insisted, "neither by laws, nor by customs, nor by education, but by *instinct* which under normal circumstances makes sexual love between the nearest kin a psychical impossibility."[1]

How, he asked, could such an aversion, as he described the incest taboo, have come about? His answer was derived from Darwin's view of evolution: an aversion to inbreeding had been "selected for" in the course of natural selection. "It is impossible to believe," he argued, "that a law which holds good for the rest of the animal kingdom, as well as for plants, does not apply to man also."[2] Early human beings, no more than animals, he acknowledged, did not necessarily recognize the evil effects of inbreeding. Rather, the aversion was to mating between individuals who had been reared together. That was the mechanism that natural selection had developed. Over time, the aversion was strengthened as those who continued to inbreed were less successful reproductively than those individuals who began to "marry out." "Thus an instinct," he concluded, "would be developed which would be powerful enough, as a rule to prevent injurious unions." Although the aversion would be to matings between those individuals which had been reared together, they would, as a matter of fact, "be blood relations, so that the result would be the survival of the fittest."[3]

It is worth noticing that Jeremy Bentham, the early nineteenth-century philosopher of utilitarianism, had also observed the behavior pattern on which Westermarck rested his explanation for the incest taboo. In his *Theory of Legislation,* Bentham had called attention to his observation that within a family the emotion he called "love," and which we would denominate more properly as "sex," did not manifest itself. "There needs to give birth to that sentiment," he contended, "a certain degree of surprise, a certain effect at novelty."[4]

Despite the early support by Bentham, which the Finnish anthropologist probably did not know about, Westermarck's theory of innate avoidance of incest came under fierce attack. Perhaps the most devastating was that delivered by the British anthropologist Sir James Frazer, the author of *The Golden Bough.* Frazer simply asked why a rule against incest was laid down by culture after culture if the resistance to the practice was innate. After all, out of their lively interest in the subject, anthropologists for a half-century or more had been identifying cultural rules in virtually every society prohibiting marriage between certain relatives, not all of whom, by any means, were closely related. Moreover, complained Frazer, by describing the incest taboo as innate, Westermarck was placing human beings and their culture in the same category as animals, something in itself that was likely to arouse opposition to any explanation for human behavior.[5]

Westermarck himself tried to answer Frazer's objections, saying that if one followed out Frazer's logic then sodomy and bestiality, against which there certainly were prohibitions or rules, would have to be considered natural. His point was that a behavioral pattern against a certain practice might be innate, yet still be consciously prohibited or restricted by laws or customs. Only an occasional student of the subject, such as Havelock Ellis and Leonard T. Hobhouse, found Westermarck's defense persuasive.

Sigmund Freud in *Totem and Taboo* (1913), for example, flatly rejected Westermarck's explanation. To Freud, as to Frazer, a social or cultural rule was unnecessary if the prohibition was already innate. Freud, of course, had an additional reason for rejecting the taboo's innateness. Psychoanalysis, after all, required the taboo to be cultural rather than innate, for his concept of the Oedipus complex assumed a conflict between a human desire for incest and the cultural need to control it. In *Totem and Taboo*, Freud rather casually offered his own explanation in the form of the myth of the "primal crime," in which the primordial sons killed their father in order to possess their mother. Their guilt caused them never to seek the mother again, but rather to look for their sexual objects (their mates) outside the family, that is, to "marry out." According to Freudian theory that promise was not easy to keep and the difficulty was measured in the recurrence of what has come to be called the Oedipus complex, the son's incompletely repressed wish to possess the mother.*

Freud's objection to Westermarck's explanation extended beyond the question of the taboo's origin. Whereas Westermarck, like Bentham before him, had discerned an absence of sexual emotion among family members, Freud postulated its presence, a presence, in fact, that became the engine of his whole psychological theory. As he himself wrote toward the end of his life, "if psychoanalysis could boast of no other achievement than the discovery of the repressed Oedipus Complex that alone would give it a claim to be included among the precious new acquisitions of mankind."[6] Yet despite his rejection of any innate or biological basis for the incest taboo, Freud, interestingly enough, and quite in line with his early biological bias, postulated a biological explanation for the *persistence* of the prohibition. Since Freud remained until his death a convinced Lamarckian, or believer in the inheritance of acquired characters, he could easily believe that the taboo was hereditary. "The incest taboo," he wrote in his *Contribution to the Theory of Sex,* "probably belongs to the historical acquisitions of humanity and, like other moral taboos, it must be fixed in many individuals through organic heredity."[7]

---

*Actually, the classic Greek story of Oedipus is not a good analogy for Freud's conception of the complex. Oedipus, after all, slept with his mother out of ignorance, rather than from incestuous desire, conscious or unconscious. He did not know her since he had been separated from her at birth. But the myth does fit well Freud's insistence that such desires would never appear undisguised, even in myths.

Bronislaw Malinowski, the well-known anthropologist of the University of London, also perceived the incest taboo to be a landmark in the evolution of humanity, a behavior pattern that delineated the transition from animal to human existence, from nature to culture, to civilization itself. Where Malinowski departed from Freud was in his rejection of the psychoanalytic assumption that the attachment between mother and child was sexual. The prohibition originated, Malinowski wrote in *Sex and Repression in Savage Society* in 1927, from the need to prevent the disruption of the family, the institution he saw as fundamental in the development of culture. "Incest, as a normal mode of behavior," he insisted, "cannot exist in humanity because it is incompatible with family life," since it would bring about the "upsetting of age distinctions, the mixing up of generations, the disorganization of sentiments and a violent exchange of roles at a time when the family is the most important educational medium. No society could exist under such conditions," he sternly warned.[8]

Malinowski, like Freud before him, discerned in human beings a powerful urge to incest. His conviction was so deep that ultimately he concluded that the urge to incestuous behavior among human beings was the foundation of social life, since it compelled men to "marry out." But, he added, such feelings were "not likely to occur in animal families governed by true instincts." Animals lacked any drive to mate with family members, a fact that in Malinowski's mind compelled us to recognize "that a social taboo does not derive its force from instinct, but that instead it always has to work against some innate impulse," in this case, the urge to incest, which was peculiar to human beings. Although that was a view quite in keeping with Freud's conception of the relation between culture and nature, it stood in clear opposition to Westermarck's explanation derived from Darwinian theory. The tendency to incest, Malinowski concluded, was "the original sin of man. This must be atoned for in all human societies by one of the most important and universal rules. Even then the taboo of incest haunts man throughout life, as psychoanalysis has revealed to us."[9]

Malinowski's explanation for the origin of the taboo is no less fanciful than Freud's, even though he rejected Freud's own fantasy of the primal crime. For Malinowski the incest barrier was as much a product of man's reason, driven by his social needs, as it was for Freud. The prohibition was necessary to preserve the family, in Malinowski's eye the foundation unit of civilization itself. Despite his failure to uncover a connection between human and animal behavior, some new information about animal behavior, acquired a few years later, apparently weakened his earlier conviction that the incest taboo was cultural in origin. Writing in the *Encyclopedia of the Social Sciences* in 1931, he suggested that "if incest could be proved to be biologically pernicious, the function of this universal taboo would be obvious," but "specialists in heredity disagree on the subject." Only four years earlier, however, he had been confident "that biologists are in agreement on the point there is no detrimental effect produced upon the species by incestuous unions."[10]

Malinowski's hint of a possible link between the biology of human beings and the incest taboo was not expanded upon. It was, nonetheless, an indication of the way biological ideas continued to influence the thinking of anthropologists during the first decades of the twentieth century. During the 1930s and 1940s, however, there would be fewer and fewer signs of that influence. Everything that was being discovered about sociality or social behavior among human groups underscored the great variety of human experience across cultures and time; increasingly it almost seemed as if the only thing that the members of the myriad of human cultures had in common was their ability to interbreed. The immense variety of social practices was apparently derived from the cultural imagination of human beings, a source of diversity that seemed almost unlimited in its influence.

Indeed, the contrast—nay, the paradox—between the universality of the taboo and the variety of cultural practices discouraged many social scientists from trying to explain it at all. Leslie White, a renowned anthropologist at the University of Michigan, writing in 1948, simply traced the origins to culture alone. Not biology, psychology, or sociology, he contended, could explain it. Anthropology, however, had accomplished the feat long before, in principle at least, with the work of Edward Tylor, who had coined the epigram to summarize it: "marry out or be killed out."

At the time that White's article appeared, anthropologists' understanding of animal behavior had already begun to shift, and the change was evident in White's remarks. Malinowski, as noted earlier, thought animals displayed no concern one way or another about the practice which human beings called incest. By 1948, however, White had concluded that "within sub-human primate families" there was a "strong inclination toward inbreeding," a view that was consistent with Freud's view of animal behavior, and presumably with the practices of proto-humans before they made the great leap to humanity through the adoption of the incest taboo. That, however, was as far as White was prepared to go in seeing a continuity between human and animal behavior.

White was just too impressed by the immense variety of marriage customs to be able to discern any single biological force underlying them. "Obviously, the kinship terms express sociological rather than biological relationships," he wrote. "Obvious also is the fact that the incest tabus follow the pattern of social ties rather than those of blood." Psychology cannot explain the taboo because from that field, he contended, "we learn that the human animal tends to unite sexually with someone close to him. The institution of exogamy is *not* explained by citing this tendency; it is contrary to it." "Incest was defined and exogamous rules were formulated," he explained, "in order to make cooperation compulsory and extensive, to the end that life be made more secure."[11] It was cultural systems that created the taboo. Hence the practice of the taboo varies from group to group according to circumstances.

White harbored none of Malinowski's functionalist worries about the future of the family, nor did he see the incest taboo arising from the individual psyche as Freud had suggested. To White, the major benefit from cooperation was economic advancement. That is why marriage, he asserted, is an economic matter in all societies, and not sexual. Primitive societies, he noted, often offer alternative means of sexual expression outside of marriage. Marriage may well be, as in monogamy, a means for limiting sexual expression. "Ideally considered," he revealingly remarked, monogamy is "the next thing to celibacy." The true cause for marriage was cultural, arising from the "exigencies of a social system that was striving to make full use of its resources for cooperative endeavor." A biological purpose, such as reproductive success, was nowhere to be seen; White's answer to the riddle was the ultimate cultural explanation.[12] Like Freud and Malinowski before him, White offered no more than a functional explanation for the origin of the incest taboo: it was culturally necessary to achieve exogamy, therefore the prohibition came into existence. Causality was replaced by function. It was like saying because people needed to eat they invented agriculture.

White was also quick to reject a suggestion from a recent follower of Westermarck who had the temerity to suggest that propinquity or familiarity inhibited sexual interest. Nonsense, retorted White, "intimacy fosters incest rather than callousness." And the evidence, he announced, was "both clinical and ethnographical," by which he meant psychoanalytical and anthropological. He rounded out his case with Freud's remark that "'the prohibition against incestuous object choice [was] perhaps the most maiming wound ever inflicted . . . on the erotic life of man.'"[13]

Just as Malinowski had let slip a suggestion that some reference to animal behavior might be pertinent in thinking about human behavior, so White, for all of his consuming commitment to cultural explanations, let fall a glimmer of recognition that human life had an animal past. In explaining why the incest taboo developed, he said that some force was required to overcome "the centripetal force of the family. This . . . was found in the definition and prohibition of incest." Persons would be forbidden to marry within the family and thus "would be compelled to marry into some other group—or remain celibate, *which is contrary to the nature of primates.*" That was one of the few references he made to a possible historical, that is, evolutionary, connection between human beings and animals. And no sooner was the reference made than we were right back with culture and human consciousness in accounting for the beginning of the taboo. "The leap was taken" White pronounced; "a way was found to unite families with one another, and social evolution as a *human* affair was launched upon its career."[14]

Throughout White's account a truly causal analysis is conspicuous by its absence; in its place is a functional explanation. Nowhere is process to be found; the incest taboo simply emerges, apparently full-blown, and for good. Like Freud, White thought it "would be difficult to exaggerate the significance of this step. Unless some way had been found to establish strong and enduring social ties between families," he emphasized, "social evolution could have gone no further on the human level than among the anthropoids."[15] Although White would undoubtedly have declared himself a believer in Darwinian evolution, it seems not to have occurred to him that an alternative explanation was available, that mating outside the group, and therefore sociality, might have come into being through natural selection and not simply as a human invention. But in order to arrive at that interpretation, he would have had to abandon his view that inbreeding was a characteristic behavior pattern of apes.

The emphasis upon cooperation or alliances among human families as the functional origin of the incest taboo reached its acme in the work of the great French anthropologist Claude Levi-Strauss. His contention, simply put, was that marriage rules in all societies could be interpreted as following the fundamental practice that men exchanged women in order to build alliances and gain supporters in defending themselves in a dangerous world. In illustration of his argument he quoted a conversation reported by Margaret Mead in *Sex and Temperament in Three Primitive Societies*. When men in a simply culture were asked why they did not marry their sisters, they pronounced the idea absurd. "'What is the matter with you?'" they responded. "'Don't you want a brother-in-law? Don't you realize that if you marry another man's sister and another man marries your sister, you will have at least two brothers-in-law, while if you marry your own sister you will have none? With whom will you hunt, with whom will you garden, whom will you visit?'" That was Levi-Strauss's way of describing marrying out; the incest taboo was simply the principle that forced the practice. In outlining his position he canvassed the previous explanations beginning with Westermarck's theory that propinquity diminished erotic feeing. Freudian psychology, he contended, successfully demonstrated just the opposite, as did his own observations of people who successfully married after having grown up together.[16]

Despite his certainty, Levi-Strauss admitted the difficulties that had been encountered in trying to account for the origins of the prohibition. "Contemporary sociology," he observed, "has often preferred to confess itself powerless than to persist in what, because of so many failures, seems to be a closed issue." He noted that the American anthropologist Robert Lowie had confessed in 1920 that it "is not a function of the ethologist but of the biologist and psychologist to explain why man has so deep-rooted a horror of incest." (That, of course, was at a time when biology still had a foothold of acceptability in social science. By 1940, in common with many other anthropologists, Lowie could no longer find a biological explanation acceptable. That year, in his *Introduction to Cultural Anthropology*, Lowie wrote that "the horror of incest is not inborn, though it is doubtless a very ancient cultural feature.")[17]

Levi-Strauss, writing in 1949, was equally sure that biology had nothing to contribute. "It is true," he wrote, "that, through its universality, the prohibition of incest touches upon nature, i.e., upon biology or psychology or both." But since the prohibition is a social rule it belongs in the realm of society, "hence to culture, and so sociology, whose study is culture." It is true, he continued, that "contemporary thought is loth to abandon the idea that the prohibition of relations between immediate consanguines or collaterals is justified by eugenic reasons," but the most recent work in biology, he remarked, does not support the view that inbreeding would be deleterious. Nor is there any support to be found in animal behavior, he added, for what uncertainties there may be regarding the sexual habits of the great apes, "it is certain that these great anthropoids practise no sexual discrimination against their near relatives."

Besides, he remarked, echoing Freud, why would a rule exist if the prohibition were innate, as Westermarck had claimed? "The origin of the prohibition of incest must be sought," Levi-Strauss insisted, "in the existence or in the assumed existence, of this danger for the group, the individuals concerned, or their descendants." In short, the origins must lie in a conscious decision at some unspecified and certainly undocumented time in the dim, prehistoric past.[18] Contrary to Freud, Levi-Strauss took the biological explanation seriously enough to explain why the social rule against suicide was not a proper analogy for the rule against incest, that is, a social rule that repeated a natural repugnance. Both rules hold among animals, he contended, that is, they neither commit suicide nor practice incest avoidance, while only human beings avoid mating with close relatives. Yet he could not dismiss biology entirely from the origins of the taboo. "The prohibition of incest is in origin neither purely cultural nor purely natural, nor is it a composite mixture of elements from both nature and culture. . . . In one sense, it belongs to nature, for it is a general condition of culture. Consequently, we should not be surprised that its formal characteristic, universality, has been taken fron nature. However, in another sense, it is already culture, exercising and imposing its rule on phenomena which initially are not subject to it. Rather, the taboo is the link between culture and nature." Not surprisingly, Levi-Strauss's careful, almost torturous delineation of the meaning of the incest taboo resulted in his raising the social rule to a transcendent level. When nature and culture are united in this taboo, "the whole situation is completely changed. . . . Before it, culture is still non-existent; with it, nature's sovereignty over man is ended. The prohibition of incest is where nature transcends itself."[19]

Despite the thoroughness—really ambivalence—with which Levi-Strauss canvassed the meanings and nature of the taboo, the origins, in his analysis, remained mentalistic, as they were in White and Freud. As he wrote in 1963, "a kinship system does not consist in the objective ties of descent or consanguinity between individuals. It exists only in human consciousness; it is an arbitrary system of representation, not the spontaneous development of a real situation."[20] The function of the taboo was to create that kinship system; the function thus becomes the explanation. Or, to put the matter another way, the explanation is completely ahistorical. Although Levi-Strauss himself might protest that he was not truly interested in the origins of the taboo, his extensive discussion of the prohibition reveals that he never quite succeeded in escaping the question. In the end, he evaded rather than ignored it. Given the centrality of the taboo to his argument, that is not surprising. Because Levi-Strauss rested his analysis, at least in part, upon certain assumptions about animal behavior, when those assumptions were shown to be dubious, his argument, as we shall see, was open to fresh criticism.

A move away from a conscious or mentalistic explanation for the origin of the incest taboo such as Levi-Strauss offered first appeared among social scientists in Mariam Slater's demographic explanation in 1959. Working on the assumption that the average life expectancy of early human beings was probably no more than thirty or thirty-five years, she contended that "by the time the children are of age to mate their parents would be most likely dead." In effect, she turned the standard argument for the origins of the taboo on its head. Rather than the taboo being a device for building connections or alliances with other families or groups, she contended that the "cooperative bonds were determined by mating patterns, not vice versa."[21] On the average, a son had no opportunity to mate with his mother because she was dead before he reached mating age. Subsequent commentators pointed out, however, that Slater had forgotten that life expectancy was not a good indicator of average life *span*, simply because the many who died in infancy pushed down the average age and thus obscured the survival of those who lived into their forties and beyond.

Demography, in short, could not explain why there would be no need for an incest taboo. Slater was, nevertheless, one of the first social scientists since the early years of the twentieth century to offer an explanation that did not assume a conscious decision or mentalistic act to bring the prohibition into being.

A much more significant departure from the mentalistic approach of Levi-Strauss and other anthropologists occurred when biologists began to take a fresh look at the effects of inbreeding. As we have seen, virtually all of the proponents of a cultural explanation for the origins of the incest taboo had rejected the possibility that matings between close relatives resulted in deleterious consequences. Those who had paid attention to Darwin, however, would have been directed to a different line of inquiry. In *The Descent of Man,* when discussing the possible origin of human marriage, Darwin had noted that most anthropologists of his day assumed that promiscuity was common among early human beings. He, however, dissented. "From the strength of the feeling of jealousy all through the animal kingdom, as well as from the analogy of the lower animals, more particularly of those which come nearest to man," he contended, "I cannot believe that absolutely promiscuous intercourse prevailed in times past, shortly before man attained to his present rank in the zoological scale." A few paragraphs later he drew a closer linkage between human and animal behavior. He speculated that, as occurred among the apes, young male hominids had probably been expelled from the group, thus being compelled to find mates elsewhere, and thus "prevent too close interbreeding within the limits of the same family."[22]

Not long after Levi-Strauss asserted the harmlessness of inbreeding, other social scientists became aware of evidence, in offspring of both animals and human beings, that supported Darwin's suggestion. In 1963, a group of social scientists who were studying animal behavior in order to shed light on the origins of the incest taboo among human beings called attention to the serious disadvantages of inbreeding in animals and therefore the likelihood that it was at least as detrimental to human beings. For that reason they speculated that the incest taboo was undoubtedly of significance in the evolution of human beings, but they could not imagine how early human beings would be aware of that danger, so they would not have been able to take it into consideration in making the decision to "marry out." Meanwhile, physiologists' and psychologists' empirical studies of incestuous matings among human beings in modern times were demonstrating that inbreeding was as genetically harmful to human offspring as it was to animals. Premature death, and psychological and physical infirmities, these investigations revealed, were strikingly higher among offspring of close relatives than among those of non-relatives of similar social background.

Despite this new evidence, however, the social scientists in their 1963 study rejected the idea that the taboo could be accounted for by reference to biology. Like so many other investigators before them, and despite their acceptance of the likelihood of possible genetic defects from inbreeding, they were brought up short by the doubt that would not go away: "It is hard to see why what is naturally repugnant would be tabooed." Moreover, they concluded, "evidence for sexual attraction among kinsmen is quite adequate for rejecting" any instinctive basis for the taboo.[23]

A quite different inference was drawn from the dangers of inbreeding by social psychologist Gardner Lindzey. In his presidential address to the American Psychological Association in 1967, he boldly announced that he was going to examine "one particular set of determinants (the biological) which I believe by itself provides a *sufficient* explanation of the origin" of the incest taboo. His point of departure was Freud's Oedipus complex, which Lindzey acknowledged as not only central to psychoanalysis but also closely connected "to the incest taboo and its consequences for psychological development." Like Freud and most anthropologists, Lindzey assumed "the givenness of human incestuous impulses," but to that point he added two significant others. He accepted the new evidence on the disadvantageous consequences of human inbreeding and "the evolutionary necessity of inhibiting them. . . ." For the first time in the long discussion since Westermarck, Darwinian evolution was being brought into the picture. Inbreeding, Lindzey explained, would be selected against in both animals and human beings, for it was "unlikely that human society would survive over long periods of time if it permitted, or encouraged, a high incidence of incest." His conclusion was "that the incest taboo (whatever other purpose it may serve) is biologically guaranteed."[24]

Lindzey, however, did not follow out the implications of the evolutionary theory with which he began. After making the point that human beings, like animals, could not have survived if inbreeding had not been inhibited in some fashion, he abandoned the evolutionary approach. For if incest was in fact selected against, as he said it must be, then it was only logical to conclude that an urge to intra-familial mating would not have persisted. But that was not the line that Lindzey followed. Instead he accepted the Freudian argument that human beings had a built-in drive to mate with family members, in support of which he cited clinical and other evidence of incestuous desires and acts among human beings. In short, he ended where Westermarck's analysis had begun.

One reason Lindzey could not see the illogic of his position was that he, like most other students of the incest taboo at that time, worked under the assumption that animals usually mated with close relatives, that incest avoidance was a peculiarly human activity. In fact, it was just that contrast in behavior that had caused Freud, Malinkowski, Levi-Strauss, and others to see the taboo as the great point of transition between animals and human beings, between nature and culture. That contrast became difficult to retain during the early 1960s as students of animal behavior, that is, psychologists and biologists, began to report on the absence of matings between close relatives in various species in the wild. Given evolutionary theory, that behavioral pattern was to be expected, especially in view of the recent recognitions of the deleterious consequences of inbreeding. One early observer of free-ranging monkeys, for example, reported in 1968 that he witnessed not a single copulation between a mother and son, even though the two were in close contact, grooming, and resting together. The more common observation was that young males usually left their natal group and thus had no occasion for intra-familial mating. Jane Goodall, who later became world-renowned for her studies of chimpanzees in the wild, reported as early as 1971 on the apparent refusal or reluctance of mothers to mate with their male offspring.[25] The evidence in support of "incest avoidance" among animals became so strong that one ethologist in 1975 remarked, with only a little exaggeration, that it is "an empirical fact that in the whole animal world with very few exceptions no species is known in which under natural conditions inbreeding occurs to any considerable degree."[26]

Although this was a statement not all ethologists could then accept with a comparable degree of assent (the author had put his assertion in italics), it was not far wrong.[27] In 1976, for example, anthropologist and primatologist Vernon Reynolds explicitly concluded on the basis of his study of various species of sub-human primates that the incest taboo could not be an invention of human beings. (Some years later, in her definitive report on the Gombe chimpanzees in 1986, Jane Goodall reinforced the conclusion on the rarity of incest, especially between mother and son. "No male has ever been observed to *try* to take his mother or sister on a consortship," she wrote. Even father-daughter copulations she thought unlikely since out of 258 consortships between 1966 and 1983 at Gombe, "in only 18 per cent was the male old enough to have been the father of the partner.")[28]

These ethological findings clearly undermined not only Lindzey's argument but Freud's and Levi-Strauss's as well. No longer could it be assumed that animals commonly mated with close relatives and that from such a behavioral pattern human beings had emerged. Nor was it any longer plausible to believe, given the evidence from animal behavior studies, that familiarity bred lustful feelings. On the contrary, familiarity bred *lack* of interest; Westermarck began to emerge from the shadows of history.

Even before the ethologists had established their position on the avoidance of incest among animals, one of their bolder colleagues, K. Kortmulder, had taken up the cudgels in behalf of Westermarck's explanation for the origin of the incest taboo. Because Levi-Strauss had explicitly ruled out a biological explanation, Kortmulder took him to task by name, arguing from ethological evidence that animals, contrary to what Levi-Strauss had written, avoided mating with close relatives. Using geese as his animal group for study, Kortmulder contended that siblings reared together did not pair because, he reasoned, familiarity reduced the emotions necessary for mating. He then looked at three different human cultures and concluded that "in humans, as in geese, familiarity is an important fact in incest avoidance."

The boldness of Kortmulder's argument from ethology to human beings was measured in the responses to his article by fellow ethologists and anthropologists. Out of the nine published responses to it, only one favored his use of ethology in accounting for this aspect of human behavior. Among the opponents was one who thought Kortmulder's position was "so simplistic as to be ludicrous"; most found his evidence lacked clear relevance to human behavior.[29]

Westermarck's theory may have been emerging from the shadows of history, but his central idea that human erotic feelings were weak within families was not yet acceptable to most anthropologists, despite Kortmulder's effort. Anthropologists did not find it easy, with their longstanding emphasis upon cultural explanations, to find a causal connection between animal and human behavioral patterns. Yet at the very time Kortmulder was making his case, the kind of anthropological evidence his argument lacked was being brought together from opposite sides of the world.

The story begins in Taiwan, where Cornell, later Stanford University anthropologist Arthur Wolf began to study the custom of *sim-pua,* in which parents of young sons brought into their household at an early age similarly aged young girls. The explicit intention was to guarantee a wife for the son. Since not all members of a community practised *sim-pua,* Wolf recognized that the phenomenon provided an opportunity to test the validity of the competing Freudian and Westermarckian hypotheses concerning the effects of familiarity upon sexual interest. For if the *sim-pua* marriages were as reproductively successful as those unions in which the partners were reared apart, then the Westermarck theory would certainly be called into question.

As Wolf soon found out, however, those couples who married in *sim-pua* showed a clear aversion to it. In his first article in 1966, Wolf concluded that "under the condition of intimate and early association propinquity does annihilate sexual desire," just as Westermarck had said. Rather than seeing the incest taboo as a human invention to enforce "marrying out," as Tylor, White, and other proponents of a cultural explanation had contended, Wolf defined exogamy as "the unintentional benefit of the very creation of the conditions of family life," namely the lack of sexual interest among siblings. Wolf speculated that the source of the lack of erotic interest was the painful socialization process experienced by the couple. He drew no connections between animal and human behavior.[30]

Five years later Wolf went back to China to study *sim-pua* further and to check his earlier results. He came away with a dramatic finding: those children who, because of changing economic and social circumstances, could escape from *sim-pua*, readily did so. Yet even in the households that continued to practise *sim-pua*, the couples hated it. Divorces were four times greater among *sim-pua* couples than among couples who were not reared together. Fertility was also much lower among *sim-pua* couples, presumably because frequency of intercourse was less. Westermarck, he concluded, was right: familiarity did breed lack of interest even if it did not necessarily breed contempt. Contrary to the conclusion in his 1966 article, he was now prepared to see some relevance between animal and human behavior. In the interim, as he noted, he had become aware of the several studies reporting incest avoidance among monkeys and other primates. Thus he now saw as the purpose of his paper "to provide another example challenging the view that man's behavior in society is largely a creation of society." He was not prepared, however, to offer an explanation for the lack of erotic feeling which he had documented.[31]

Across the globe in Israel, another "natural experiment" was providing additional "back-handed" support for Westermarck's hypothesis. In 1964, Yonina Talmon, an Israeli sociologist, reported the result of her study of 125 couples who had been brought up togther in Israeli kibbutzim. Within that group there was not a single couple in which both mates were reared from birth in the same peer group or children's house. As she noted, parents offered no objections to such marriages; in fact, the parents preferred such unions. The children, however, clearly did not.[32]

In subsequent years, several additional studies were conducted, involving hundreds of kibbutz young people. Anthropologist Seymour Parker summarized the findings in 1976. No marriages and very few extra-marital sexual liaisons occurred among the members of the same peer group. The avoidance of such relations, he added, "was completely voluntary." The behavior of the kibbutz-niks was especially relevant not only because it was voluntary, but also because, as Talmon had found, it was generally contrary to the behavior encouraged by the children's parents. There was, therefore, no reason, as there may have been in the case of the *sim-pua* marriages, to think that any extraneous influence was suppressing erotic feelings among the young people. Israeli sociologist Joseph Shepher, a longtime student of the subject, also ruled out sexual repression as a explanation inasmuch as the social climate in the kibbutzim actually favored the elimination of embarrassment about sexual matters. The kibbutz cases provided powerful support for the Westermarck hypothesis even though in those instances there was no genetic danger inasmuch as the young people were not related. That is why sociologist Pierre van den Berghe referred to the evidence from the kibbutzim as "culture fooling Mother-Nature."[33] As the young people themselves often complained, to marry someone from your own children's house would be like marrying your sister or brother.

"Human beings, like birds and other mammals," wrote psychologist Ray Bixler in 1981, "are *generally disposed* to lack sexual interest in each other if they *experienced*, to a degree not yet well defined, close and extensive contact with each other during early prepubescence."[34] In 1981 that was no longer a bold or rare conclusion to be heard among social scientists.

One anthropologist in 1977 had frankly taken up Westermarck's cause not only from human and primate evidence but from a neurophysiological standpoint as well. Drawing on the work of Karl Primbram in neuropsychology, anthropologist William Demarest argued that "the capacity for incest avoidance did not have to be selected for . . . but was always present in the structure of the mammalian mind, . . . and only when long-maturing mammals were organized into social groups," he argued, "did the capacity become behaviorally apparent." He had no doubt that incest avoidance was embedded in the biological history of human beings and other mammals.[35]

New anthropological findings in support of the Westermarck hypothesis continued to be published in the 1980s. Justine McCabe, an anthropologist at the University of California at Davis, reported in 1983 on a practice in Lebanon in which parents encouraged marriages between a son of one brother with the daughter of the other. The consequences, measured by divorce rates and fertility, were similar to those discovered in the *sim-pua* cases: divorces were four times more frequent and number of children was 23 percent fewer than in marriages between unrelated persons. Familiarity, not genetic closeness, was the operative ingredient. Marriages between cousins who did *not* grow up with each other showed neither the divorce nor fertility characteristics that distinguished marriages between cousins who had been reared close by.[36]

This latest example of the working out of the Westermarck thesis—backhandedly, to be sure—was especially pertinent since only a few years earlier a group of philosophers and anthropologists studying the biosocial explanation for the incest taboo had used marriage between cousins as a refutation of the argument that proximity inhibited sexual interest. Their argument was that "if the proximity hypothesis were correct, and such an inhibition did in fact serve the evolutionary function, then sex and marriage between cousins raised together should be significantly different from sex and marriage between cousins raised separately. No such luck," they concluded. "Marriage between cousins is common and even encouraged in many instances of *both* situations."[37] In the Lebanon example, as in *sim-pua*, marriages were indeed encouraged, but the reactions of the couples who entered them differed markedly from those found in marriages between persons who were not raised together, a contrast that was consistent with the Westermarck hypothesis.

Support for Westermarck's theory may have been strong by the 1980s, but that still did not mean most social scientists were prepared to view it as a plausible example of how evolutionary theory and ethology might provide insight into the origins of the incest taboo. That the thesis undermined a basic assumption of psychoanalysis almost guaranteed that it would be resisted when not denied outright. And even when psychologist Ray Bixler tried to make a case for incest avoidance as a function of *both* environment and heredity, he found himself criticized for not being able to show a genetic basis for incest avoidance among animals, much less among human beings.[38] University of Michigan anthropologist Frank Livingstone as late as 1969 flatly denied that incest avoidance among animals could be accepted on the basis of who mates with whom since mating was a function of the animals' social structure. And for similar structural reasons, he added, one could not draw valid conclusions about the causes of human behavior from the *sim-pua* or kibbutz experiences. "Incest could have developed," he insisted, "only after the development of language," since human kinship patterns and incest avoidance depended upon symbols.[39]

Other critics complained that "the incest taboo has not been shown to be universal in any sufficiently well-defined sense such that biological accounts are appropriate." Therefore, given the great variety of social practices throughout the world's cultures, "cultural explanations on a local scale seem far more persuasive."[40] The depth of the disagreement by this time among social scientists concerning the role of biology in accounting for the origins of the prohibition is captured in the comment of anthropologist Melvin Ember in 1975. He noted the resistance of anthropologists to "the possibility that natural selection may have favored the incest taboo for biological rather than social reasons." Why this should be so, he concluded with some exasperation, "is a question that psychohistorians may be called upon to answer."[41]

Yet a biological explanation for the taboo against sex with close relatives did leave some elements unaccounted for. The Westermarck explanation, for example, dealt only with sex among family members; it said nothing about prohibitions against sex with or marriage to other persons, yet such anti-incest rules occurred in many cultures. And even within the family, none of the anthropological evidence in support of the Westermarck theory—the *sim-pua* and kibbutz examples—reached the question of father-daughter incest, though in the

modern world that incestuous relation seems to show by far the highest incidence.* Several critics doubted that a taboo, that is, a social rule, could develop
from a biologically rooted aversion, as the biosocial explanation assumes. If that
were true, philosopher John R. Searle pointed out, then we would have to
expect that the Israelis would "evolve moral norms forbidding sexual relations
among members of groups that grew up together on kibbutzim." The idea he
thought "preposterous," yet implicit in the logic of sociobiological explanation.[42] On the other hand, a sociobiologist would undoubtedly reply that thirty
years is hardly a period of time sufficient for a moral norm to evolve.

Perhaps the most persistent criticism of the sociobiological explanation is that
which Frazer and Freud raised against Westermarck many years before. As philosopher Philip Kitcher, the author of a book-length denunciation of sociobiology, complained, sociobiology has done "nothing to help explain why so
many societies have incest taboos. Why do we need taboos to prohibit what we
naturally resist?" he wanted to know.[43]

Actually, modern supporters of the Westermarck view have tried to meet
that precise argument. As early as 1973, Roger V. Burton saw the taboo as a
reinforcement of the natural avoidance, an argument elaborated upon subsequently by other social scientists. "The function of cultural norms," wrote sociologist Joseph Shepher in his biosocial study of the incest taboo, "is to safeguard
those propensities which motivate rather than determine human behavior."
Another sociologist, Joseph Lopreato of the University of Texas, who thought
"incest avoidance is a biological feature of all human populations," agreed with
Shepher that the taboo itself "is a reinforcer of the biological predisposition and,
thus, a form of insurance against inbreeding depression," that is, the deleterious
effects of mating between close relatives. Shepher's essential argument was that
culture and biology go hand in hand, "a process that has rendered committed
incest a rare phenomenon."[44]

Richard Alexander, a University of Michigan biologist and vocal advocate of
sociobiology, responded to the objection of the redundancy of the taboo by
offering an ingenious if somewhat forced analogy between the adaptive value
of incest avoidance on the one hand and that of the inhibition of ovulation
during lactation on the other. Both biological functions, he noted, have
spawned social rules: one against sexual intercourse after childbirth or during
nursing and the other against marrying close relatives. The incest taboo, however, it is relevant to note, is much more widespread across cultures than the
restrictions on sexual relations. A closer analogy would be with suicide, which

---

*Shepher, *Incest,* p. 127, reports that mother-son incest occurs about four times out of 10,000
population; sister-brother relations four times out of a 1000 population, and father-daughter incest
1.6 times out of 100.

is widely prohibited yet, at the same time, presumably selected against. Anthropologist Robin Fox has pointed to murder as a nearly universal taboo, and likely to be selected against, yet there, too, the taboo is supported by apparently redundant social rules.[45]

The most extended response to the old objection to Westermarck's conception of incest avoidance was made in 1976 by Seymour Parker, an anthropologist. Like some other writers on the subject, Parker saw culture and biology working in tandem to avoid the bad effects of inbreeding. For one of the recognized advantages of incest avoidance, he pointed out, was the development of cooperation beyond family lines. It reduced "what might have been 'crippling' sexual competition within familistic units requiring cooperation. As the cultural way of life became established," he argued, "additional adaptive pressures arose for this biopsychological tendency to become institutionalized as the incest taboo, because it increased the stability of the family unit, assured wide alliances, and reduced the number of births to economically immature individuals." He doubted that incest avoidance alone was a sufficient condition to account for the development of the taboo, and it may not even have been necessary, but it surely was "a facilitating condition," in his judgment. Yet insofar as the taboo was based upon biological needs—to avoid inbreeding depression— the taboo was easier to learn because of the reinforcement from what he called "intraorganismic sources," that is, biology.[46]

A few years later psychologist Ray Bixler moved the argument of biocultural cooperation a little further along by observing that cultural prohibitions alone could not account for the low incidence of incest if only because few, if any, cultural prohibitions work as effectively as the incest taboo. He called attention to the widespread incidence in our own time of adultery, fornication, and alcohol consumption, despite strong social prohibitions. But those restrictions, he stressed, lack reinforcement from inbreeding depression. It was an argument that Westermarck himself had advanced in slightly different form many years earlier. Bixler also countered the objection that the Westermarck thesis did not account for the extension of the taboo to persons outside the immediate or nuclear family. The prohibition was extended beyond family, Bixler asserted, by human reflection, imagination, and perhaps by the observation that reproduction with close relatives could result in defective offspring.

The view that there might be a conscious element in the establishment of the taboo was also advanced by anthropologist Melvin Ember, who noted that one-third of the incest myths in the Human Relations Areas Files mentioned deformed offspring from incestuous marriages. "Such a substantial degree of association suggests," wrote psychologist Gardner Lindzey after reporting the same finding, "that on occasion some degree of biological insight may well have accompanied the origin of the taboo."[47] Anthropologist William Durham made a similar point when he doubted that incest avoidance was genetically based, even in animals, though he had no doubt that avoidance of sexual relations with close relatives carried a clear reproductive advantage. Like Bixler, Ember, and Lindzey, Durham thought primitive peoples might well have recognized the consequences of such behavior and thus prohibited it. And he agreed with Parker that it was to be expected that culture and biology worked in the same direction, namely, to enhance reproductive success.

That there has been a growing and often enthusiastic interest among social scientists in looking at the history of the incest taboo from a biological point of view seems clear. Is there anything else that may be concluded from this revival of interest in the biological roots of human nature?* Has anything been concluded or established? What aspects of the ancient issue still remain unaddressed, despite all the new information and insights?

To start with the more or less settled aspects, it now seems dubious to insist, as Freud, Levi-Strauss, and other anthropologists certainly did, that the development of incest avoidance marked the beginnings of humanity. The ethological evidence, mounting each year, renders the avoidance of incest within the family or basic social unit no longer a peculiarly human activity, whatever the subtle and complex workings out of it that have undoubtedly been added by culture. Animals may avoid incest, but there is no reason to believe that they abhor it. Yet what a group of social scientists wrote in 1963 seems accurate: "In no animal group are there restrictions on inbreeding except for the family unit, whereas in no human group are incest taboos limited to the nuclear family. Once the familial taboo is in existence, extension of the taboo to other categories of kin becomes a simple evolutionary step."[48]

When such evidence is combined with that from the kibbutzim and *sim-pua*, Westermarck's conception of the lack of intra-familial erotic attraction seems to be more accurate than the assumption that among human beings there is a

---

* Among ethologists, as might be expected, the shift has been notable. John Maynard Smith, the eminent English zoologist, announced his conversion in 1978 despite an earlier conviction that the incest taboo was "entirely cultural in origin. Three things," he testified, "have made me change my mind. The first was the realization that if the deleterious effects of inbreeding are sufficiently serious to have been responsible for evolution of self-incompatibility in plants, they would also provide a strong selection pressure for reducing inbreeding in animals. The second was the discovery by a number of workers that, in the wild and the laboratory, many animals, including primates, have behavior patterns which reduce the frequency of incestuous matings. The third was the work of Shepher on Israeli kibbutzim, suggesting very strongly that if given a choice human beings avoid incestuous matings. Of course, the details of the incest taboo in different societies may have a cultural basis. But it seems very likely that our ancestors avoided incest long before they could talk." J. Maynard Smith, "The Concepts of Sociobiology," in G. S. Stent, ed., *Morality as a Biological Phenomenon* (Berlin: Abakon, Verlagsgesellschaft, 1978), 32–33.

natural urge to incest. It is also worth repeating that Westermarck's theory, as he himself recognized, is consonant with what evolutionary theory would predict: avoid inbreeding because it leads not to reproductive success but to gene extinction. Or, as University of Michigan sociobiologist Richard Alexander pointed out, the kibbutzniks' and *sim-pua* experiences suggest that culture evolved in accordance with genetic advantage. Moreover, to discern a continuity between animal and human behavior fits well the underlying principle of Darwinian evolution. Rather than assuming a sharp discontinuity between the behavior of animals and human beings, as the cultural explanations do, a biosocial explanation for the taboo includes human beings in the continuum of natural beings that Darwin taught us to see and appreciate.

Yet even if one accepts a continuity between human and animal behavior and also accepts the idea that incest avoidance is built into animal behavior, the puzzle is still not quite solved. One of the things that recent ethological investigations seem to have settled is that the family, which we take for granted among human beings, is quite uncommon among apes and monkeys. In only a few species of primates, principally gibbons and siamangs, do both parents rear the young. Chimpanzees, who are our closest relatives, genetically as well as morphologically, in the animal world, live in groups with the mother-child dyad as the primary reproductive unit. Males are present, but they mate promiscuously and play almost no role in rearing young. If this behavior pattern is taken as the historic behavior of early human beings, then the question of marrying out would not have arisen since that was the established practice. Thus the question posed by anthropologists about the origins of marrying out may be largely irrelevant in the light of primate ethology. But by the same token, that same evidence also leaves a problem for the biosocial approach, since there is no obvious continuity between the child-rearing behavior of chimpanzees and that of human beings. "Heterosexual pair bonding for reproductive purposes is well known in fishes, birds, and some types of mammals," psychologist Frank Beach has observed, "but it is rare in primates. . . . Seen against this background of anthropoid behavior," Beach emphasized, "*Homo sapiens* stand out as a striking exception. There is not today and probably has never been a human society which lacked the essential elements of family structure," he added.[49]

Beach's remarks suggest that the real issue that still awaits explanation is the origin of the nuclear family, which is so widespread among human cultures. Could it have evolved independently, unrelated to the behavior of our animal ancestors, that is, as an adaptation to the special life of a mobile, ground-dwelling, bipedal, omnivorous mammal with a rapidly growing brain? A genetic basis for the rearing of offspring by both parents, as Beach noted, is common among birds and other animals, but the phylogenetic or evolutionary distance between them and *Homo sapiens* leaves us with no more than the clue that the behavior pattern of dual parenting is capable of being rooted in biology. So far as I can tell, no sociobiologists have ventured into the thicket of the biological basis of human families.

For those who harbor doubts about the validity of the Freudian view of human nature, the new support for the Westermarckian hypothesis about intra-familial erotic attraction provides fresh ammunition. For if family members do not have a strong sexual attraction, then the biological basis of Freudian theory, particularly the Oedipus complex, is brought into serious question. As anthropologist Parker has written, Freud held the belief that "man's humanity and civilization have emerged with a concomitant repression of his instinctual demands."[50] Indeed, the threat is so severe that Robin Fox, a leading anthropologist, has sought to salvage Freud by depicting the incest taboo as relatively ineffective. Although Fox is one of the earliest proponents among anthropologists of a biosocial approach to human nature, he dissents from certain propositions advanced by sociobiologists. For example, he wrote in 1976 that "incest is generally avoided rather than actively prevented," by which he meant it came out of our animal ancestry.[51] But later, in his book *The Red Lamp of Incest* (1981), he went along with Freud in maintaining that there is indeed an urge toward incest among human beings, especially by sons in rivalry with fathers, or as he preferred to put it, between the old and the young males. The object of their rivalry, he argues, is access to females, not merely to the mother, as Freud had maintained.

Undoubtedly, from the standpoint of sociobiologists, the principal gain from a biosocial interpretation of the incest taboo is that it provides an *explanation,* while alternative interpretations, such as those offered by anthropologists in general and by Freud and Levi-Strauss in particular, offer no explanation at all. Rather, they set forth an ahistorical functional *assertion* of why the taboo ought to have developed; the means by which it was brought about being no more than a shift in human consciousness at some remote and undesignated point in the prehistoric past. The biosocial interpretation links the taboo to evolutionary theory in a plausible way, by seeing its origins, though not its cultural elaborations, as a product of natural selection.

A frequently presented argument against a biological basis for the incest taboo has been that incest is much more common than most people know, or at least like to admit. After all, so the argument goes, biology is much more regular in its impact upon behavior than culture, so that if incest occurs, even as rarely as it does, the taboo is most likely, by that fact alone, to be cultural in origin. Proponents of a sociobiological outlook have usually answered the objection as sociologist Pierre van den Berghe does: "Genetic predispositions are not rigidly deterministic; they allow for flexible adaptations to a wide range of ecological situations." Or as the historian of science Stephen Jay Gould had written: "in our vernacular 'inherited' often means 'inevitable.' But not to a biologist. Genes do not make specific bits and pieces of a body; they code for a range of forms under an array of environmental conditions."[52] Others have simply contended that examples of incest among the Egyptian and other royal families were not truly reproductive, that such incestuous behavior was essentially symbolic and, in any case, relatively rare in occurrence.

The response of rarity, however, fails to meet the challenge that came to light in 1980. It is evidence of a rather widespread incidence of brother-sister marriages among the common people of Roman Egypt. These cases from ancient Egypt are indeed distressing to any believer in the reality of the incest taboo, sociobiologist or not. As one such troubled student of the subject has written, the Egyptian evidence is a "particularly bedeviling instance," but since it appears to be "a unique class," he thought perhaps it could be dismissed, at least for now, as only an "ethnographic and historical oddity."[53] The Egyptian evidence, as reported by Keith Hopkins in 1980, is thus far unique. Hopkins discovered 113 instances of a brother marrying a sister, a number which constituted about a third of the families he studied who had sons and daughters who reached marriageable age. (Many children died earlier than that.) So far as the records reveal, the marriages were voluntary and, according to Hopkins, were not occasioned by a concern to conserve family property, or because of legal or parental requirements. At that period in Egyptian history, Hopkins further observed, women held a relatively high social status and therefore, presumably, were in a position to make more or less independent decisions about choice of marriage partners.[54]

As noted already, these findings raise troubling questions not only about the Westermarck hypothesis but also about other interpretations as well. Supporters of the Westermarck hypothesis would probably like to know the degree of reproductive success from such unions. Supporters of a cultural interpretation cannot fall back upon Leslie White's acceptance of exceptions to the taboo among certain royal families, for these marriages are relatively numerous and among commoners. Perhaps the most comforting route would be to follow White's definition of incest as "something criminal and prohibited."[55]

Meanwhile, William Arens, the anthropologist who described the discovery of incestuous marriages in Roman Egypt as "particularly bedeviling," has accepted the biological or ethological roots of the taboo, only to move beyond it. In his book *The Original Sin*—his name for incest—he follows Robin Fox in being impressed by the evidence that incestuous acts are far from rare, despite the prohibition and despite his personal conviction that the taboo is innate. Like Westermarck, Arens wrote in 1986 that "the source of aversion to the familiar is an innate proclivity, while the prohibition of others as sexual or marriage partners is social or cultural in origin." In substance he follows those who see a combination of biology and culture standing behind the prohibition. Contrary to Freud and Levi-Strauss, however, he sees no flowering of human culture either in the onset of the prohibition or in its meaning. It is true, he admits, that the concept of a biological basis for the taboo rests on circumstantial evidence, but the idea of a cultural origin rests on no evidence at all, circumstantial or otherwise! It is not culture, he argues, that suppresses or limits sexual expression among human beings as compared to animals. On the contrary, "culture allows for human sexuality to flourish" and to move far beyond mere reproduction, to take "expressive directions inconceivable and thus unavailable to lower order primates, which lack our imagination." Arens's view would be quite in line with Kinsey's observation: "In matters of sex, everything you can possibly imagine has occurred and much that you cannot imagine."[56]

Having said that, Arens then uses his conception of human sexuality to account for the paradox of a fairly wide practice of incest in the midst of a general taboo against it. In arriving at his answer he stands Levi-Strauss on his head. Incest, Arens asserts, is "a product of our unique capacity to supersede natural inclinations." Thus incest itself, "rather than its absence, is the unsightly, but nonetheless true mark of humanity, culture and civilization." Just because incest is prohibited by nature, its practice is a sign and a measure of human transcendence over nature or the animal. That is the meaning of its practice by certain royal families in Egypt, Hawaii, and Equatorial Africa. The practice is also associated, Arens points out, following the suggestion of the English anthropologist Edmund Leach, with the creation of a society or a new social order. Leach's examples of such new societies being created out of incest are the children of Adam and Eve, the sons and daughters of Noah after the Flood, and Lot and his daughters after the destruction of Sodom and Gomorrah. Arens's fundamental argument "is that incest is a product of both human imagination and capability. From this it follows that the main concern of the deed is the transmission of profound cultural messages about what it means to be human."[57]

Provocative and even unsettling as Arens's theory may be, it nevertheless forcefully illustrates the impact that biological ideas have had on social science over the last few years. The acceptance of the findings of ethology and the application of the Darwinian idea of reproductive success by some social scientists have caused them to come full circle in accounting for or interpreting the incest taboo. At one time, virtually all social scientists were agreed upon the proposition that the phenomenon—the incest taboo—was a measure of humanity. Now many have found it to be almost the opposite: a behavior we have in common with animals. And for some, like Arens, that which can be said to set us apart from animals is our ability to defy nature—that is, to accept incest under certain circumstances as a uniquely human act.* If that means a return to the old Freudian view that culture and biology are at odds, that would not be the first time social scientists have retraced their steps in seeking the key to the nature of human nature.

Seeking an answer to social science's most enduring enigma was not the only use that has been made of the insights of sociobiology, ethology, and Darwinian evolution. An increasing number of social scientists in a variety of disciplines have brought the new insights to bear upon a variety of new and old problems. It is time now to look at some of them.

---

* It is perhaps necessary to point out that the acts of incest to which Arens refers are not those committed by ordinary individuals and which are legal crimes, for they are, in Arens's view also "crimes against nature." Arens is dealing only with those instances of incest which have been institutionalized or accepted by a culture, like the royal marriages in Ptolemaic Egypt.

11

The Uses of Biology

We are fond of saying that society exists, and that hence it inevitably exerts a constraint on its members. . . . But in the first place, for society to exist at all the individual must bring into it a whole group of inborn tendencies; society therefore is not self-explanatory. So we must search below the social accretion.

Henri Bergson, 1932

Economics has no near kinship with any physical science. It is a branch of biology broadly interpreted.

Alfred Marshall, 1920

The conundrum of the roots of the incest taboo among human beings is surely the most elaborate example of the way biological knowledge has been drawn upon by social scientists. That was but one use, admittedly a dramatic one, that social scientists have made of biological ideas in bettering their understanding of human nature. Some of their efforts have been no more than suggestions; others have specified clues for understanding human actions which have long baffled investigators and social theorists. Still others gained fresh insights into human nature from Darwinian evolutionary theory.

The impetus for these new approaches, as we have seen already, frequently came from ethological investigations, though not all of them recent. At least one of the relevant findings actually went back to the 1920s, when it was first discovered that among some animal species individuals in a group situation arranged themselves in a hierarchical pattern of relations. This identification of what came to be called dominance/submission behavior was first reported among chickens, where the pattern soon was colloquially called a "pecking order." Few social scientists at that time, however, were prepared to draw any analogies between the behavior of chickens and human beings. For as sociologist Allan Mazur explained in 1973, if animal behavior is to be drawn upon in analyzing human behavior, there has to be some closeness in what he called the "phylogenetic tree," which is a shorthand way of saying that the animals must be close to human beings in evolutionary development.[1] He further pointed out that if dominance/submission patterns were being studied, then the character

270

of them also needed to be similar to those of human beings. Most of the dominant/submissive behavior among chickens, for example, involved violent attacks (pecking) upon one animal by another in order to establish dominance. That was a much more demonstrative—and violent—form of dominance assertion than prevailed in human relations in small, face-to-face groups. Among primates, however, as Mazur pointed out, dominance behavior was subtle, as among human beings, sometimes not even requiring overt threats. Observations of dominant/submissive behavior among primates in the wild were not made until the 1960s when primatologists began to study systematically social structures among the apes (chimpanzees and gorillas, primarily).

Indeed, the similarity between primate and human social behavior in face-to-face relations was so striking that primatologists could not help commenting upon it. After describing the dominant individual in a troop of apes, one primatologist in 1970 was moved to draw analogies with the character of human leadership. The social order of apes, he observed, was largely organized around a dominant figure, usually a male, who served as the focal point of the group for its protection and internal order. The process of dominance, the primatologist contended, may well offer fresh ways of understanding or interpreting human behavior. In particular, he argued, the difference between "power" and "influence" in human society may well "rest on features indistinguishable in other primate societies." For in the societies of many sub-human primates, "agonistic" (threatening) behavior overtly coerces individuals to perform social functions for the protection and service of the group. At the same time some primate groups express a "cognitive set of relations reinforced by patterns of information exchange based primarily on the hedonic mode of behavior," by which he meant play behavior, tool use, and other mannerisms which capture the friendly attention of other individuals in the group. Such behavior leads toward "flexible social relations which act as the medium for the dissemination of information with the society."[2] That latter kind of behavior, he suggested, was a recent finding and would reshape the way ape societies would be studied in the future. Both kinds of dominant behavior, he emphasized, were found in human social groups.

As suggested already in referring to the difference between "influence" and "power," Max Weber's ideas concerning leadership seemed obviously related to what was being observed among sub-human primates. Weber's conception of charismatic leadership, which involves a person who "demands attention by appearing in the presence of or before his followers," the primatologists suggested, was similar to the behavior of the dominant individual in primate societies. "Therefore in attempting to explain charismatic movements," they advised, "we should look for the operation of an ethologically based attention binding mechanism."[3]

Sociologists who followed the suggestions of primatologists were less optimistic or more cautious in drawing analogies between primate and human behavior, but some did make the inter-species connection. Allan Mazur, for example, writing in 1973 in a sociological journal, concluded that the closer on the evolutionary "tree" a primate species came to human beings the more similar their submissive behavior was to that of human beings. Among baboons and macaques, for example, the submission took the form of what Mazur called "overt gestures," while among chimpanzees and gorillas, who are biologically more closely related to human beings, the gestures were much less overt and almost entirely absent, as usually happened in comparable relations between human beings. Though obviously impressed with the merit of the analogy, Mazur was still cautious. (A decade later he would be less so.) He would say only that among human beings submissive behavior "has some noncultural basis," thereby avoiding saying it was "biological." His caution is also evident in his reminder that to identify a noncultural influence did "not mean learning was not involved."[4] Another sociologist, Donald Ball, writing about the same time, was less circumspect: "Dominance order, roughly, is to animals what social stratification is to man . . . ," he forthrightly asserted.[5]

Political scientists were close behind sociologists in identifying connections between the recent findings in primate behavior and some of the classic concerns in their field. In 1971, Fred Willhoite, then a young political scientist at Coe College in Iowa, called the attention of his colleagues to ethological findings that he thought were relevant to the ideas of the seventeenth-century political philosopher Thomas Hobbes. Among human beings Hobbes had perceived a need for dominance and submission if peace and a social order were to be maintained. To Hobbes, Willhoite remarked, coercion was necessary to ensure submission, but now modern ethological studies suggest otherwise. Dominance, of course, did not imply "dictatorial domination," he emphasized, nor did it imply any particular means of selecting leaders. All that was required was "some kind of hierarchical distinction, however minimal, between leaders and followers." That requirement, Willhoite concluded, referring to the ethological findings, "may well be an ineradicable feature of the social and political life of mankind."[6]

Five years later, Willhoite expanded upon the idea, this time in his discipline's principal journal, the *American Political Science Review*. There he cited and quoted from a wealth of studies on primate behavior to clinch his argument that among the animals most closely related to human beings patterns of dominance and submission were as ubiquitous and as subtle as in human social relations. Like the primatologists from whom he quoted, Willhoite took for granted that dominance/submission patterns had adaptive value to human beings in the course of their evolution. Among other things, it helped to maintain order in the group, a structure that was necessary for reproductive success. He was also sympathetic to the suggestion of one primatologist that, from our common ancestry with apes, human beings may well have inherited the rudiments of self-control. For primates, he noted, "show signs of this, such as waiting for the group before drinking, or holding back from copulating with another's harem member." He called attention, as well, to the ability of primates to adapt quickly to new hierarchical patterns, much as human beings do. He told of a female savanna baboon who, when placed in a group of hamadryas baboons, learned within an hour to remain near and to follow the male which had threatened or attacked her. That is the usual behavior of a hamadryas female, but the "wrong" behavior pattern for a savanna female when in her own group. Hamadryas females when placed among savanna baboons, he further noted, also soon learned not to follow individual males.

To discover connections between primate and human dominance/submission behavior raised questions about some of the long-held assumptions of classical political theorists like John Locke and Jean-Jacques Rousseau. Both of them had postulated a mythical time, before the institution of government, when each human being lived alone. Indeed, to Locke, Rousseau, and Hobbes, the problem had been how to account for the establishment of government or authority, that is, the institutionalization of dominance and submission. Willhoite's point was that primate ethology in conjunction with evolution implied that no such time had ever occurred in the human past. Contrary to the assumption of a modern radical thinker like Robert Paul Wolff, Willhoite observed, there probably was never a time when the human individual was either independent or without a built-in need to submit to authority. In the transition from animal to human being, culture without a doubt elaborated the character and details of the relation between ruler and subject. Nevertheless, the core of social behavior, including a pattern of dominance and submission, was probably present long before culture had emerged.

Willhoite was highly sensitive to the criticism that he was ignoring the role of culture in human development. "I am not arguing that any single *type* of dominance hierarchy is the only 'natural' pattern," he explained. Nor did he intend to argue that an animal origin for authority precluded "attempts to construct or reconstruct orders of political stratification. . . . The last thing I desire to do," he insisted, "is to revivify the justly discredited ideas of 'Social Darwinism.'"[7]

Roger Masters, a Dartmouth political scientist and a leading authority on Rousseau's political thought, drew a sophisticated connection between human and primate dominance/submission behavior. Politics, he contended, was more than simple competition for control, as competition in sports, on the school playground, and in business make evident. None of these activities constitutes politics, he pointed out. Political behavior, rather, involved those "actions in which the rivalry for and perpetuation of social dominance impinges on the legal or customary rules governing a group." Thus he placed political science "at the intersection of ethology and anthropology—or more broadly, at the point where the social and natural sciences meet." And just because that is the intellectual locus of political science, political thinkers from Aristotle to Hobbes and Rousseau "were almost always concerned with the definition of human nature and the relationship between nature and society."[8]

Political scientists were not the only social scientists to discover that ethology offered clues to the question of why human beings accept authority. Psychologists have at least as deep an interest in the issue. Perhaps the most famous experiment by a psychologist on the roots of authority was that designed and carried out by Stanley Milgrim at Yale University in the early 1970s. The experiment was performed with a large number of subjects and in various forms, though the details do not really concern us here. The important conclusion of the experiment was that ordinary people proved to be quite capable of harming strangers who had themselves not harmed anyone *providing* a recognized "authority" condoned the harmful acts.

The "ordinary people" or subjects were asked to apply electrical shocks to a person they could see and hear protesting in an alarming way to the administration of the voltage. When a subject would object to what he or she was being asked to do because of the protests or pleas of the victim, a white-coated researcher, that is, a recognized "authority," would instruct the troubled subject to proceed nonetheless. Twenty-five out of 40 subjects continued the shocks as instructed until a level of 450 volts had apparently been reached, and the person being shocked had lapsed into silence after heart-rending protests. (The whole experiment, of course, was a trick on the subjects. No real shock was administered and the "victim" was actually an actor playing a role.) The question Milgrim confronted was: Why did these subjects—all ordinary citizens recruited through newspaper advertisements—obey their instructions so slavishly despite what appeared to be the harmful effects of their behavior?

At first Milgrim thought that perhaps an innate sense of aggression on the part of those administering the "shocks" was at the root, that the subjects simply liked to punish strangers. This possible explanation, however, was rejected because, whenever the subjects were given the opportunity to select the level of electrical shock they preferred to administer, almost all of them chose the lowest level. "The key to the behavior of subjects," Milgrim concluded, "lies not in pent-up anger or aggression, but in the nature of their relationship to authority." Thus, the question remained: Why would these people give themselves over so completely to an authority?

At this point Milgrim turned to animal behavior. We know, he wrote, that "in birds, amphibians, and mammals we find dominance structures, and in human beings, structures of authority mediated by symbols rather than direct contests of physical strength." Since organization, or a pattern of submission and dominance, aids in survival, he suggested it was likely that this behavior had been shaped "through successive generations ... by the requirements of survival. Behavior patterns that did not enhance the chances of survival," he thought, "were successively bred out of the organism because they led to the eventual extinction of the groups that displayed them." The potential for obedience, he concluded in Darwinian fashion, had survival value and thus "was bred into the organism through the extended operation of evolutionary processes." The emphasis, he warned, must be placed on "potential," not on obedience as such. He rejected any suggestion that human beings possess an "instinct" for obedience, but, he added, "we are born with a *potential* for obedience, which then interacts with the influence of society to produce the obedient man," he concluded. Whether nature or culture was more important in the process, he was not prepared to say. But "from the standpoint of evolutionary survival," he was sure "all that matters is that we end up with organisms that can function in hierarchies."⁹

The relevance of dominance/submission patterns among primates to the study of human behavior also captured the attention of a young anthropologist, Jerome Barkow. The problem to be explained by primate behavior for Barlow was not authority but the prevalence of concerns about prestige in human social relations. Anthropology, he pointed out, has long recognized the importance of prestige among human beings, as the study of potlatch accumulation, the counting of coups, and other rituals, not to mention social stratification, all make evident. In effect, Barkow interpreted the seeking of prestige or the advance of self-esteem as a kind of human "universal," an explanation for which he thought could be found in the study of primate behavior. "The zoological concept closest to that of prestige," he contended, "is *social rank* or *dominance*." Barkow noted that overt threats and abject appeasement tend to moderate or disappear entirely from the behavior of animals as one "ascends" the phylogenetic tree, that is, comes closer to human beings. Chimpanzees and gorillas win "prestige," Barkow said, by attracting the attention of their fellows. He then cited a well-known example from Jane Goodall's observation of chimpanzees in the wild. She described how a chimpanzee she named "Mike" managed to become the dominant member of the group she was studying by banging together a pair of tin cans he had picked up from human beings. Barkow's argument, in short, was that "natural selection has transformed our ancestors' general tendency to strive for high social rank into a need to maintain self-esteem," which is achieved "by seeking prestige."

Like other social scientists who looked to ethology, Barkow was careful not to draw simple analogies between primate and human hierarchies. After all, he warned, among human beings "abstract principle and cognitive evaluations" are heavily involved, as they are not among animals. His aim, rather, was to show "how our distinctively human behavior may have evolved from the general primate trait of social dominance." Since the ancestors of human beings were primates and today's primates seek social rank, it seems likely that "our ancestors . . . were socially ranked." Human beings evolved from those ancestors while organized in bands of hunters and gatherers, a form of organization that implies at least some form of social ranking. "It would seem," he continued, "that at no point during our phylogenesis did we lack social hierarchy." Barkow's point was that the similarity in behavior between primates and human beings was more than simply analogous; it was "homologous" or continuous, as he said, which meant that it had been inherited from the common ancestor from which both species descended. Among modern human beings, he added, "cultural systems of social rank are based on, not opposed to, our primate heritage."[10]

The appearance of Barkow's argument in the journal *Current Anthropology* precipitated a vehement attack by a prominent anthropologist, Eliot Chapple, who, ironically enough, had been himself active in urging biological explanations upon his fellows. Among Chapple's complaints was the objection that Barkow had seemed to play down the evolutionary closeness of primates and human beings. In his defense, Barkow pointed out that his primary interest in ethological studies was to clarify "what kind of animal *Homo sapiens* is and how we got to be that way." To accomplish that goal, he acknowledged, we must recognize the continuities between human beings and other primates since we have shared much of our evolutionary history with them. "But it also requires recognition," he concluded, "that we have been diverging from them for several million years and have acquired some rather unique adaptations of our own."[11] In short, culture was not irrelevant in accounting for human behavior; it just was not the whole explanation.

More recently, sociologist Joseph Lopreato of the University of Texas at Austin has carried much further the emphasis upon what he calls "the climbing maneuver," that is, the tendency of human beings to seek to improve their ability to gain access to mates, resources, and prestige. It is true, he concedes, "that the working of dominance orders are to an extent the result of environmental, including for us, sociocultural factors. But the wide diffusion of the institution [of ranking] in human societies as well as in the animal kingdom leaves little doubt about its biological, evolutionary origin." Indeed, Lopreato considers the presence of a drive to dominance to be the root of class conflict among individuals. He also sees it as the reason egalitarian institutions like the Israeli kibbutz have not spread. It is, he contends, a behavior force "with which egalitarianism must in fact contend hopelessly, at least in the foreseeable future." It is not surprising that he concludes with the pessimistic thought that "people do not really wish equality for all." On the other hand, and with considerably more caution, he does not believe that the predisposition determines "any particular form that a system of institutionalized inequalities may take in given times and places."[12]

Awareness that dominance/submission behavior is present in a number of species is a relatively old story, even if its application to an understanding of human behavior is fairly recent. Another and more widely applied ethological principle—reproductive success—is both more recent and less easily applied to human behavior. On one level, the concept of reproductive success is the basic principle of the modern theory of Darwinian evolution. It argues that the goal or purpose of all behavior is to maximize the survival and reproduction of an individual organism's genetic material, that is, its genes. The working out of the concept varies greatly across and within species. The number of offspring, for example, is not necessarily a proper or adequate measure of success since survival and reproduction may actually be enhanced or furthered by fewer, rather than more offspring, depending on the ecological or environmental situation into which the young are born. A fundamental principle of ethologists and evolutionary biologists is that reproductive success is measured for each individual, not for a group or a species. Individual organisms compete with members of their own species as well as with those from other species, whether predatory or not. As we shall see a little later, this emphasis upon individual survival presents theoretical problems for ethologists—as it did for Darwin. For the moment let us look at the way in which social scientists have drawn upon the concept of reproductive success in trying to account for human behavior.

Reproductive success has figured prominently in the theory of co-evolution of Stanford anthropologist William Durham. As he wrote in 1978, *cultural* features have evolved "to promote the success of an individual human being in his or her natural and sociocultural environment." The measure of that success is the same for biological success, namely, "the extent to which the attribute permits individuals to survive and reproduce and thereby contribute genes to later generations. . . ." In support of his argument, he cited examples of societies that have declined or died out as a result of their maladaptive behaviors, such as a nineteenth-century Russian religious sect that banned sexual intercourse.[13] Durham's point is similar to Jerome Barkow's observation that cultural practices tend to be in line with biological requirements, rather than to work in contrary directions. Durham's principle also opens up a fresh line of inquiry into the origins or purposes of a cultural phenomenon by asking if it may have arisen as a means of enhancing reproductive success.

That precise point was made by psychologist N. G. Blurton Jones in 1983 in the course of a discussion of childrearing among the !Kung people of the Kalahari desert. Once a researcher is cognizant of the significance of reproductive success as a human goal then questions that otherwise might not have been even considered as germane now come to the fore, such as the back load borne by the mother, the spacing of children, and so forth. To think in this evolutionary manner, Blurton Jones contends, "differs radically . . . from most approaches to human population regulation." One soon finds that people act to leave the largest number of descendants given the social and economic conditions in which they live, which may mean that *fewer* babies actually result in *more* descendants who reproduce. An additional advantage in searching for the relation between reproductive success and cultural practices is that one often discovers an explanation for behavior that otherwise "leaves us wondering where people's ideas come from." For if people are not influenced "by the necessity of survival and the consequences of reproduction," he wondered, "what are they influenced by?"[14]

Blurton Jones's text does not make clear if he is seeing reproductive success as a principle followed consciously by the Kalahari people or one that has been worked out over time in a dimly conscious or perhaps even wholly unconscious manner. Joseph Shepher, a sociologist, left no doubt that he thought the drive for reproductive success among human beings is largely unconscious or innate. Drawing on sociobiological theory, he suggested, for example, that one can explain the relative rarity of polyandrous societies in the world by reference to the principle of reproductive success. There is very little advantage, Shepher argued, for a woman to have more than one mate since her reproductive capacity is not enhanced at all by such a practice. On the other hand, a man expands his reproductive capacity by having more than one wife, hence the large proportion of polygynist societies among the hundreds of social orders identified and studied by anthropologists.[15]

Even critics of the use of evolutionary theory and biology in accounting for human behavior have, at times, given grudging acknowledgment of the usefulness of reproductive success as a principle in accounting for some aspects of human behavior. Earlier students of human nature who looked to biological explanations lacked such a principle, sociologist Kenneth Bock pointed out in 1980 in his critique of sociobiology, yet the principle has a plausibility that is "striking and persuasive." But, he cautioned, in the absence of hard evidence linking human behavior and a concern for reproductive success, he thought the connection could be viewed as no more than speculation.[16] But that is precisely the meaning of the principle to Shepher, who recognized that there are other ways in which the predominance of polygynous societies might be explained. Most proponents of the use of evolutionary theory, however, think that it provides a plausible as well a fresh, though not necessarily a conclusive explanation. It might be said to be of the same order of explanation as provided by Freudian psychoanalysis or Marxism, that is, a theory that accounts for a broad range of social behavior though without the possibility of experimental or incontrovertible proof. After all, definitive proof in accounting for human behavior is extremely difficult to establish; the best that can be expected are plausible explanations derived from broad theories of human behavior. Among such theories, neo-Darwinian evolution and ethology have the signal advantage of being highly respected and powerful explanations for animal behavior, a fact that surely enhances their value to an understanding of at least some of the behavior of *Homo sapiens,* who also is an animal.

Reproductive success is a fundamental concept of ethology, but the real theoretical breakthrough came with a concept that has come to be called "kin selection" or "inclusive fitness." Only through it has one of the most difficult puzzles of Darwinian theory been solved. As noted earlier, reproductive success assumes that each individual organism strives against all others for that goal. But if that is the case, then how does one explain the cooperative behavior or "mutual aid," as the late nineteenth-century political thinker Prince Peter Kropotkin called it, that we observe among many species of animals? It is well known, for example, that in some herd or social animals one member or another will risk danger from a predator in order to warn the group or distract the predator. Obviously, such risky behavior can hardly be explained as single-minded pursuit of success in reproduction. Dead animals do not increase their chances to reproduce.

Perhaps the most striking example of such behavior, which has been denominated "altruistic," is found among certain social insects: ants, wasps, and bees. Among them the impairment of reproductive success seems total in that the workers are sterile females, having lost entirely the ability to reproduce. In the context of the individual drive to reproduce, such a situation amounts to the height of self-sacrifice. Indeed, this practice of eusocial insects was so threatening to Darwin's original theory of natural selection that he spent several pages in *The Origin of Species* seeking an explanation that could reconcile that form of behavior with his theory, which depended upon an individual organism's struggle against all other organisms for reproductive success. The solution of the problem was so central to Darwin's theory of natural selection, according to historian of science Robert Richards, that Darwin delayed the publication of the *Origin* for almost twenty years while seeking an answer to the puzzle.[17] Considering that Mendelian genetics was then unknown, it is remarkable that Darwin's own solution came as close as it did to the modern solution, which was provided only in the 1960s by zoologist William D. Hamilton.

Hamilton's theory of kin selection, or "inclusive fitness," as it is sometimes also called, postulated that individual organisms supported other organisms in direct proportion to their relatedness. The concept rested on a fundamental principle of genetics, namely, that offspring share genes with their relatives in varying degrees. Among human beings, for example, brothers and sisters share, on the average, half the genes of their parents. Uncles and nephews share three-eighths, and cousins share one-eighth. It was this distribution of degrees of relatedness that stood behind the only half facetious response of the English geneticist J. B. S. Haldane when asked for whom he would sacrifice his life. "For three brothers or nine cousins," he is reputed to have replied. In that calculus, his genetic material would have had a better chance of being carried forward to reproduce than if he himself had lived.

Kin selection has been highly productive in accounting for the behavior of many animal species, especially in regard to the paradox already mentioned of sterile female workers among the social insects. Among ants and bees (though not among human beings) a female receives all the genes of her father and, on the average, half those of her mother. Genetically, this means that if an individual female ant works to support her sisters, as she in fact does, she is, on the average, working to help individuals who share three-quarters of her genes. On the other hand, if she laid eggs herself, only half of her genes would go to the offspring, the other half being those of the father of the offspring. In short, rather than having given up reproductive opportunities by becoming infertile, the female workers have actually enhanced the likelihood that their genes will be passed on to offspring.

The process has been expressed here in our language of consciousness, but the source of the behavior among the social insects is not conscious. Rather, it is the result of natural selection; the theory being that the "ant" mode of reproduction has proved to be more advantageous in the circumstances than individual reproduction. Ethological studies among a number of mammal species have shown that relatedness does offer an explanation for what otherwise seems "altruistic" or selfless behavior. George Schaller, in his study of lions in the wild, for example, has reported that females in a pride are usually all related; they are sisters, mothers, and even grandmothers, hence their cooperation in the kills. A study of the behavior of a species of ground squirrels found that cooperation between certain females and the lack of it with others can be explained by kinship. Sisters will be fed; non-related females will be chased off. A student of female animal behavior concluded that with few exceptions "cooperation among females in the animal world depends on kinship ties. Females will help their relatives where they would not help strangers."[18]

Although the behavior which kin selection explains is called "altruism" by ethologists, a full understanding of the way it works makes clear that the term does not carry the meaning it usually holds for human beings. To us altruism means sustaining a disadvantage in behalf of another. Kin selection is nothing more than what in human affairs is called "enlightened self-interest," since the individual organism that appears to be sacrificing itself for another is actually gaining an advantage through that behavior.

The theory of kin selection has had considerable influence on the thinking of social scientists interested in accounting for human behavior. Cornell University anthropologist Davydd Greenwood, for example, has called the idea a major breakthrough in thinking about social behavior, even though he also recognizes that the concept is not easy to use in actual social science research. Rather than seeing its application to human beings as "a diabolical plot," as some critics apparently have, Greenwood advises social scientists to recognize that if the concept is "relevant to at least some social species it is reasonable to entertain possible applications to humans." This is true even if, as he admits, some applications that have been set forth so far "cannot be taken seriously."[19]

Two other young anthropologists, Peter Richerson and Robert Boyd, take a more positive position in defense of the use of kin selection theory in explaining human behavior. They recognize that a purely genetic approach to kinship cannot account for the patterns found in some societies where a male will show favoritism toward offspring of a brother over the children of a sister (though the genetic connection is the same). Or in a situation where a sister's child is deemed "closer" than his own child. They recognize that kinship can be highly diverse across cultures and where it is not obviously derived from biology it should be considered cultural in origin. But, they add, all kinship patterns derive in the end from a foundation of genetic relations. For if they are not so derived, they ask, why do "humans almost invariably organize their societies around altruism between genetic kin?" If genes play no role through natural selection, why do people care about kinship at all, why do they develop a system of relatives? Where does the idea of relation come from if not from biology? Moreover, are not genes the ultimate explanation as to why parents rear their children? "If the possibility that such bonds are genetic is excluded," Richerson and Boyd contend, "it is difficult to explain their universality." Even a biologist as opposed to sociobiology as the Harvard geneticist Richard Lewontin concedes that "kin selection has operated in some instances to establish some traits of organisms. It is undoubtedly true that human behavior like human anatomy is not impervious to natural selection and that some aspects of human social existence owe their historical manifestations to limitations and initial conditions placed upon them by our evolutionary history."[20]

One of the earliest and most detailed criticisms of kin selection as a shaper of human social behavior was mounted by the eminent anthropologist Marshall Sahlins in his book *Use and Abuse of Biology,* published in 1976. The principal objection was one already mentioned by Richerson and Boyd, namely, that many kinship patterns which anthropologists have studied do not conform to the pattern predicted by kin selection theory. Genetically close relatives are not always deemed "close" by the kinship rules of a given society. Fellow anthropologist Martin Etter forcefully responded to Sahlins's objection. According to kin selection theory, there should be, on the average, a correspondence between genetic closeness and the likelihood of an altruistic act. "All kinship terminologies of which I am aware," Etter contended, "generally satisfy this . . . expectation." He concluded, along with Richerson and Boyd, that "what is remarkable is that kinship—this paramount human system for behavioral orientation—has any relation whatever to genealogical categories. I believe it can be fairly said that no purely anthropological theory, but only biological theory, has ever suggested why this should be so."

Anthropologists Davydd Greenwood and William A. Stini made a similar point in general terms when they noted that the complexity of human social orders results from the unique human capacity for culture. Unfortunately, they quickly added, this culturally induced complexity "has led some anthropologists [read Marshall Sahlins] to abandon the study of human biology as irrelevant to the study of kinship and marriage." No error could be greater, they insisted. "Kinship and marriage alliances are built up over a common primate biological heritage in which sexual differences, the long infancy and socialization of human children, and the maintenance of genetic variability within the population are fundamental."[21]

Sociologist Pierre van den Berghe and psychologist David Barash corroborated the argument even more directly by reminding us that in "practically all societies, people know that they have two parents, four grandparents, and so on, and behave preferentially toward not only the members of their own clan and lineage, but also toward other relatives belonging to different lineages." Or as van den Berghe expressed the point in another place, "Why do human societies organize so much around kinship and marriage?"[22]

All these commentators were simply contending that genetic relations are so important, so fundamental, that they are taken for granted in identifying connections among human beings. Since kinship is biological in origin and concept, to emphasize those patterns of kinship that are not biological, as some cultural anthropologists like Sahlins do, is to ignore the fundamental in favor of the exceptional.

Sociologist Joseph Lopreato illustrated this tendency in his analysis of an anthropological example advanced by a vigorous opponent of sociobiology, Stephen Jay Gould, the brilliant Harvard University paleontologist and historian of science. The example concerns a practice among an Eskimo tribe in which, during times of food shortages, the family groups must migrate. The grandparents offer to stay behind (and die) in order to relieve the group of the burden they would be in making a long and difficult migration. Gould admitted that a genetic explanation was as plausible as a cultural one, but that since neither had any explanatory advantage over the other, he preferred to opt for the cultural one. Lopreato pointed out, however, that although both the cultural and biological explanations were adaptive—"Families with no tradition of sacrifice do not survive for many generations," Gould himself remarked—only the biological would offer an explanation or theory, namely reproductive success, to account for it. Even the cultural explanation—the good will of the oldsters— as Gould's comment showed, depended upon reproductive success as a motive. Lopreato pressed his point by adding that one ought to expect culture and biology to work toward the same ends, for culture, which depends on biological roots, would thus also be driven by a need for reproductive success.[23]

Anthropologist Donald Symons suggested how kin selection theory helped to identify a novel interpretation of kinship. An anthropologist named Hanes discovered in 1979 among the Ye'Kwann Indians in Venezuela that the "degree of genetic relatedness is a good predictor of the frequency of interaction between individuals, a much better predictor," Symons emphasized, "than Ye'Kwann kin terms. Although evolutionary theory in no sense 'predicted' this finding," Symons noted, "it would never have been discovered had Hanes not been inspired to analyze his data" according to Hamilton's kin selection theory.[24]

For many social scientists who have been attracted to biological explanations for human behavior, kin selection theory has a glaring weakness. The theory is certainly helpful in explaining why human beings are supportive of people related to them, but it does not account for behavior that is supportive of nonrelatives, or what we usually refer to as truly altruistic or selfless behavior. Yet that kind of behavior is hardly rare in human societies. In an effort to meet that objection, sociobiologist Robert Trivers has introduced another form of altruistic behavior, to which he has given the name "reciprocal altruism." As the name suggests, the behavior pattern is one in which an individual is supportive of a non-relative on the assumption that at a future time that non-relative will reciprocate.

Two things are worth noticing about this form of so-called altruistic behavior. First of all, it is no more selfless than kin selection; it explains apparently selfless behavior by identifying expectations of a return in the future, which is to say, the action is not selfless as we would usually define that term. The second point is that though there is no obvious biological mechanism fostering the behavior (the individuals are not related, as they are in kin selection), Trivers's assumption is that the behavior has been selected for in the course of evolution because it advanced reproductive success.

At least until recently, ethological evidence in support of Trivers's assumption has been thin, especially when one tries to imagine how animals are programmed to distinguish between those animals who are likely to reciprocate and those who are "deadbeats," that is, will take a favor but not return it. There is some evidence that monkeys in grooming one another do draw distinctions between those who reciprocate and those who do not, but that may or may not be genetically based.

Among primatologists, however, Jane Goodall, for one, is convinced of the validity of reciprocal altruism. "Chimpanzees can readily remember who helped them (as well as who sided against them) from one occasion to the next, even when the occasions are separated by long intervals," she reports. Sometimes, she added, failure to reciprocate was punished. "Thus, the old adage 'one good turn deserves another,'" she remarked, "is, without doubt, deeply rooted in our primate heritage." She was even convinced that altruism toward kin, over the course of time, would be extended to non-kin. In higher animals, like the chimpanzees she studied, "attitudes toward kin are shaped, to a large extent, by the degree of familiarity. It is very logical, therefore, that helping behaviors will on occasion be extended to familiar individuals even when they are *not* close kin. . . . Because altruistic behavior toward near kin is dependent, genetically, upon firmly established helping behaviors among close kin," she suggested, "we do not have to argue a separate evolutionary mechanism for its selection."[25]

Curiously enough, Goodall's defense of reciprocal altruism has received important support from a social scientist. Political scientist Robert Axelrod of the University of Michigan has not drawn upon evolutionary theory so much as he has contributed to it. In his imaginative book *The Evolution of Cooperation* (1984), he has shown how the application of the theory of games (the Prisoners' Dilemma) explains how cooperation can evolve between individuals even when the participants are neither relatives nor conscious of the game, that is, how reciprocal altruism could develop among animals or human beings. His intent was to demonstrate how in political or economic rivalry a particular strategy of reactions to another's behavior, which he called Tit for Tat, could evolve into a stable situation of cooperation. Evolutionary biologists have found his analysis particularly valuable because in Darwinian theory individuals within a species compete for survival one against the others, yet they also cooperate with each other. The origin of that cooperation, Axelrod argues, lies in inclusive fitness among related individuals, but in time, through the strategy of Tit for Tat, the cooperation moves beyond that to a reciprocal altruism that no longer depends on inclusive fitness, or relatives.[26]

Despite the enthusiasm of Goodall and Axelrod's support, social scientists, even those otherwise favorably disposed toward biological explanations for human behavior, have been less supportive of Trivers's theory. They tend to consider reciprocal altruism as "problematic," as two sociologists phrased it, dismissing Trivers's idea as being "of little use because, conceptually, it does not seem to be distinct from more general notions of sociality."[27]

As many of the examples already discussed show, the question of why human beings are social in their behavior drew some social scientists to evolutionary theory even before the publication of Edward Wilson's provocative book *Sociobiology*. One such pioneering social scientist was the psychologist Donald Campbell. In his 1975 presidential address to his fellow psychologists, Campbell told of his interest in the recent literature on evolution, which argued that competition among individuals, including human beings, was the driving force of evolution. That literature, he correctly pointed out, asserted that no behavior was directed toward helping the group or species. Yet he also noticed, and therein he identified the central problem: among vertebrates, human beings have a higher degree of sociality and altruism than any other species. Only the social insects, Campbell observed, exceeded the social cohesion or cooperation of human societies, an achievement that suggested to him "that some complex forms can be achieved on a genetic basis." But the additional fact that intelligent animals like "wolves and chimpanzees have never achieved it" he thought resulted from the "evolutionary trap or conflict produced by genetic competition among the cooperators. Man," he warned, "is in the same genetic predicament."

Campbell either did not know of kin selection or chose to ignore it even though he clearly recognized the problem it sought to resolve. To Campbell, sociality derived from culture, which he clearly saw as working in the opposite direction from biologically rooted competition between individuals. Three years later in an elaboration of his thesis his basic argument remained unchanged: only cultural forces like religion contained or restrained the otherwise selfish proclivities of human beings. In his presidential address earlier he had gone so far as to suggest that his own discipline of psychology in teaching self-interest may be unwittingly undermining a needed "social-evolutionary inhibitory" system.[28] Although there is no doubt that Campbell was in the forefront in drawing on evolutionary theory in psychology, his views differ sharply from more recent social scientists who have been proponents of evolutionary theory; they generally see cultural and biological evolution working in the same rather than opposing directions. For them, as for Darwin, sociality is no less grounded in biology than is competition. For that reason alone, Campbell's denial of a biological basis for cooperation has failed to attract supporters.

By recognizing the biological roots of human sociality or social behavior, some students of human society have found a basis for a human nature, that is, a nature that underpins cultures; anthropologists refer to such characteristics as "universals." "For far too long, sociologists have left the question of human nature to 'social philosophy,'" complained sociologist Pierre van den Berghe. Some have even denied there was any such thing as human nature given the great diversity of cultures in the world. For such thinkers, van den Berghe continued, "to all intents and purposes, man was man-made," created, as it were, by culture. But as philosopher Mary Midgley pointed out, "if we were genuinely plastic and indeterminate at birth, there could be no reason why society should not stamp us into any shape that might suit it." Acceptance of a human nature born within us, Midgley rightly maintained, does not threaten freedom as some have feared. On the contrary, "the early architects of our current notion of freedom made human nature their cornerstone." Echoing Roger Masters's observation, she noted that "Rousseau's trumpet call, 'Man is born free, but everywhere he is in chains,' makes sense only as a description of our innate constitution as something positive, already determined, and conflicting with what society does to us." Anthropologist Robin Fox acknowledged an innate human nature that "is intensely social" and which, by its very innateness, "gives the lie to all sanctimonious manipulators from Mill through Stalin."[29]

Economists who have drawn upon sociobiological ideas have not been concerned with human nature as such, but they have certainly been more in tune with modern ethological theory than psychologist Campbell seems to have been. (That may not be surprising since it was classical economist Thomas Malthus whose demographic work gave Charles Darwin the clue that led to his theory of natural selection through competition!) Jack Hirschliefer of the University of California at Los Angeles, for example, has reported that a number of his fellow economists have been struck by the similarity between the behavior of business firms and the behavior patterns that ethologists study and theorize about. The economists have shown that firms act "as if" they are maximizing profits, just as animals act "as if" they are seeking reproductive success. From that analogy Hirschliefer concluded that biological theory might be applied to the behavior of such a "highly specialized and consciously contrived 'cultural' grouping," for the principles of evolutionary theory "all play roles in explaining observed patterns of survivorship and activity." His contention is that "the social processes studied by economics, or rather by the social sciences collectively, are not mere analogs but are rather *instances* of sociobiological mechanisms."*

An imaginative group of economists designed an experiment that, in effect, tested Hirschliefer's conception. "Ever since Darwin," the economists pointed out, "it has been widely recognized that behavior and structure vary continuously across species and that behavioral principles do not stop suddenly at the boundary separating humans from other animals." Thus, if it can be established, they continued, "that analogous economic processes exist across species, our chances of finding out more about these processes are vastly increased by experimenting with laboratory animals." So they designed an experiment to see if rats would respond to price and budgets as economic theory said human beings did.

They set up a cage in which Collins mix and root beer were made available to the rats when they pressed a lever. By setting a limit of 300 presses being available, a "budget" was created. By varying the size of the cup from which the liquid was drunk they set a "price." They then found that by enlarging the cup (that is, lowering the price), but keeping the budget the same, the consumption increased. When they kept the budget steady, but lowered the price for Collins mix as against root beer, the rats' consumption moved strongly to Collins mix. To avoid possible associations with certain levers, they shifted the Collins mix at a lower price to a different lever, but the lower price invariably attracted the rats. To control for possible preference for one liquid over the other, despite price, they put root beer in both cups but had one cup larger than the other. Virtually all of the pressings were for the larger cup. When the cups were shifted from one lever to another, the rats shifted their preferences almost immediately. The economists thought the experiment opened up opportunities for research since the controls and monitoring could be more closely achieved

---

*Economist Gary Becker was similarly impressed by the congruency between sociobiological theory and economics. The new sociobiology, he wrote in 1976, is congenial to economists "since both rely on competition, allocation and limited resources—food and energy—efficient adaptation to the environment." But sociobiology does not use models with rational actors, as he thinks economics does. Gary S. Becker, "Altruism, Egoism, and Genetic Fitness: Economics and Sociobiology," *Journal of Economic Literature* 14 (1976), 818.

than with human subjects.[30] Whether there was any follow-up on this experiment is not known.

Hirschliefer, in addition to transporting theories of competition from evolutionary biology to economics, envisions a use for sociobiological theory since it can account for "tastes," something which economists either dismiss as trivial or deem incapable of being accounted for. Reproductive success, Hirschliefer stressed, is a basic biological principle that economists might look to in trying to understand why certain human tastes exist. For example, economists usually assume, as psychologist Campbell did, too, that human individuals are ultimately self-concerned. Yet economists also recognize that these same individuals have "tastes," to use the economist's jargon, for family and relatives, tastes that belie a self-centered conception of human nature. Such altruism, he concluded, is not simply a "taste" if one looks to sociobiological principles. It can, instead, be identified as rooted in biological evolution. "That a parent is more benevolent to his own child than to a stranger's," Hirschliefer thought "is surely capable of explanation." Moreover, he continued, "what is true for the specific 'taste for altruism' holds in considerable degree for preferences in general—that these are not arbitrary or accidental, but rather the resultants of systematic evolutionary processes." He was careful to add that such a conclusion did not preclude changes in preference in the future. "On the contrary, the inbuilt drives themselves contain the capability of expressing themselves in diverse ways depending upon environmental circumstances, which will in turn be modified by cultural evolution." It was not, he explained, that human preferences are "biologically determined in any complete way—but rather, that they are scientifically analyzable and even in principle predictable . . . [from] past genetic and cultural adaptations."[31]

Hirschliefer's conviction that sociobiology could help economics did not wane with the years. In 1987 he thought the extension of economic principles to "study the human interactions involved in phenomena like mate choice and crime and charity and politics" would be advanced by drawing on sociobiology. "Interaction via market exchange under the rules of the game . . . called *political economy* is only a part, often a small part, of the economic picture. Not only plants and animals, but human beings as well," he contended, "interact economically to a very large degree under *natural* economy rather than political economy, without benefit of law or property or contract. Economists," he warned, "have been studying only a chapter of the book of economic life. By following in the direction pointed out by sociobiology, we will be able for the first time to take cognizance of the book as a whole."[32]

Political scientist Roger Masters was especially impressed by the usefulness of kin selection in his field. Although he doubted that inclusive fitness theory resolved the question of whether human beings are basically selfish or altruistic, it could explain altruism toward relatives as the theories of certain political thinkers like the Sophists and pre-Socratics, who depicted human beings as fundamentally self-interested, could not. He quoted from Thrasymachus's speech in Plato's *Republic* in which an egoistic basis of behavior was set forth: "Men draw life from the things that are advantageous to them; they incur death from the things that are disadvantageous to them," he quoted Thrasymachus as saying. To the Greek, altruism seemed "inimical to nature." But, like Hirschliefer, Masters recognized that human beings more than occasionally showed themselves to be truly unselfish. A large part of that genuine altruism, Masters asserted, was likely to be cultural in origin, rather than biological. At the same time, he called attention to that tradition of political and social theory to which Aristotle, Plato, Hegel, and Marx belonged, in which social behavior rested not on self-interest, as it did with Hobbes, Locke, and Rousseau, but upon an innate sociality. "Inclusive fitness theory," Masters concluded, "thus shares a formal characteristic with one of the two major traditions in western social and political thought."[33] Or, as he would phrase it some years later, economics emphasizes the competition; sociology the whole. Both need to be taken seriously, if only because they are both in our biology. Aristotle, he thought, transcended the dichotomy when he denominated man as a "political animal." A careful reading of Aristotle's *Politics,* Masters was convinced, shows that the philosopher's views are "strikingly consistent with modern biology."[34]

In light of these social scientists' remarks, it comes as no surprise that kin selection theory has also been suggested as a means of accounting for the widespread incidence of nepotism throughout history. "Favoritism toward real or putative kin," wrote sociologist Pierre van den Berghe and psychologist David Barash in 1977, "has been observed in practically all societies, and most societies seem to take it for granted. . . . Nepotism triumphs in the end," they concluded, "and most societies have been realistic enough not to try to stamp it out."[35]

Political scientists, as might be anticipated, have been stimulated by the possibility that kin selection theory offers novel insights into the origins and development of political and administrative organizations that compete with family loyalties. Stephen Balch, for example, was impressed by the light that sociobiological theory might throw on the problem of institutional or bureaucratic loyalty. He took it as axiomatic (as well as being sociobiologically sound) that the lives of most people center around their families and that the cohesion and integrity of all organizations are likely to be "constantly imperilled," as he phrased it, by a generally accorded priority to family. To validate his contention, he surveyed the strenuous and persistent efforts made in the Chinese, Turkish, and Byzantine empires to reduce or weaken that primary loyalty to family in the course of developing bureaucracies. One of the advantages of eunuchs, which all three of these empires employed, he emphasized, was that they lacked progeny, whose interest could compete with that of the ruler's house. The extremes to which these empires went in seeking to counter familial loyalty, he suggested, measured the importance of the problem, "while the wide historical provenance nicely displays its rather universal character."[36]

Legal scholar John Beckstrom has suggested several ways in which the sociobiological theory of kin selection might be helpful to lawyers. He observes that in recent years empirical studies have consistently shown that the risk of child abuse is higher in homes with step-parents than in homes with biological parents. Yet there has been no satisfactory explanation for the finding, though, in Beckstrom's eyes, that sociobiological explanation is obvious: a parent tends to favor his or her own biological children over those who are "genetic strangers."[37]

Kin selection, or the principle that the degree of aid given to another person is generally commensurate with the degree of genetic relationship, is also relevant, Beckstrom suggests, in regard to the law of intestates. Most people die without a will and courts (and lawmakers) are charged with making rules to distribute their assets as closely as possible to what the deceased would want. The theory and empirical data behind such decisions, Beckstrom notes, are very weak and are usually derived from evidence taken from those who die with wills, a quite different population. He suggested that lawmakers and judges might look into the degree of genetic relation as a principle to apply in such cases. If they do, he notices parenthetically, they will find a surprising degree of agreement with present practice.[38]

Finally, Beckstrom observes that in 1985 eleven states and the District of Columbia prohibit marriage between people related only by marriage. Whatever purpose those prohibitions were once expected to serve, genetic danger could not have been one of them since, by definition, the relation of the parties is legal, not genetic. Furthermore, Beckstrom continued, sociobiological theory suggests that such barriers to marriage run counter to the general social policy of favoring measures that advance the welfare of children. Since the sister of a deceased mother of small children shares one-quarter of the genes (the same as a grandparent) of those children, for her to marry her brother-in-law would give her a decided genetic interest in the care of his children, a fact that would make her an especially desirable step-mother. Thus it would seem good social policy to repeal prohibitions against such marriages.[39]

If, on the one hand, sociobiological theory emphasizes the likelihood of favorable behavior toward kin, on the other hand, it sets a limit on that behavior because of its fundamental principle that natural selection works through individuals, not groups. Even families are not exempt from the rule. Among mammals mothers clearly subordinate their short-term interest to take care of their young, for, according to ethological theory, that behavior serves the mother's long-term interest in ensuring her reproductive success. The mother, nevertheless, will nurse only so long, depending on the environmental circumstances, for an excessively lengthy nursing period will interfere with the proper care of the new offspring that inevitably come along. In sum, a time arrives in the life of the mother/offspring relation when a conflict of individual interest arises between them.

Robert Trivers, the originator of the principle of reciprocal altruism, has systematically explored this idea of mother/offspring conflict among animals. Some social scientists have found it a useful concept even if they reject the application of other aspects of ethological theory to human behavior. One of these critics, political scientist Joseph Losco, concedes that Trivers's view of child/parent conflict may well serve as a "guide to the study of human behavior by generating new lines of research and by acting as a guide against which some . . . explanations can be assessed." Child psychologist Peter J. Smith in practice found sociobiological predictions concerning conflicts over weaning useful in his work among the Kalahari San of southern Africa. Child care occupies a good deal of attention there as it does among western parents, though in both societies cultural influences can interfere with what can be objectively viewed as the optimum care. He saw wet-nursing in the West, for example, as just such a negative cultural influence since in practice chances of survival were generally reduced under such a regimen.[40]

Even a historian of child rearing has found sociobiological theory useful. Linda Pollack, a historian of child care in the United States and Britain from the sixteenth to the end of the nineteenth century, noted that "despite the enormous differences among cultures, and although there are obviously cultural influences on child care, a good deal of similarity has been found in child-rearing practices." She drew upon sociobiological studies on the rearings of offspring among animals to justify her controversial conclusion. Since "good parental care is adaptive," she thought it likely that human parents would be inclined to provide for their children, contrary to the view of many historians that before the nineteenth century indifferent care was the common practice among parents. "Sociobiological theory predicts," she noted, "that animals should invest in reproduction as much as possible consistent with the maximisation of inclusive fitness." And though human behavior is not to be viewed as genetically determined, human beings "are at least genetically influenced," and thus are likely to behave toward their offspring as animals do, that is, to strive for reproductive success.[41]

Reproduction among mammals means sex, to be sure, but it also means that each of the sexes involved play quite different roles in bringing reproduction about. One member of the pair bears the brunt of the job, carrying the offspring within its body and then supplying the food for the first phase of life. That difference between the sexes, which is as characteristic of other mammals as it is of human beings, quickly leads to a consideration of how that difference in the reproductive process affects behavior. Even with the triumph of culture, that difference was never entirely removed from scholarly consciousness however much it may have been played down or pushed aside. With the revival in the 1970s of social scientists' interest in ethology and Darwinian evolutionary theory, the old question resurfaced once again, sometimes with old responses, but also with some novel ones, as will appear in the next chapter.

12

Biology and the Nature
of Females

Many naturalists doubt, or deny, that female animals ever exert any choice,
so as to select certain males in preferance to others. It would, however, to be
more correct to speak of the females being excited or attracted in an especial
degree by the appearance, voice, etc., of certain males, rather than of deliberately selecting them.

Charles Darwin, 1882

Given the indisputable physiological fact that women differ from men, it was
almost inevitable that, once social scientists began to re-examine the place of
biological influences in human action, they would again be alert to the ways in
which those physiological differences might affect behavior. That second look
produced an effect upon the well-known feminist and sociologist Alice Rossi
which nicely measures the impact that biological ideas have had upon social
scientists.

In 1965 Rossi published a powerful defense of equal rights for women in
Daedalus, the journal of the American Academy of Arts and Sciences. The
underlying assumption of the article was that, though women and men were
different physiologically, socially they were interchangeable. What a man could
do a woman could do if only the social order did not raise barriers against her
intentions. Twelve years later, she published a second article in *Daedalus;* this
time the emphasis was on the differences between men and women. The title
of her second article captured the new place biology had obtained in her thinking about women: "A Biosocial Perspective on Parenting." Her point was that
men and women differed biologically in their ability to be parents. She frankly
acknowledged her reluctance to adopt the new posture. As a feminist, she
admitted, her outlook changed only "after a long period of personal and intellectual concern," in the course of which she had to confront her "own previous
analysis of sexual equality" in her 1965 article. In 1977 she found that earlier
view "wanting." Resolute in continuing to adhere to her feminist outlook, she
argued that men and women do not have to do the same things or be the same
"in order to be socially, economically, and politically equal." She regretted that

the current tendency among many feminists was to "confuse equality with identity and diversity with inequality." That tendency induced her to remind them that "diversity is a biological fact, while equality is a political, ethical, and social concept." A biological approach, or what she preferred to call "a biosocial perspective[,] does not argue that there is a genetic determination of what men can do compared to women," she insisted. "It suggests that the biological contributions shape what is learned, and that there are differences in the ease with which the sexes can learn certain things." (Rossi, it is only fair to report, explicitly dissociated her views from Edward Wilson's sociobiology.)[1]

Seven years later, in her presidential address to the American Association of Sociologists, Rossi explained at length why she had arrived at her biosocial point of view. Present sociological theories of families she had concluded were inadequate. This was true, she explained, whether exchange theory, general systems, conflict, feminist, or developmental theories were used, because none of them seeks "an integration of biological and social constructs." No one would think of ignoring biology in discussing aging or geriatrics, she reminded her audience, "but research on gender has studiously avoided efforts in this direction." Yet the differences between men and women, she continued, "are not simply a function of socialization, capitalist production, or patriarchy." They are grounded in a diversity between the sexes "that serves the fundamental purpose of reproducing the species. Hence sociological units of analysis such as roles, groups, networks, and classes divert attention from the fact that the subjects of our work are male and female animals with genes, glands, bone and flesh occupying an ecological niche of a particular kind in a tiny fragment of time." Those differences between the sexes, she emphasized, "emerged from the long pre-history of mammalian and primate evolution." To neglect those differences shaped by time and natural selection, she warned her fellow sociologists, was to run "a high risk of eventual irrelevance."

Sex, Rossi carefully pointed out, differs from other social categories of analysis such as race. "The attributes of mothering and fathering," she affirmed, "are inherent parts of sex differentiation that paves the way to reproduction. This is where the sociological analogy so often drawn between race and sex breaks down in the most fundamental sense. Genetic assimilation is possible through interracial mating," she recognized, "and we can envisage a society that is color-blind. But genetic assimilation of male and female is impossible, and no society will be sex-blind. Except for a small minority, awareness of and attraction to differences between male and females are essential features of the species," she contended.[2]

Women differ from men in another way: they, she noted in her *Daedalus* article, have been concentrated in child care in all societies, while men have not.

(A point also emphasized by Margaret Mead ten years before.) Evolution suggests why this division of labor prevailed: "The female is more closely involved in the reproductive process than the male. Biologically males have only one innate orientation, a sexual one that draws them to women, while women have two such orientations, a sexual one toward men and a reproductive one toward the young." Fathering is socially learned, she concluded, since such behavior is "often non-existent among other primates." Bonding between mother and child she insisted is even more important among human beings than among animals because evolution has caused babies to be born "early" thanks to the narrowness of the human female pelvic girdle relative to the size of an infant's head.

Rossi was careful to distinguish her conception of mother-child bonding from what she considered a mere cultural phenomenon dependent upon the mother's behavior immediately after a birth. So-called "bonding" at birth she thought not significant on evolutionary grounds. Evolutionary theory suggests that "it is highly unlikely that small variations in early contact could be critical to human attachment to infants." In a complex organism like a human being, the fixing of such a bond is hardly "to be dependent on a brief period or specific experience following childbirth." True bonding between mother and child, she argued, took place "long before a first pregnancy," that is, through evolution. Although she acknowledged a great variety of social forms into which human societies organized themselves, nevertheless "no society replaces the mother as the primary infant-tender except in cases of small and special categories of women." On the other hand, she made clear that she was *not* arguing for a "maternal instinct" in women, but rather for a pattern of behavior that was triggered by pregnancy.[3]

Rossi was quite willing to live with the social implications of her conviction that women were better fitted by evolution to be parents than men. "If society wishes to create shared parental roles," she warned, "it must either accept the high probability that the mother-infant relationship will continue to have greater emotional depth than the father-infant relationship, or institutionalize the means for providing men with compensatory exposure and training in infant and child care in order to close the gap produced by the physiological experience of pregnancy, birth, and nursing." Persistently she pressed home the point that, unless "biosocial facts are confronted, allowed for, and, if desired, compensated for, the current cultural drive toward sexual equality in marriage and the workplace and shared child-rearing may show the same episodic history that so many social experiments have demonstrated in the past." As she pointed out in her 1984 presentation to her fellow sociologists, studies of "solo fathering" and "solo mothering" show traditional differences between the sexes in parenting behavior, all of which suggested to her that "there is more involved

than a need to unlearn old habits and learn new ones specific for parenting."
Changes in child-care practices in Israeli kibbutzim, where mothers now want
their children at home at night, also suggested to her that something more than
socialization lies at the root of differential parenting behavior.

Other scholars have been taking similar positions on the question. The
English feminist and philosopher Janet Radcliffe Richards also looked to biological evolution for the tendency that women are generally the prime rearers
of children. "There is no need to presume," Richards maintained, "that this
feeling is entirely culturally induced; it is hard to see how evolution could have
produced a female mammal without a strong inbuilt disposition to care for its
young." American sociologist Allan Mazur drew explicitly upon Rossi's work
in asserting the cross-cultural similarities in women's role in child care. He
thought "many of the nonverbal aspects of 'human' mother-infant interaction
have evolved along with the physical character of the species."[4]

Other social scientists placed different emphases upon biology's role from
Rossi's, but they did not doubt that biology had a part in shaping women's
nature. Anthropologist Melvin Konner, for example, was not as sure as Rossi
that women were naturally better child carers than men, though he did not rule
it out. On the other hand, psychologist Daniel Freedman of the University of
Chicago went considerably beyond Rossi in suggesting that genetic influences
on child rearing were sufficiently strong to make a difference between natural
and adoptive parents' care of children. His argument depended on the sociobiological principle that natural children carried the parents' genes and adopted
children did not. "For although a parent may be scrupulously fair with his
adopted children or stepchildren," Freedman asserted on the basis of his ongoing work, "he or she may not be as all-forgiving or as ready to step beyond
the 'call of duty' as with blood children, which, in turn, would differentially
affect a child's self-concept."[5]

Freedman is a confirmed believer in the application of sociobiological principles to the study of human behavior, so it is not surprising that he should
conclude that studies of adopted children also show that they seek to locate or
know their biological parents, for that implies the importance of genes or
genetic connections in human beings. Yet it is difficult to know how such an
urge, which is familiar enough in everyday life, would be explicable by sociobiological theory. What biological gain would a grown child or adolescent
obtain from being reunited with its biological parents? According to sociological
theory, it is the *parents* who are interested in passing on their genes; children
have no interest in their parents' genes. Insofar as such urges do manifest themselves in adopted children, they would be better seen as cultural rather than

biological in origin.* At the same time, the undeniable presence of such an interest among adopted children provides a good measure of the enduring role that biological relatedness plays, for whatever reason, in the thinking of human beings. Most people, after all, consider it quite "natural" that adopted children, even when sympathetically reared by others, should be interested in finding and meeting their "real" parents, as, significantly enough, their natural parents are generally called.

To argue, as Alice Rossi and other social scientists have, that evolution has left men and women with different parenting skills opens up the much larger and more controversial issue of other differences in the behavior of the sexes. That controversial idea, of course, was not new in the 1960s, when an interest in the biological roots of human behavior resurfaced among social scientists. Indeed, it is probably true to say that the concept of behavioral differences between the sexes never disappeared from the minds of social investigators. Like everyone else, social scientists arrange their lives around the awareness that there are two sexes and that the connections between them are central to life and reproduction. Yet until the 1970s, most social scientists, with the notable exception of Margaret Mead, displayed only moderate professional interest in studying the differences in relation to biological influences. After all, did not culture or socialization explain whatever behavioral differences were commonly observed? By the 1970s that answer seemed less than satisfactory to some social scientists.

The shift began with a well-known fact of gender differences, namely, that the division of labor between the sexes is universal in human societies. Recognition of that fact caused one anthropologist in 1970 to suggest that perhaps something more enduring than culture might lie at its source. Judith Brown, the author of the article, noted that planned societies like those of Israel and Mao Tse-tung's China, where men and women were officially considered to be interchangeable in the work place, had only limited success in achieving that interchangeability in practice. Seven years before Alice Rossi made the point, Brown observed that nowhere in the world was the rearing of children primarily the responsibility of men. After examining the work of women in five societies, she concluded that the kind of work women performed depended on the

* More recently, legal scholar John Oakley speculated in 1985 that an argument might be developed to the effect that legally closed adoption records could be opened on the grounds that an adopted child had a *biological* need to know its natural parents. Another legal scholar, however, pointed out, along the lines offered in the text, that sociobiological theory offers no support for such a case. "Culturally (environmentally) induced impulses rather than genetically programmed impulses operating in the adoptees are likely to be the main factors behind any such needs," he contended. See John J. Beckstrom, *Evolutionary Jurisprudence: Prospects and Limitations on the Use of Modern Darwinism throughout the Legal Process* (Urbana: Univ. of Illinois Press, 1989), 56–57.

opportunities available for combining work and what she called "child watching."

Five years later, Brown returned to the same subject from a slightly different angle in the pages of the scholarly journal *Current Anthropology*. This time she asked why women are almost universally the preparers and processors of food. She was merely raising the question, rather than providing an answer, but it was apparent that she was seeing the child-rearing role of women as influential, if not formative.[6]

Brown's approach was biological in that she suggested that the physiology of woman—namely, her bearing and feeding a child from her body—helped to shape the noticeably different relations a woman and a man had with children. Brown said nothing, however, about different patterns of behavior between the sexes that were unrelated to child rearing, or possible differences in outlook on the world between men and women. By the mid-seventies, however, social scientists and other students of human behavior were beginning to acknowledge that those kinds of differences between the sexes might be traced to biology as well.

Undoubtedly the most authoritative study identifying a role for biology in the behavioral differentiation of the two sexes was published in 1974 by Stanford psychologists Eleanor Maccoby and Carol Jacklin. Their book, *The Psychology of Sex Differences*, reviewed a large literature that had examined differences between the sexes over the previous ten years. As they informed their readers, the aim was "to sift the evidence to determine which of the many beliefs about sex differences have a solid basis in fact, and which do not." Given their field of expertise, they understandably focused their attention on psychological differences. Nevertheless, they were well aware that sex is both a "biological and a social fact"; they did not expect to escape biology. And they did not. For when they came to summarize their analysis of the more than 1400 studies they had reviewed, they forthrightly concluded that "biological factors have been most implicated" in two of the "fairly well established" differences between the sexes, namely, the "differences in aggression and visual-spatial ability."[7]

Maccoby and Jacklin set forth three general reasons for seeing aggressiveness as rooted in biology: "(1) the sex difference manifests itself in similar ways in man and subhuman primates; (2) it is cross-culturally universal; and (3) levels of aggression are responsive to sex hormones." Additionally, they found that cultural influences could be largely ruled out since "there is no good evidence that adults reinforce boys' aggression more than girls' aggression; in fact," they acknowledged, "the contrary may be true." In another place they reinforced their conclusion of the greater aggressiveness of the male by referring to it as "one of the best established, and most pervasive of all psychological differences," for which they had reason "to believe there is a biological component underlying this difference."[8]

They were even more convinced of the biological basis of the visual-spatial ability, for it could be traced to genetics. "There is evidence," they reported, "of a recessive sex-linked gene that contributes an element to high spatial ability. Present estimates are that approximately 50 percent of men and 25 percent of women show this element phenotypically [that is, in their bodies], although of course more women than this are 'carriers.'"[9]

Maccoby and Jacklin recognized that their conclusions might well be used to limit women's opportunities in the everyday world. They rejected, for example, the argument that the greater aggressiveness of males precluded a significant leadership role in society or, say, the job of engineer for women. In the modern world, they observed, aggression was no longer important. And besides, they properly pointed out, the differences between the sexes lay in the *distribution* of aggression and visual-spatial abilities between the sexes, not in the absence of those traits from one sex and their presence in the other.

Rossi, Maccoby, and Jacklin found a sort of follower in sociologist Steven Goldberg in regard to the different physiologies of women and men in accounting for differences in their social behavior, but in regard to little else. In a book provocatively entitled *The Inevitability of Patriarchy,* first published in 1972 and expanded in a 1973 edition, Goldberg drew upon biology to explain the presence "of patriarchy, male dominance, and male attainments" in every known society in the past and present. *"The thesis put forth here,"* he emphasized, *"is that the hormonal renders the social inevitable."*[10]

The differences in the hormones of males and females, he concluded from the work of biopsychologist John Money, and others, gives "males a greater capacity (or lower threshold for the release of) for 'aggression'; this is an advantage," Goldberg explained, "in that the 'aggression' can be involved in any area for which it will lead to success." He explicitly denied making any case for an aggressive or a killer instinct, as some ethologically inclined scholars had argued.[11] Basically he was postulating no more than the "aggression" Maccoby and Jacklin had identified in their studies. Maccoby and Jacklin thought aggression was less important as a trait in modern society and therefore its relatively lower level in females was no longer a constraint on women's activities. Goldberg, however, contended that the difference in aggression precluded the existence of a non-patriarchal society so long as only women bore children, for the difference in degree of "aggression" in the two sexes Goldberg explained by reference to their different reproductive (hormonal) machinery. His point was that simply for reproductive reasons, males, on the average, were inherently more aggressive than females.

It is only fair to Goldberg to note that he was not aiming to limit women's opportunities or to suggest that any social policy be based on his conception of the hormonal roots of patriarchy. His purpose was to *explain* the absence of matriarchal societies among human beings, an absence now generally acknowledged by modern anthropologists.[12] Goldberg then propounded what he called the "environmentalist dilemma." If it is agreed that all known societies are male-dominated, he asked, what is the "universal cultural-environmental factor" that explains or accounts for that social phenomenon as effectively as "the biological factor" (male aggressiveness) which he has proposed?[13]

Goldberg recognized that biological explanations can be misused, but thought the "dangers hardly justified a denial of the determinative effect of biology. . . . Biology should never be used as an excuse for discrimination," he insisted. His central point, however, remained: Males were more likely to be hormonally disposed to be "aggressive" than females.[14] Hence, for that physiological reason, all societies had been and are today male-dominated and so they would be in the future.

The differences in aggressiveness of the two sexes figure, too, in explanations by two Harvard social scientists for the disparities in the crime rates of men and women. "Just as the gender gap [they really mean sex gap] in crime has survived the changes of recent times," James Q. Wilson and Richard Herrnstein wrote, "so also have the major sex roles. . . . The underpinnings of the sexual division of labor in human societies, from the family to commerce and industry to government may not be rigidly fixed in the genes," they concede, "but their roots go . . . deep into the biological substratum."[15]

Some other social scientists who accepted biological sources for the differences between the sexes interpreted the differences to be less constraining on women. When Jerome Barkow suggested in the pages of *Current Anthropology* in 1975 that dominance/submission theory might have some relevance to human behavior, Mary Knudsen of the University of Texas at Galveston took the occasion to suggest that there might be a male and a female "preferred strategy" in dominant behavior. Males, she suggested, would be inclined to follow a "more individual-oriented strategy," while females would be likely "to use more group-oriented strategies." She speculated that international politics "might change if traditional female strategies like 'Make love, not war' were utilized."

Knudsen's idea, which, as we have seen, had a long history in American social thought extending well back into the nineteenth century, sounded more than an echo in the minds of other social scientists in the late 1970s. Political scientist

Glendon Schubert, for instance, argued in 1976 for the deliberate study of the biology of sex differences since he thought there was now a good deal of evidence of hormonal and neural differences between the sexes. He was following Knudsen's lead when he observed that the behavior of the few female national heads of government then available for study—namely, Golda Meir and Indira Gandhi—comprised too small (and perhaps too war-like) a sample from which to draw conclusions about a possible difference in the quality of leadership of men and women. In a larger sample, he speculated, the character of women leaders would probably differ from that of men. Those few women presently in positions of leadership had been compelled to act like men in order to achieve their offices, so little could be learned about sex differences in general from them. He thought it worthwhile for political scientists to pursue the theory "that a substantial increase of female participation in positions of national political leadership at the highest levels would enhance chances for international peace."[16]

After reviewing the experimental, clinical, and observational work on differences between the sexes in hormones, brain development, and aggression, anthropologist Melvin Konner in 1982 came down on the same side as Schubert, though somewhat more tentatively. Konner thought that, even with more examples to draw from in 1982 than Schubert had had, female rulers of states in past and present were not sufficiently numerous to provide a good representative sample from which to draw conclusions about the nature of female versus male leadership. He conceded that "some women are, of course, as violent as almost any man. But speaking of averages . . . we can have little doubt that we would all be safer if the world's weapons systems were controlled by average women instead of by average men."[17] It is worth noting that Konner based his conception of women's relative lack of aggression on the earlier work of Maccoby and Jacklin.

The diverse ideological interpretations that can be made of biology in thinking about human behavior is well illustrated by the divergent meaning that Konner, on the one hand, and Maccoby and Jacklin, on the other, ascribe to women's relative lack of aggressiveness, even though they agree on the biological "facts." In Maccoby and Jacklin's hands, the trait required some interpretive qualification in order to avoid placing limits on opportunities for women, while in Konner's hands it became an attribute that the world needed in order to improve itself. As frequently happens, the social values that are attached to biological findings largely determine how important or useful those findings turn out to be.

Other social scientists in the 1970s and 1980s were not as sanguine as Konner that differences between the sexes could result in an improved world. Nor were they necessarily convinced that behavioral differences between women and men were biologically based. Perhaps the most striking example of the ambivalence in the thinking of some social scientists is provided by the work of sociologist Melford E. Spiro. In 1955 Spiro had published a study on the success of the Israeli kibbutz in breaking down the differences in the traditional behavior of the sexes. All jobs in the kibbutz community were open to women, and child rearing was a public not a private function with the children reared in "Children's Houses" in order to separate parenthood from child care. As Spiro observed, the original intention behind the kibbutz experiment had been to restructure the social order in a deliberate effort to equalize opportunities for the two sexes. It was, in modern parlance, a concerted assault on gender, or the socially constructed differences between men and women. Only their physiological differences, that is, those that were sexual and reproductive, were to be acknowledged. Twenty years after his first visit, Spiro returned to the kibbutzim to see what had happened in the interim.

What he found and what disturbed him he reported in his lectures *Gender and Culture: Kibbutz Women Revisited,* published in 1980. The burden of his report and the source of his disturbance were that the women, despite the social engineering inherent in the kibbutz idea, had reverted to what he described as "traditional" roles. Not only were the women congregated in conventionally female jobs, but they were also the primary child rearers in the Children's Houses. Moreover, almost uniformly, mothers wanted their children to sleep at home, rather than in the Children's Houses. The original kibbutz ideal had intended that the Children's Houses should be surrogate parents. In the interim since Spiro's first visit they had become little more than child-care centers while the mothers were at work.

As Spiro explained with some puzzlement, "it is a basic axiom of the social sciences that human behavior and motive are primarily, if not exclusively, culturally programmed." Yet the changes he had found in the kibbutzim constituted, in his judgment, "a rather unequivocal exception to our social science axiom." The piece of evidence that struck him as most telling was the resistance of girls, once they had passed through puberty, to dressing and showering with boys, as the kibbutz required. "In sum," he concluded, "the original kibbutz belief, that in the proper learning environment children would be sex-blind, was proven to be false even in the sexually enlightened conditions in which these children were raised." In view of this development, he suggested that "it seems likely that these sex differences (like their sense of sexual shame) were brought about not by culture, but by the triumph of human nature over culture, that is, by motivational dispositions based on sex differences in precultural rather than culturally constituted, needs." And if these needs are in fact precultural, as he believed, then, "all things being equal, it is probable that they are shared by females in any society."[18] As fellow sociologist Alice Rossi pointed out, Spiro "gives no detail on what he thinks those 'precultural' factors might be." Apparently, he was reluctant to ascribe them to biology, though that would seem to be a logical and obvious explanation for such influences.

Rossi has been even more critical of Harvard psychologist Carol Gilligan, who, like Spiro, has been identifying differences between men and women while clearly reluctant to connect them to biological sources. Gilligan, Rossi pointedly observed, proposed no theory "to explain *why* intimacy is threatening to men and impersonality to women: or *why* she finds women's mode of thinking to be contextual and narrative while men's is formal, linear and abstract. She merely argues that theories of human development have used males' lives as a norm and tried to fashion out of it a masculine cloth that does not fit." Rossi also complained that a study of how sex roles are inverted in old age in four different societies similarly lacked any theory to account for the finding. "The lack of explanatory specificity in all three examples—Spiro, Gilligan, and Gutmann—is based," Rossi concluded, "on the entrenched but erroneous view that biology is properly left outside the ken of the social sciences."[19]

It is worth noting that, though the identification of sex differences has an ancient history, these more recent identifications, unlike many of those in the past, lack any implication that the differences justify confining women's place or opportunities in society. Maccoby and Jacklin, for example, emphasize that societies, after all, have "the option of minimizing, rather than maximizing sex differences through their socialization process. A society could, for example, devote its energies more toward moderating male aggression than toward preparing women to submit to male aggression, or encouraging rather than discouraging male nurturance activities." Social institutions and practices need not follow biology. "A variety of social institutions are viable within the framework set by biology," they advised. "It is up to human beings to select those that foster the life styles they most value."[20] Even Steven Goldberg, who thought biology kept women in general from being dominant over men, acknowledged that biology does not "give us the right to judge any individual by the characteristics of the group to which that individual belongs. . . . Biology can never justify refusing any particular woman any option," he held. And, like Maccoby and Jacklin, he recognized that "both males and females are capable of nearly every type of sexually differentiated behavior," including dominating others.[21]

Political scientist Schubert and anthropologist Konner, both of whom think women's lesser aggression is a social gain, clearly look forward to a day when the special quality of female leadership will have greater prominence than it does today. Moreover, as we shall see a little later, some modern women feminists agree that differences between the sexes do not necessarily limit or confine women. Instead, the special nature of women opens opportunities to them that society needs taken up, but which few men can adequately fulfill.

Doris and David Jonas, writing in *Current Anthropology* in 1975, did not proclaim themselves to be feminists, but they freely acknowledged a difference between the sexes that they clearly considered a source of human achievement. Modern studies demonstrate, they asserted, that girls are more verbal than boys. "We must assume," they argued, drawing on ethological theory, "that this superiority is not without a biological history and that it must have arisen in connection with the reproductive success of the species." From that they ascribed to females nothing less than the origin of human speech. It came about, they suggested, from mothers' communicating with their children, for even among primates mothers vocalize to their offspring. Among the bushmen of the Kalahari Desert, mothers talk frequently to their infants, while communication among the males during the hunt is by silent signals, probably to avoid frightening the game, though the Jonases do not say that. Their theory was heavily criticized in subsequent issues of the journal, but even the criticism provided fresh testimony of the burgeoning interest among anthropologists in drawing upon ethological research to gain insights into the workings of human nature.[22]

Anthropologist Donald Symons of the University of California at Santa Barbara has drawn quite explicitly upon ethological theory. In his book *The Evolution of Human Sexuality,* published in 1979, Symons began his argument by referring to the sociobiological principle of "differential reproductive strategies" of males and females. Simply put, the principle states that males and females, especially among mammals, have different approaches to reproduction because of the nature of sexual reproduction. Among female mammals the cost of reproduction is high, entailing in sociobiological language a "high investment" of energy in the form of "large" eggs (as compared with sperm), and, most important, long periods of gestation and lactation. For males, on the other hand, reproduction is quick, easy, and above all, cheap.

Symons draws on this ethological principle to uncover the biological basis of the common observation that women are generally less eager to engage in sex than men. Like other female mammals, women have a higher personal investment in offspring than men and hence are more cautious in engaging in sexual behavior. "Selection favored the basic male tendency to be aroused sexually by the sight of females," Symons asserted. "A human female, on the other hand, incurred an immense risk, in terms of time and energy, by becoming pregnant, hence selection favored the basic female tendency to discriminate with respect both to sexual partners and to the circumstances in which copulation occurred." A possible physiological explanation for this difference is that sexual arousal is necessary for reproduction by males while in females sexual arousal is not necessary for reproduction. That difference enhances male sexuality by tying it to nature's drive for reproduction. Although it might be objected that the less insistent urge to copulation in women is the result of social repression of their sexuality, Symons's response was that, even in societies in which female sexual activity is virtually without social constraints, men exhibit a greater sexual drive than women, by the testimony of both men and women.[22]

The differences in sexual interest of males and females, which Symons relates to different reproductive strategies, receives some confirmation from the investigations of University of Michigan psychologist David Buss. His research sought to ascertain if the varying sexual strategies derived from evolution might throw light on sources of marital conflicts. His findings suggested that they did inasmuch as "the strongest predictors of sexual dissatisfaction for males" he found to be "sexual withholding by the wife." On the other hand, for women the strongest predictor is "sexual aggressiveness by the husband."[23]

Symons has connected another difference in sex drives to biology. It is the disparity in orgasmic experience. Modern studies of human sexuality often report that women are much less likely to experience orgasm than men. Symons, for example, quotes the well-known comment of Margaret Mead that "comparative cultural material gives no grounds for assuming that an orgasm is an integral and unlearned part of women's sexual response as it is of men's sexual response."

At least two biological facts seem relevant here. The first has already been made, namely, that an orgasm is not necessary in order for a woman to produce. The second is that the existence of orgasms among female primates other than *Homo sapiens* seems at best problematical.* For his part, Symons flatly pronounced the "female orgasm . . . a characteristic essentially restricted to our own species."[24]

Certainly the mode of copulation among almost all primates, from the rear, rather than face to face, renders unlikely the stimulation of the clitoris, which is a primary source of female sexual excitation. In fact, among human beings it is not clear that "standard" face-to-face intercourse results in female orgasm without additional stimulation. On the face of the available biological evidence, the sexual behavior of human females, at least as measured by orgasmic experience, would differ from that of males. Or, put another way, there are biolog-

---

* Primatologist Frans de Waal in his recent book *Peacemaking Among Primates* (Cambridge, Mass.: Harvard Univ. Press, 1989) stresses the human-like sexual behavior, including what seem to be orgasmic responses, of female bonomos, or pygmy chimpanzees. Unlike other apes, but like human beings, bonomo females have frontal vulvas, thus permitting face-to-face sexual intercourse. See the section entitled "Kama Sutra Primates," pp. 198–206.

ical reasons why the oft observed and commented upon differences in the sexual drives and responses of the two sexes are to be anticipated. Men's sexuality is tightly linked to the basic evolutionary drive to maximize reproduction while women's is to control it to maximize quality.

An implication that might be drawn from Symons's explication of the difference in orgasmic experiences is that the female orgasm is cultural, rather than biological. It is biological, to be sure, in that it is a physiological phenomenon just as the male's is. Unlike the male orgasm, however, it is not necessary for reproduction. In that sense it is not a part of the natural reproductive system of the species any more than it is for other female primates. They have clitorises, but provide almost no evidence of orgasms.

At the same time, history and contemporary reports more than assure us that women do indeed experience orgasms. And since the experience is as pleasurable as a male orgasm, if not more so, it follows that women would encourage the experience. But inasmuch as the orgasm is not built into the reproductive arrangements, the impetus behind that encouragement or fostering is cultural rather than biological. It is an experience that women have cultivated for their own pleasure, just as human beings, males and females alike, have made sexuality into a lengthy and pleasurable ritual far beyond that which is necessary for "natural" reproduction. Simple and quick copulation works as well reproductively for us as it does for other primates. (Controlling premature ejaculation would be a human male version of cultural influence that reshapes the reproductive function in order to enhance sexuality among human beings.)

Viewing the human female orgasm as a culturally induced experience would have the advantage of reversing the usual misogynistic habit of perceiving the male as culture (and therefore superior) and the female as nature and therefore baser. The male orgasm is "natural" because it is reproductive, while the female climax is a creation of human imagination, or, to phrase it another way: it is peculiarly human.*

Accounting biologically for the differences between male and female sexuality would also help to explain certain historically recognized differences in the behavior of men and women. The presence through a good part of recorded

---

*Primatologist Frans de Waal's emphasis upon the abundant sexual playfulness of bonomos, including the possibility of female orgasms, which he has observed at the San Diego Zoo, lends some support to the supposition in the text. In a zoo, animals have much more time and many more occasions to elaborate upon sexual activity than they do in the wild where the search for food is demanding, offspring must be cared for, and enemies evaded. Early human social organization would provide opportunities even superior to those of a zoo. No one knows, de Waal tells us, how the bonomos behave sexually in their natural habitat since they are very elusive. Frans de Waal, *Peacemaking Among Primates*, 198–206.

history of pornography directed at men, and of female prostitution in service to men, could be accounted for by the theory. Since, according to the theory, males must be ready at any time to copulate, their sexual arousal will be much more closely attuned to potential stimulation than females'. Given, too, the high visual acuity of primates, and of human beings in particular, it comes as no surprise that pictorial pornography, almost all of which has been produced for males, is common in a variety of cultures. Similarly, given the males' propensity for more frequent and more indiscriminate copulation, recourse to prostitutes would be much more common, as it clearly is, among males than among females.

None of this, of course, implies or defines what ought to be viewed as normative sexual behavior for females today. Nor, as a general statement, does it tell us anything about the normative sexual behavior or attitudes of individual women in the past. History, after all, has recorded a number of highly sexed women. The interpretation offered here merely suggests that a fit exists between the predictions of the sociobiological theory of male-female differences in sexuality and the general sexual behavior of women in history. (The counter argument that, in a social order in which men were no longer dominant, women's sexual behavior would *not* differ from men's really begs the question: it assumes as an alternative the very conclusion that is at issue, and for which historical and contemporary evidence is presently lacking.)

Although social scientists' revived interest in sex differences occasionally implied limitations on women's activities, on the whole, the revival was more pro-feminist than not. As we have seen, feminism and the recognition of sex differences could find common ground on which to stand. Thus in 1978 psychologist John Fuller endorsed both feminism and biological influences on human behavior even as he was noticing that biology was reclaiming the attention of social scientists. "Currently in the western world," he observed at a conference on sociobiology held in San Francisco, "there is a challenge to the traditional differences in the social roles of males and females." This widening "of role choices for both sexes," he thought "a great advance," but then confessed that he could not go along with those who contended "that the differences between the interests of men and women are solely the product of exposure to sex stereotypes in the nursery and in their school readers," a reference, presumably, to the assumption of some feminists that socialization was the only basis for any identifiable differences between the sexes. Whether his disagreement was justified, he circumspectly concluded, only time would tell. But he could not resist adding that "testosterone and estrogen are potent compounds that affect the brain as well as the reproductive system."[25]

Fuller's carefully phrased views of sex differences were expanded upon with greater force by Elizabeth Wolgast in her book *Equality and Rights of Women*, published in 1980. She had no doubt that the bearing of children separated a woman fundamentally from a man. Indeed, the principal point of her forcefully argued book was that sex differences needed to be recognized, and those characteristics peculiar to women should not be seen as simply a burden to be borne and then ignored. She denied the implicit argument of many feminists that there was a kind of androgynous person or single human type into which both sexes could be fitted if only enough energy and imagination were expended to bring it about. That "is the wrong order of things," she protested. "Being female is a disadvantage in a society that makes it one." Why not envision a different kind of society, one that puts "egalitarian androgynous models aside and look[s] to a social form fitting a species of two sexes, both having their own strengths, virtues, distinctive tendencies, and weaknesses, neither being fully assimilable to the other?" Along with Rossi and other writers who pressed for a recognition of sex differences, Wolgast drew upon ethological researches to support her argument, calling attention to the differential behavior of the sexes among other primate species.

"If, as the present studies suggest, the sexes are different in dimensions not connected with reproduction," Wolgast argued from a feminist premise, "we have reason for concluding that women should make their own distinctive contribution to the culture and society." Such studies further suggested, she thought, that a woman's perspective was needed, and therefore women ought to be included in all areas of thought, art, and science that touch upon the lives of both sexes. Interestingly enough, this argument was almost a restatment of feminist Charlotte Perkins Gilman's contention of almost eighty years earlier that the time was at hand when society required the contributions of females to civilized life as well as those of males. Wolgast did not refer to Gilman's argument, but she did point out that other women, like anthropologist Margaret Mead and philosopher Mary Midgley, had similarly emphasized the differences between the sexes without believing that such an acknowledgment denigrated women's activities.[26]

Radical feminist Lynda Birke is neither as committed nor as forthright in her recognition of the role of biology in separating women from men as Mead and Midgley. Yet after a good deal of qualification of the place of biology, she concluded that "our concepts of gender should not be seen as purely social phenomena, but should admit biological experience." Birke's ambivalence toward, even fear of, biology as a casual factor was obvious, yet that very ambivalence captured the way in which even a radical feminist has been affected by the renewed interest among social theorists in biological differences between the sexes.[27]*

Questions of social theory have not been confined by any means to issues of sexual differences. Indeed, the whole question of the place of human beings in nature has been invigorated by the return of Darwinian evolution to social theory. The discussion has always been controversial, sometimes even acrimonious. In the next chapter we look at some of the theoretical gains social scientists think they have achieved from evolutionary theory, and we ask why some scholars have found evolution and sociobiology useless, if not dangerous, in the study of human nature.

---

* Her ambivalence comes through, too, in the words she added to those quoted in the text: "Our biology, however, does not determine anything," and in the remark with which she opened the book, p. x : "This book is founded upon my belief in the relevance of biology to feminist thought." Her primary fear is what she calls biological determinism.

13

The Uses and Misuses
of Evolutionary Theory
in Social Science

The history of man is a real part of the history of nature. . . . The sciences of
nature will integrate the sciences of man, just as the sciences of man will integrate the sciences of nature, and there will be only one science.

Karl Marx, 1844

There was a time when sociobiology was seen by friends as well as critics as
imperialistic, seeking to subsume all of human experience under the Darwinian
Imperative. But that one-sided approach to the relation between evolutionary
theory and ethology, on the one hand, and the social sciences, on the other, has
been replaced, even in the minds of its biologically oriented proponents, by a
full recognition of the power and influence of environment or culture. Even
Edward Wilson by the early 1980s was talking interaction. "In sum," he wrote
in 1983 with his co-author Charles Lumsden, "culture is created and shaped by
biological processes while the biological processes are simultaneously altered in
response to cultural change."[1]

Part of this shift in emphasis undoubtedly stemmed from the criticism to
which both biologists and social scientists had subjected the last chapter in Wilson's book *Sociobiology*. But much also derived from the old battle within social
science during the 1920s and early 1930s over nature versus nurture. That bruis-
ing struggle had been resolved, more or less, as we have seen, by a general agreement that both were inextricably connected in accounting for human behavior.
In the years since the 1930s, however, many social scientists had increasingly
forgotten about the biological side of the compromise. As the anthropologist
James Silverberg remarked in 1980 to members attending a conference on sociobiology, everyone says they believe in epigenesis, or the interaction of culture
and biology, but when they come to discuss particular behaviors they usually
come down in favor of one or the other. He correctly and sadly remarked that

practice was nothing less than a throwback to the old nature/nurture dichotomy. Actually, virtually all of those social scientists who called for a recognition of the role of biology in human behavior had long emphasized the interaction of nature and nurture. "Biologists now agree," wrote anthropologist Alexander Alland in 1967, "that the argument over the primacy of environment or heredity in the development of organism is a dead issue. It is now generally accepted that the function and form of organisms can be understood only as the result of a highly complicated process of interaction." Political scientist Roger Masters echoed this view in 1973 when he denominated any attempt to identify the cause of a behavior pattern as "either purely genetic or purely environmental . . . as a biological absurdity," for all biological phenomena "presuppose an interaction between genotype and environment." In his 1978 critique of Marshall Sahlins's objections to sociobiology, Martin Etter took the well-known anthropologist to task for failing to appreciate that "this hard-won comprehension, between the rock of genetic determinism and the whirlpool of Lamarckian environmentalism, is assumed by sociobiology." Or as Cornell anthropologist Davydd Greenwood has simply phrased the point: "Nature without culture or culture without nature cannot be conceived."

Sociologist Alice Rossi offered a concrete example of the reality of interaction when she reminded her readers that not only do hormones affect behavior but behavior affects the secretion of hormones as well. "It makes no sense," she stressed in 1984, "to view biology and social experience as separate domains contesting for election as 'primary causes.' Biological processes unfold in a cultural context, and are themselves malleable, not stable and inevitable. So, too, cultural processes take place within and through biological organism; they do not take place in a biological vacuum." Even confirmed proponents of human sociobiology Pierre van den Berghe and David Barash acknowledged that "there is no question that human behavior must always be understood as the product of the interplay between biological, ethological, and sociocultural conditions." Their disagreement with colleagues in the social sciences, they insisted, is confined to those "who dogmatically assert the irrelevance of biology to our social behavior."[2]

After anthropologist Victor Turner, a long-time student of rituals and myth, came to recognize a role for biology in his study of human rituals, he began to emphasize interaction between biology and culture. "Exciting new findings [that] were coming from genetics, ethology, and neurology, particularly the neurology of the brain," had raised pointed questions for him. "Can we enlarge understanding of the ritual process by relating it to some of these findings? After all," he continued, "can we escape from something like animal ritualization without escaping our own bodies and psyches, the rhythms and structures of which arise on their own?" His answer was that "I am at least half convinced that there can be genuine dialogue between neurology and culturology, since both take into account the capacity of the upper brain for adaptability, resilience, learning and symbolizing."[3]

Some social scientists who are convinced that biology as well as culture must be taken into consideration in trying to understand human behavior have moved beyond interaction or epigenesis. Impressed by the power of evolutionary theory, they have worked to create a theory of cultural evolution. Foremost among them has been William Durham, presently a Stanford anthropologist, who began his work at the University of Michigan with a decidedly sociobiological bias to his work. By 1978, however, he was complaining that "somehow it escapes a number of sociobiologists that human behaviors can be highly adaptive for entirely cultural reasons." As noted in an earlier chapter, he soon began to work toward a co-evolutionary theory, one in which culture, like biology, would be seen as evolving under pressure from the natural drive for reproductive success. "However culture changed and evolved," he explained, "and whatever meaning was given by people to their cultural attributes, the net effect of those attributes was to enhance human survival and reproduction." Cultural traits, he argued, could be selected for their usefulness to survival and reproduction just as natural selection might select a particular behavior pattern for the same end. After all, Durham recalled, not all new ideas or cultural innovations are ultimately adopted. Child-rearing practices, for example, were particularly responsive to ecological circumstances.

Durham identified two cultural practices which could not be easily explained except by seeing culture as an evolving process that sought to maximize reproductive success along with biological evolution. The first related to a practice among certain South American cultures of introducing lime into their preparation of maize. A modern anthropologist had discovered that only with the introduction of lime could that food release its full nutrition for human consumption. Presumably, the native peoples had discovered the need for lime without knowing the chemical or nutritional explanation and had then integrated it into their tradition because it was conducive to what sociobiologists would call reproductive success. The other cultural practice was head hunting among the Mundurucu, a tribe of Brazilian Indians. A traditional ethnologist would be likely to provide a functional explanation: head hunting was a way to preserve social order by providing an outlet for the expression of animosities toward others. But a co-evolutionary interpretation contends that head hunting was actually related to maximizing the food supply by killing rival hunters of tapirs, the Mundurucu's main food supply. (The Mundurucu believe that accumulating human heads encourages an increase in the number of tapirs. Killing rival hunters, of course, does that, too, but indirectly.)

Durham's aim, in short, was to show concretely how evolution functioned biologically and culturally. His aim was "to persuade biologists that theories of human behavioral adaptation must make *explicit allowance* for non-genetic mechanism behind the transmission of traits in a population." By the same token, he wanted "to persuade anthropologists that human survival and reproduction remains an important influence behind the ongoing evolution of cultural practices."[4]

At about the same time, a young political scientist, John Langton, had arrived at a similar theory of co-evolution, based upon psychological "reinforcement." Langton drew an analogy between Darwin's random and infinite variations in the biological world and the random and infinite variety of social traits or practices in the cultural world. Like Darwin's traits, cultural practices can be "inherited," that is, passed on to successive generations; many, however, are not passed down simply because they serve no selective purpose. Those that do soon spread through the population because they are advantageous. Just as there is a struggle for existence in Darwin's world, so in the world of culture there is a struggle for "reinforcement," as Langton described the process.[5]

Sociologist Marion Blute's development of a conception of co-evolution was similarly stimulated by the explanatory power of evolutionary theory. Students of cultural change, she contended, could learn from biological evolution by thinking of individual cultural innovations as competing with one another for survival or endurance just as biological variations in living organisms struggle for survival in changing environments. The true Darwinian approach to cultural evolution, she stressed, had nothing to do with "stages" of cultural change, as some of the earlier theorists of social evolution like Montesquieu, Comte, Marx, and Lewis Henry Morgan had contended. The engine of change in both cultural and biological evolution was individual competition. As she cogently phrased it, biology is a historical science and the competition that accounts for the changes in nature can also account for the changes in culture.[6]

More recently, two anthropologists, Peter Richerson and Robert Boyd, have attempted to reduce the idea of co-evolution to a model the underlying assumption of which was similar to that suggested by Marion Blute: both nature and culture change under the impact of competition, or natural selection. Like most co-evolutionists, Richerson and Boyd worked on the assumption that culture and biology pushed in the same direction since both seek, ultimately, to maximize reproductive success. It is unlikely, they argue, "that an ordinary organism would evolve a second system" of behavior—that is, culture—"unless the new system increased genetic fitness." If it can be accepted that culture is a form of inheritance, they maintained, then Darwin's approach will be as helpful with culture as with biology. "Today's cultural traditions are the result of cumulative changes made by past and present bearers of them. To understand why cultural traditions have the form they do, we need to account for the processes that increase the frequency of some cultural variants and reduce that of others." Or, to put the idea in Darwinian biological terms, why do some animal forms flourish and others die out?

Evolutionary theory appealed to economists Richard Nelson and Sidney Winter for similar reasons: it would bring a breath of fresh air to a discipline too long rooted in narrow and static theory. Traditional economic theory of business organization, they complained, was overly committed "to unbounded rationalization and optimization" and thus was "inherently inflexible," unable to handle process or change. An evolutionary approach, Nelson and Winter maintained, was inherently dynamic. In their theory, business firms were assumed to have certain capabilities and rules, which over time "are modified as a result of both deliberate problem-solving efforts and random events. And over time the economic analogue of natural selection operates as the market determines which firms are profitable and which are unprofitable, and tends to winnow out the latter." In their system equilibrium is never achieved, no more than it is in nature. Quite candidly they acknowledged the theoretical influence of Edward Wilson's *Sociobiology* on economists, and particularly on "our work," they added, which "may be viewed as a specialized branch of such a theory." A final advantage they saw in an evolutionary approach was that "it would be a step toward freer trade in ideas" with the other social sciences.[7]

University of Texas sociologist Joseph Lopreato shaped his book *Human Nature and Biocultural Evolution,* which he published in 1984, around the proposition that a theory of cultural evolution could usefully draw upon Darwinian evolutionary theory. He hoped to be able, like Boyd and Richerson and William Durham, "to make a contribution to the emerging but still embryonic theory of biocultural evolution. My commitment is to the idea that such a theory can be constructed along the lines, roughly, of the variation-and-selective-retention model found in Darwin's theory of evolution by natural selection." Co-evolution, he rather extravagantly concluded, "is the beginning of a major breakthrough in the history of science."[8]

A final theoretical inspiration that some social scientists have discovered in evolutionary theory has been its potentiality to unify the social sciences, to provide an umbrella of large theory that would subsume the increasingly diverse fields that make up the once celebrated and hoped for "science of man." In one sense, of course, Darwinian theory as set forth in the *Origin of Species* and *The Descent of Man* was a catalyst to that dream. But things did not quite work out as hoped. The reasons were several, but notable among them was the misinterpretation or misunderstanding of Darwinism by nineteenth-century social Darwinists, eugenicists, and others. By the second half of the twentieth century, however, Darwinian theory had been put on a much firmer basis, thanks to the "modern synthesis" of genetics and population theory. By that time, too, the new science of ethology had begun to provide the observational data necessary to put flesh upon the skeleton of theory. Out of those developments emerged in 1975 Edward Wilson's landmark book, which boldly asserted sociobiology's ability to unite all the human sciences. Some social scientists, however, had come to that conclusion even before Wilson sounded the clarion.

Few advocates preceded or were more persistent in their advocacy than Columbia University anthropologist Alexander Alland, Jr. In his 1967 book, *Evolution and Human Behavior* he laid out a case for combining biology and culture in a single system of analysis. "Evolution is a process through which systems develop and are modified in relation to specific environmental backgrounds," he began. All that a theory of evolution demands is that variations occur and that there be ways in which the survival of variations can occur in the system, and, finally, "that these systems be subject to environmental selections." The mechanisms do not have to be biological, he pointed out in anticipation of later co-evolutionists, for the theory is about process. When that is understood, he emphasized, then the old distinctions between biological and cultural evolution disappear. "There is only one evolutionary process—*adaptation,* and one measure, comparative population size."

He rightly observed that most American social scientists follow some particular first cause in their social explanations. "Some are economic determinists, others psychological determinists, still others cultural determinists." But if somatic and behavioral traits are studied in connection with their adaptive values in particular environments, he pointed out, "we may find that some variables in a system are psychologically adaptive, others economically adaptive and so on." Traits that are biologically adaptive for a population as a whole are also biologically adaptive in the sense usual in evolutionary theory. "Hence," he concluded, "the biological model provides an umbrella for other deterministic theories." Such a theoretical approach, he predicted, would bring together "two rather disparate interests of American anthropology: enthnohistory and behavioral science."[9]

Three years later, political scientist Thomas Landon Thorson offered his justification for the wedding of social and biological sciences. Addressing his fellow professionals in his book *Biopolitics,* published in 1970, Thorson declared it "altogether sound to grant the existence of a biologically ordained social nature to man," as Freud and Jung had suggested when they looked "behind the

sweetly rational picture of man presented by Enlightenment *culture* into his
biological nature." Thorson left no doubt that his model was Darwinian evolution, for it was only with Darwin, he maintained, that biology became
dynamic, that is, concerned with change over time. A similar transformation
he hoped to see in political science, which, for too long, he thought, had erroneously taken physics as its model of science. Unlike physics, which is static,
biology is like politics or history; change over time is its hallmark. "What I am
saying," he summarized, "is simply that taking time and evolution seriously can
open the door to a genuine science of society, one which is capable of grasping
the whole phenomenon of man. . . . Whatever deserves the name political science must start with a conception of the human animal in nature and must
conclude with an account which will deal with the whole phenomenon of politics, the choosing and creating as well as the reacting and responding."[10]

During the 1970s, probably the most active political scientist pressing his
professional colleagues to pick up the challenge thrown down by Thorson was
Roger Masters of Dartmouth College. In recent years, Masters complained in
1975, "sociology, economics, political science, anthropology and psychology
have become so highly specialized that it is difficult to conceive of a single generally accepted theory of social life. . . ." But thanks to recent gains in biology,
"the time has come . . . to reconsider the conventional assumptions underlying
twentieth century social and political history." Biology is the appropriate link,
he insisted, because we are, first of all, living beings. Evolutionary theory would
offer a comprehensive approach to human behavior, "bridging the existing gaps
between the various social sciences and linking them to the natural sciences."
Political behavior, he declared, can "most fruitfully" be understood as a biological subject because doing so will break the "simplistic dichotomy" between
nature and culture. Six years later he called sociobiology "a bridge between the
natural and social (or human) sciences, placing social life in a Darwinian theoretical context without reducing humans to unthinking beasts."[11]

Since then other political scientists have followed Masters's lead in making,
as one of them phrased it, "the social sciences into biologically sensitive disciplines." Other social scientists have not been far behind in identifying evolutionary theory as a unifier of disciplines concerned with human social life. One
economist suggested that, in terms of research strategy, a sociobiological
approach "holds out great hope for breaking down not only the 'vertical' discontinuity between the sciences of human behavior and the more fundamental
studies of life, but also the 'horizontal' barriers among the various social studies
themselves." To one enthusiastic psychologist, "sociobiology in its many ver-
sions" seemed to be "the most important theoretical contribution of the second
half of this century." "Whether one's motivation as a sociologist is rooted in

passionate commitment to social change or passionate commitment to scientific advance, or both," sociologist Alice Rossi told her colleagues at a professional meeting, "the goals we seek are best approached through an integrated biosocial science."[12]

Simply because the social sciences are "rich in studies of reciprocity, social exchange, and related behaviors"—the very subjects with which sociobiology deals—sociologist Joseph Lopreato pointed out, "collaborative effort is both inevitable and very promising." For much too long the social and biological sciences have been on "entirely separate courses." Each, he recognized, has developed its own "unique and respective contributions to behavioral science," but convergence now seems to be taking place. "The gradual development of behavioral biology, underscored of late by the emergence of sociobiology," he argued, "is but one major sign of the convergences now being forged on many fronts, even if in some cases without awareness of the fact by the forgers themselves." Anthropologist Vernon Reynolds, noting that "there is no holistic science of man," but only "many human sciences," suggested that perhaps sociobiology would provide the answer to the question of "how, in any meaningful scientific sense, are flesh and spirit to be reconciled?"

The accomplishment of that task, warned psychologists Julian Jaynes and Marvin Bressler as long ago as 1971, will not be easy, however desirable it undoubtedly is. "An aspiration of this magnitude will require a revision in the modus operandi of the social sciences. It will demand sociologists who are "more receptive to cross-societal analysis, anthropologists who are more attentive to complex civilizations, and historians who can discern common patterns in discrete events." Social scientists, in short, who at present are in the distinct minority in their respective fields. At the same time, they warned, "ethologists will be obliged to show more respect for the subtlety and complexity of evolutionary processes."[13]

In all of these sometimes overly optimistic statements of how biology in one way or another would unify the deplorably disunited social sciences can be discerned once again a powerful motive behind social scientists' interest in biology. With it they would be able to draw upon a dynamic, comprehensive evolutionary theory that had already won its spurs among natural science. The prospect it offered was nothing less than dazzling: the integration of the many fields dedicated to the study of human behavior into a single science of human nature. For those social scientists, evolutionary theory promised to revive and invigorate an old, and now long tarnished, cherished dream.

Such idealistic aspirations, however, were not the motives their opponents perceived. And so it is time to look at some of the objections to this new interest among social scientists in the role of biology in defining human nature. Some of the critics, many of them social scientists themselves, could find only the most sinister of explanations for the revival of interest in biological theories. The distinguished physical anthropologist Sherwood Washburn of the University of California at Berkeley, for example, accused sociobiologists of not being adequately "worried by their repetition of the errors of the eugenicists, social Darwinists, or racists." In calmer moments, Washburn zeroed in on the inappropriateness of seeing any analogies, much less any homologies, between the behavior of animals and human beings. Social science might be defined, he wrote in 1978, as the study of "the nature, complexity, and effectiveness of linguistically mediated behaviors." Armed with that definition, he thought it obvious that the roots of human behavior could not be deduced from the activities of animals since only human beings had developed language. He scornfully identified human ethology as "the science that pretends humans cannot speak."

Even more hostile was the assessment of Stephen Chorover in his deliberately provocative book *From Genesis to Genocide,* published in 1979. "In the last analysis," Chorover concluded, "it was sociobiological scholarship . . . that provided the conceptual framework by which eugenic theory was transformed into genocidal practice" in Nazi Germany. Chorover would discern in "sociobiological falsehoods" nothing more than an attempt "to justify social inequality." In a book published in 1983 on sociological theory, Randall Collins was only slightly more circumspect. He described sociobiology as "antihumanistic and antiliberal" in tone, and fitting in "with the swing of the political pendulum to the right in the 1970s." Other social scientists, too, feared the political fallout from an application of ethological principles to human behavior. Historian of science Nancy Stepan, for instance, was especially troubled by the use of sociobiological principles in human affairs because such "simplistic speculations . . . about human behaviour do a disservice to society by falsely making hereditarianism popular."[14]

It is true that an application of evolutionary theory to human behavior, as sociobiology seeks to do, flies in the face of some deeply seated beliefs of social scientists. As Harvard historian of science Gerald Holton observed in 1978, "sociobiology violates Durkheim's injunction—which is bedrock in the training of social scientists in this country—that social phenomena can only be explained in terms of social variables." By rooting human behavior in genetics, added biologist George Barlow, sociobiology sets off an alarm. "When behavior is left out, there is no voiced objection to the field of human genetics." Thus genetically caused diseases or defects, he pointed out, are readily acknowledged, for they are not behavioral.[15]

The political and ideological origins of some of the American opposition to sociobiology was noticed, too, and not without irony, by the eminent French anthropologist Claude Levi-Strauss, though he himself was no proponent of human sociobiology. In France, he observed in 1983, sociobiology has been taken up by the political left, out of a *"neo-rousseauiste* inspiration in an effort to integrate man in nature." In the United States, at the same time, he observed, it has been denounced "as neo-fascist doctrine" and around it liberals have thrown a "veritable prohibition over all such research," a development Levi-Strauss deplored.

Frequent as the charge of a conservative bias to biological explanations may have been, it was both inaccurate and irrelevant. "One need not be a fascist or a social Darwinist," Arthur Caplan pointed out in 1980, "to be committed to the view that if the methodology of evolutionary biology is valid for analyzing bones, muscles, and morphology, it should also be true for analyzing brains, minds, cultures, and languages."[16] Furthermore, as seen earlier in this book, many of the early twentieth-century social scientists were reformers rather than conservatives, yet many of them looked to biology in some fashion in seeking to explain human behavior.

Even in modern times, political radicals have been proponents of a biosocial perspective, as Marxist anthropologist Paul Heyer observed in 1982. He instanced political philosopher Herbert Marcuse and Noam Chomsky, the highly regarded student of linguistics, as well-known radicals who "have worked within a conceptual framework that assumes innate parameters to human nature. ..." Many of the best-known proponents of a biosocial approach to social science, as we have seen, like Alice Rossi, Melvin Konner, Allan Mazur, Davydd Greenwood, and Fred Willhoite, Jr., to name only the most vocal, are clearly political liberals. Heyer went on to describe as misguided any attempt to denominate biosocial or sociobiological approaches as "social Darwinism resurrected. ... The belief that human nature has innate parameters does not invariably lead to a justification of the status quo." Political theorist Roger Masters's answer to the charge of the inevitability of a conservative outlook was that the cost/benefit analysis or individualistic approach to social behavior, which inclusive fitness, or kin selection, seems to promote, actually permeates a number of interpretations of human behavior, running from pre-Socratic political thinkers, to neo-classical economics, to the ideas of Hobbes, Locke, and Rousseau. It is, Masters emphasized, an interpretation that has been drawn upon by a variety of political philosophies, from right to left. On the other hand, Masters noted, sociological theories that look to whole systems, like those of Durkheim or Parsons, can be conservative in social and political outlook. After all, Masters correctly observed, cost/benefit analysis in the hands of a Locke or a Rousseau turned out to be quite revolutionary.[17]

Some social scientists met the charge that the use of biological ideas gave an ideologically conservative bias to social inquiries by noting how a biological orientation can "provide the means for modification and change," as sociologist Alice Rossi told her colleagues at a professional meeting. "Ignorance of biological processes may doom efforts at social change to failure," she warned, "because we misidentify what the targets for change should be, and hence what our means should be to attain the change we desire." Anthropologist Melvin Konner recalled that "countless eyeglass wearers and insulin-takers" might have something to say about the relevance of biology to social improvement. He was convinced that "genetic analysis of behavior can lead to an increase, not only in human welfare, but in human freedom."[18]

Some of the doubts about the use of biological concepts, it needs to be said, did not really confront the central purposes of the new approach. For example, Sherwood Washburn in 1983 laid out in some detail his several reasons for doubting that sociobiology had anything to offer in explaining human nature. Upon inspection, however, virtually all of his objections referred to the inability of sociobiology to account for the *differences* among human groups or societies. Since very few of those who draw upon biology in discussing human behavior try to account for differences within the human species, Washburn's criticisms missed their target.* As Edward Wilson himself wrote in 1978, "the evidence is strong that almost all differences between human societies are based on learning and social conditioning rather than on heredity." A little later, he rephrased the same point: "Because we are a single species, not two or more, . . . mankind viewed over many generations shares a single human nature within which relatively minor hereditary influences recycle through ever changing patterns, between the sexes and across families and entire populations." Washburn was quite correct, therefore, in implying there was no danger sociobiology would eliminate the need for anthropology or any other social science that studied the *differences* among human social groups.[19]

Generally speaking, sociobiologists and other supporters of a biosocial approach seek to explain the behavior of the species, hence their interest in so-called universals across cultures. That is also why few, if any, biosocially inclined social scientists concern themselves with the question of racial differ-

---

*Sociologist Marion Blute, in her article "Sociocultural Evolutionism," and psychologist Daniel Freedman, in his book *Human Sociobiology*, are among the very few social scientists who have pointed to biology in accounting for differences *among* human social groups. Psychologist Arthur Jensen is perhaps the best-known advocate among social scientists of the existence of racial differences in intelligence. See his article, "How Much Can We Boost I.Q. and Scholastic Achievement?," *Harvard Educational Review* 39 (Winter 1969), 1–23. Unlike Blute and Freedman, Jensen does not draw upon either sociobiology or evolutionary theory. His work is more properly seen as following in the tradition of psychological testing of the 1920s.

ences. As political scientist Elliott White has written in this connection, "the only overall policy that is warranted from a population-oriented perspective is a program of equality of opportunity allowing and encouraging each individual to develop to the maximum of whatever capacities he or she possesses."[20] A further reason why racial differences are generally not discussed by social scientists who find value in a biosocial perspective is that many of them, as noticed already, are political liberals, for whom the investigation of racial differences is still fraught with disturbing memories of the evil misuse of such inquiries in the recent past.

Charges of a necessary conservative bias to a biosocial orientation may be relatively easy to counter, but other objections are not. Stephen Jay Gould, the well-known historian of science and paleontologist, who also happens to be a departmental colleague of Edward Wilson, is a persistent and outspoken opponent of almost everything sociobiologists support, despite his own fervent commitment to Darwinian evolutionary theory. The essential objection he sets forth in his book *Mismeasure of Man* is that biology, as opposed to culture, limits human opportunities. To see a biological influence in human behavior, he contended, is to offer a reason why something cannot be done. And it is true that in the past that is the way in which biological influences have been interpreted. Those were the kinds of conclusions arrived at by eugenicists in the early part of this century and they are often the conclusions pointed to even today in regard to opportunities for women. As Gould phrases the issue, sociobiology "is fundamentally a theory about limits. It takes current ranges in modern environments as expression of direct genetic programming, rather than a limited display of much broader potentialities." (But, by the same token, the logic of Gould's position often pushes cultural determinists, though not Gould himself, to the untenable conclusion that there are no biological or genetic limits on human nature.) Anthropologist Davydd Greenwood advances Gould's essential argument, but without the untenable implication. Greenwood takes a view that Lester Frank Ward advanced a century earlier. He sees culture or environment limiting biological potentialities. And it is quite true that today we know that better medical care, more food, and better nutrition, among other things, have greatly improved and lengthened the lives of people, especially in western Europe, North America, and Japan, from what they were a century ago. Life expectancy at birth in this country, to take another example, has almost doubled since 1900. The average size of Japanese and Americans has increased as well, and in the case of the Japanese the gain has come about in less than half a century. The limiting factor in those cases was not biology, but environment or culture. The biological potential was there all along; an improved environment was required to realize it.

The above examples and the comments of Gould and Greenwood illustrate once again the need to consider environment and biology and especially their interaction. An opponent of sociobiology like Gould does indeed emphasize that interaction, yet at the same time, he persistently resists investigations of the role of each of the interacting elements. "We cannot factor a complex social situation into so much biology on one side, and so much culture on the other," Gould insisted in 1984. Instead, we must try to understand the consequences "arising from an inextricable interpenetration of genes and environments."

For Gould the operative word seems to be "inextricable." Social scientists interested in the role of biology in human nature also recognize the interaction, as we have seen, of nature and nurture, but they also want to look more closely at the role of biology since it has been so frequently ignored. "Organisms are not passive objects, acted upon by internal genetic forces, as some sociobiologists claim," wrote Alice Rossi in 1984, "nor are they passive objects acted upon by external environmental forces as some social scientists claim." Instead, they are the product of both forces, which "interpenetrate and mutually determine each other." But rather than leave the matter at that level, as Gould recommends, Rossi urges study of the role of biological processes "in the same way sociologists try to specify social processes."[21]

It is true, as Gould and others have often reminded us, that culture is both quicker and easier to alter than biology, which, after all, is embedded in genes and has to work through generations. Biological evolution teaches that when environment shifts, and there is genetic diversity, those organisms best able genetically to adjust to the changed environment will survive, while those less well fitted genetically to the changed environment will gradually die out. The process is extremely lengthy by human standards. Culture, on the other hand, we like to think, is a matter of mind, and, as we all know, we can change our minds at will.

Socially and historically speaking, however, things are not that simple. For one thing, the cliché that history is a combination of continuity and change should alert us to the tendency of culture to persist, rather than to shift easily and quickly. Indeed, the very justification for the study of history as a form of understanding of human nature rests on the assumption that the past is difficult to escape. As Marx so graphically pointed out, history lies like a nightmare on the brains of the living. Friedrich Nietzsche in his essay "The Uses and Abuses of History" offered a similar view when he contended that the influence of the past on the present is so burdensome as to inhibit action or change.

To acknowledge the tendency of culture to endure, even in the face of efforts to alter it, should not be taken to mean that cultural and biological changes are equally slow. The point is merely to remind ourselves that both cultural change and biological change are historical and often slow, for as Gould himself has frequently emphasized, Darwin's explanation of the evolutionary process was actually a demonstration of the power of history to explain how nature, in which human beings are included, came to be as it is today. In Darwin's case it was history, that is, a changing environment, working upon nature. Human history, in turn, results from the interaction of human beings with their biological and cultural history and the present environment. In both cases, the past sets the framework within which the future evolves; in neither case can the process be reversed.

To see evolution and human cultural life as but two aspects of an integrated historical process does not dispose of all the conceptual doubts that have been raised against a biosocial approach to human experience. Even someone like Arthur Caplan, who defended sociobiology against the charge of social Darwinism and potential fascism, could still find a "conceptual pitfall" in drawing upon evolutionary biology in accounting for human behavior. There are "problems concerning causation, comparison, adaptation, and determinism," he warned, which cannot be resolved "by professions of faith in the utility of evolutionary analysis." And the underlying assumption of many sociobiologists that, because "a trait common in one culture is common across the cultural board" it must be genetically based, caused anthropologist James Silverberg a good deal of trouble. How about the controlled use of fire across cultures? he asked. That trait or skill is certainly universal among human societies and adaptive as well, but fortunately no sociobiologist has been foolish enough to claim it as biologically based.[22]

William Durham, an anthropologist much more committed to a co-evolutionary and biosocial approach to human behavior than Silverberg, nonetheless has similar reservations about some of the propensities of sociobiologists. "Correlations between theories from evolutionary biology and observed human behavior," he warns, "in no way constitute evidence of causation." Such correlations "may equally result from the *cultural* inheritance mechanism and a complementary process of *cultural* evolution." In the end, Durham's worry was the intellectual imperialism of some of the proponents of sociobiology rather than their use of biological concepts.

A similar reservation was expressed by Roger Trigg, a philosopher, who at other times has indicated his support of the use of biological concepts in analyzing human behavior. "Sociobiology must learn a certain humility," he warned. "It may well cast light on human nature by showing some of the constraints at work, but it cannot explain everything about human culture or the products of the human mind." More specifically, he directed attention to a limitation on the explanatory power of kin selection and reciprocal altruism, two key concepts of sociobiology theory. "Reciprocal altruism may explain our apparent concern for some beyond our family," he conceded, "but it does nothing to explain why we should care for those who are never likely to be able to benefit us in return."[23]

One of the most controversial aspects of Edward Wilson's conception of sociobiology was his suggestion that evolution—or natural selection—would favor one kind of morality or system of values over another. The contention, of course, struck at the heart of the concept of culture, since values were defined as the special creations of culture. Cornell anthropologist Davydd Greenwood, as we have seen, is certainly a committed believer in the importance of biological influences on human behavior, but he backed off sharply from Wilson's assertions concerning morality. "Evolutionary theory produces no clear moral imperatives," Greenwood flatly asserted. "Evolutionary theory must argue that the difference between 'is' and 'ought' cannot be bridged by science." That has been the view of the great majority of social scientists.

Not all students of human nature, however, have been so certain. Over thirty years ago, biologist-philosopher C. Judson Herrick in *Evolution of Human Nature* boldly announced that our biological history shaped values at least to the extent of making life a good thing. Much more recently, Peter Singer, the Australian philosopher, in his 1981 book *The Expanding Circle,* has argued that kin selection and other sociobiological concepts should be seen as the biological roots of human altruism and social concern, though he, too, in the end, believes human beings are free to choose their values, thanks to culture. Interestingly enough, and undoubtedly influential in Edward Wilson's own thinking, Darwin himself suggested that the content of human morality—not simply the tendency to be moral—was "selected" by evolution.

Darwin set forth his position in *The Descent of Man* at some length. A high standard of morality, he admitted, did not seem to give an individual in a tribe an advantage over others, yet an "increase in the number of well-endowed men and an advancement in the standard of morality will certainly give an immense advantage to one tribe over another," he contended. "A tribe including many members, who, from possessing in a high degree the spirit of patriotism, fidelity, obedience, courage, and sympathy, were always ready to aid one another, and sacrifice themselves for the common good, would be victorious over most other tribes." Since throughout history tribes have supplanted other tribes and "morality is one important element in their success," Darwin maintained, "the standard of morality and the number of well-endowed men will thus everywhere tend to rise and increase."[24] In sum, certain human moral values would be inherited and survive because they were advantageous.

Harvard child development psychologist Jerome Kagan is no sociobiologist, but he has long been interested in rediscovering the biological bases of child behavior. One result is that he comes closer to Wilson's, Singer's, and Darwin's conception of the biological sources of human morality than almost any other social scientist.* His work with young children has convinced him that "beneath the extraordinary variety in surface behavior and consciously articulated ideals, there is a set of emotional states that form the bases for a limited number of universal moral categories that transcend time and locality." Sometime before a child's third birthday, Kagan discovered, "we can count on the appearance of empathy and an appreciation of right and wrong. . . . Thus, there are both biological as well as cultural influences on the growth of morality," he concluded.

Kagan's striking conclusion derived from his work with children. Soon after the middle of the second year, he noted, "children become aware of standards. They will point to broken objects, torn cloth, and missing buttons revealing in voice and face a mood of concern." When fourteen- and nineteen-month-old children were in a room with deliberately broken toys, not one of them showed any interest in them; but over half of children over two became much concerned with the broken toys, taking them to their mothers with requests for repair. On the other hand, none of the children showed any interest in coats with extra buttons or a brush with too many bristles. It was the presence of defects, or missing parts that attracted their attention. A three-year-old, Kagan concluded, "is biologically prepared to acquire standards." Parents certainly contribute to that acquisition, he recognized, but even without that assistance, he believes, "all children have a capacity to generate ideas about good and bad states, actions, and outcomes." Empathy toward others and standards "are a part of a child's development," he explained, "because they are necessary for the socialization of aggression and destructive behavior."

In most of the world in the past and in the present, Kagan remarked in explanation, mothers have a number of children and it would be "adaptive if all three-year-olds appreciated that certain behaviors are wrong and were aware of their ability to inhibit such actions. Without this fundamental human capacity, which nineteenth-century observers called a *moral sense* the child could not be socialized." Or as he expressed the same point elsewhere, "the child does not

*A notable exception is Stanford anthropologist George Spindler, who has suggested that students of human nature may have to consider seriously the possibility that certain values like "conformity to social pressure, suggestibility to prestige figures, need for moral norms, bravery, and other tendencies of this kind are a part of human nature and a product of biocultural evolution." See his Introduction to George D. Spindler, ed., *The Making of Psychological Anthropology* (Berkeley, Calif.: Univ. of California Press, 1978), 28–29.

have to be taught that hurting others is bad; that insight accompanies growth."
Kagan did not flinch in following out the logic of his case and connecting his
finding with man's animal past. "Humans are driven to invent moral criteria,"
he concluded, "as newly hatched turtles move toward water and moths toward
light."[25]

As noted already, Kagan is unusual among social scientists, as Singer is
among philosophers, in being willing to see human morality in any way
indebted to animal origins or evolution. Indeed, the more common attitude
among critics is that biologically oriented social scientists are faced with a contradiction when they try to find liberal and humane values emerging out of
evolution and natural selection. Sociologist Howard Kaye in *The Social Meaning of Modern Biology* has been especially unrelenting in making this criticism.

After dismissing several proponents of such an approach as crude and unconvincing in their efforts to locate liberal values in the biological past of man, Kaye
turned to Melvin Konner's *Tangled Wing*, which he described as an "admirable
and finely balanced 'treatise on the biology of the emotions.'" Despite these
words of praise, however, Kaye concluded that "even Melvin Konner . . . fails
in his own, far more subtle attempt to 'cannibalize' the 'soul,' by giving a physiological, biochemical, and evolutionary account of its structure and function,"
while seeking to save it with a sense of awe. "Konner cannot succeed," Kaye
insisted, "because his reductionist account of the human mind . . . also 'chips
away at the lofty soul' he so desires to preserve."

In Kaye's view, liberal sociobiologists like Konner want to have their biology
play a controlling role in the development of human—and humane—values.
To do so, however, he argues, they must disconnect themselves from a biological system that is, in his judgment, clearly not humane. For "the scientific nat-
uralism with which he [Konner] hopes to 'modulate social chaos' and to design
a 'workable world' conflict with his moral longing for peace, equality, and
'some sacred social symbol that can make us whole again.'"

Perhaps the last, though certainly not the final, word on the application of
biological concepts to understanding human nature should be a little more balanced and positive. There is no requirement, writes physician-biologist Stuart
Kauffman, that "Sociobiology assert that every behavior is biologically heritable,
only that some are in some respects. Nor need Sociobiology assert that every
behavior is adapted for maximal reproductive success. If some are, that is interesting and important." Nor is it wrong or unscientific, he continued, to use
"our own cultural knowledge to aid in understanding the behavior of another
species close to our own. If there is no neutral way to understand another
human culture" except by importing "our own culture to the investigation of
the second culture" we generally then "come to feel we understand the second

when we have obtained a coherent, consistent picture of its functioning. The same happens when we attempt to understand the behavior of higher primates. That we succeed is prima facia evidence of strong homologies in our behavior patterns, not our inventiveness in imposing our categories on other organisms. Similar efforts would fail with an analysis of a hydra colony," he concluded. Or as political scientist Roger Masters remarked, "the first requisite for a rigorously scientific approach to human nature is . . . willingness to abandon the belief that answers are either/or: our behavior can be both innate and acquired; both selfish and cooperative; both similar to that of other species and uniquely human."[26]

The return of biology to the social sciences is still in its infancy, as some of the uses make clear. The process, however, is likely to continue if only because the biological sciences continue to throw fresh light on the nature of human beings in their relation with the remainder of the animal world. Already some ethological findings are reopening old questions. Is consciousness and awareness confined to human beings alone? Is it proper to see culture as man's identifying badge, as so many social scientists have insisted, or is this achievement, too, one we must share with our fellow animals? Are we, because of the uniqueness of our language, the only beings who think? The growth of an animal rights movement has no obvious or direct connection with the revival of biology in the social sciences, but are not its intellectual springs to be found in the fateful continuity that Darwin discerned between animals and us? These and other tenuous ramifications of the Darwinian Imperative that loom above and well beyond the present scholarly concerns of social scientists are the subjects of the Epilogue.

Epilogue: Beyond Social Science

The popular mind has always been in advance of the metaphysicians with reference to the mental endowments of animals. For some reason, there has been a perpetual hesitation among many of the latter to recognize in the manifestations of the animal mind, the same characteristics that are displayed by the human intellect: less the high position of man should be shaken or impaired.

Lewis Henry Morgan, 1868

Drawing analogies "between people and animals" is, on the face of it, rather like drawing them "between foreigners and people" or "between people and intelligent beings."

Mary Midgley, 1978

The return to biology in thinking about human nature necessarily meant revisiting Darwin and his insistence upon the continuity between human and animal experience. Many of the social scientists who argued for that continuity drew their inspiration from Darwin. Indeed, by the middle years of the twentieth century the great majority of American social scientists considered themselves Darwinians, just as they thought of themselves as heirs of Franz Boas and Alfred Kroeber, who saw culture as a peculiarly human mode of social expression. Even the committed proponents of a role for biology in human nature readily acknowledged the centrality of cultures. Their fundamental contention was that culture and biology worked together in the evolution of human nature. Their complaint against those who emphasized the place of culture was not that culture was peripheral but that in the search for human nature its advocates had pushed biology out of the picture.

From the outset, placing human beings within the evolutionary process had been problematic for many students of the nature of humanity. The difficulty was symbolized by the dispute between Darwin and Alfred Russel Wallace soon after their joint discovery of natural selection. Darwin to the day of his death never wavered from his conviction that human beings were included in evolution, that they were a product of natural selection. Wallace, as we have seen, retreated from that view in 1869 to espouse a non-materialist Spiritualist interpretation, which placed human beings outside the process that had molded all other living things.

Wallace's supernatural explanation gained few followers among social scientists in the second half of the twentieth century, but his assertion of the special, indeed unique, nature of man, because of his brain, continued to influence many, directly or indirectly. The eminent modern American anthropologist Loren Eiseley, for example, was among them. His sympathetic response to Wallace reflects the views of many other American social scientists today. Eiseley did not doubt that Wallace has a better understanding of the roots of human nature than Darwin. In his book *Darwin's Century,* Eiseley contrasted Darwin's conception with that of Wallace. "The mind of man, by indetermination, by the power of choice and cultural communication," he wrote, "is on the verge of escape from the blind control of that deterministic world with which the Darwinists had unconsciously shackled man. The inborn characteristics laid upon him by the biological extremists have crumbled away," he was relieved to report. In Eiseley's judgement, Wallace stood out among evolutionists of his own time because he recognized even then that human beings had escaped from biological evolution. "Wallace saw and saw correctly, that with the rise of man the evolution of parts was to a marked degree outmoded, that mind was now the arbiter of human destiny."[1]

It is not clear if "mind," as Eiseley used it, meant culture or if it meant the ability of human beings to be aware of themselves. In either case, Eiseley could not escape recognizing that he came up against Darwin's own view of man's evolution. For Darwin had stressed not only the evolutionary continuity in physical form of animals and human beings, he had also taken great pains to demonstrate continuity in mentality as well. Darwin had not hesitated to ascribe goals and purposes to the actions of wild and domestic animals. Some of his most committed followers worked even harder than he in collecting evidence that would demonstrate the continuity between man and beast. None was more vigorous in pursuit of that goal than George Romanes, a young English naturalist whom Darwin himself, in his declining years, selected to carry on his work. In the very year of Darwin's death in 1882 Romanes published *Animal Intelligence,* which was followed two years later by his *Mental Evolution in Animals.*

Romanes's efforts to probe the animal mind, however, soon came under attack, along with Darwin's own efforts. The rising discipline of professional psychology was too intent upon turning itself into a science to be able to accommodate "subjective" studies like comparative psychology, that is, the comparison of the minds of human beings and animals. The sin of anthropomorphism was created to rid psychology of just such comparisons. Ironically enough,

Romanes's best student, the Welsh psychologist C. Lloyd Morgan, became the leader of the attack upon animal consciousness. As Lloyd Morgan wrote in 1898, "the evidence now before us is not, in my opinion, sufficient to justify the hypothesis that any animals have reached that stage of mental evolution at which they are even incipiently rational." Animals, in short, contrary to Darwin and Romanes, could not think. In time, Morgan produced a highly regarded "canon," as it came to be called, which required psychologists to assume, until proved otherwise, that animal action lacked awareness.[2]

The culmination in the denial of animal consciousness by American social scientists came in the early twentieth century with the conquest of psychology by John Watson's behaviorism. Watson excoriated the study of consciousness not only in animals, but in human beings as well. Neither giants of biology, like Jacques Loeb, nor giants of the social sciences like Luther Bernard, Franz Boas, and Alfred Kroeber, would have anything to do with the "animal mind." As the president of the American Psychological Association, James Angell warned in 1913, "there are all kinds of dangers in assuming in animals consciousness of human beings even if the behavior is the same." Therefore, he was pleased to say, "most students of animal behavior shun every reference to consciousness."[3]

A few die-hards, it is true, continued to work under the assumptions of the old Darwinian continuity between the minds of animals and human beings. Robert Yerkes and Wolfgang Köhler, for example, doggedly pursued their innovative experiments with chimpanzees, setting tasks for them that required thought and decision making, seeking signs and measures of animal intelligence and awareness. And Teachers College psychologists Edward Thorndike, who had begun his experiments with animal behavior at the opening of the century, never abandoned his early belief in the biological roots of human behavior and in the mental continuity between man and beast.

The extent of the hostility among most psychologists toward any serious consideration of animal awareness is evident from the complaint of one psychologist in 1939. Writing in *Psychological Review,* the author denounced the profession's excessive fear of anthropomorphism. Some of the recent important work in learning theory, he contended, was actually anthropomorphic, but none of the researchers would admit it. The author, on the other hand, was convinced that the time had come to repudiate behaviorism's emphasis upon the nervous system in studying animals. "I consider it inevitable," he boldly announced, "that anthropomorphism must be used. We can make intelligible some principles of animal behavior only by describing it in anthropomorphic terms," he argued. But few of his fellow professionals were prepared to be so bold. It was left to popular writers, like journalist-naturalist Joseph Wood Krutch, to make a case for animal consciousness, even if in a low key. "Animal consciousness," he wrote in 1957, is not like ours. "But if we really are animals, then the difference," he thought, "is hardly likely to be as great as the difference between sentience and automatism." He then made a connection with Darwin that few social scientists were then prepared to acknowledge. "If our consciousness 'evolved' it must have evolved from something in some degree like it. If we have thoughts and feelings," Krutch maintained, "it seems at least probable that something analogous exists in these from whom we are descended."[4]

Krutch's probability matured into a state close to certainty with the flood of new ethological investigations in the 1960s and after. In effect, though perhaps without any direct knowledge of the connection, these mid-twentieth-century researchers were picking up where Darwin and George Romanes had left off. By turning to the observation of animals in the wild—in their natural habitat—these new ethologists added valuable insights to the Darwinian quest. By demonstrating the important place learning held in animal behavior under the pressure of diverse *natural* circumstances, the ethologists moved the study of the animal mind far beyond what Thorndike with his primitive psychological puzzles and mazes had begun, or what Köhler and Yerkes had been able to discover even with their more sophisticated tests of ape intelligence. Increasingly the new ethology discerned consciousness or "mind" in animals. In the process, the gap between human and animal nature narrowed just a little further.

Natural scientists were quick to link the recent ethological findings to Darwin. Paleontologist George Gaylord Simpson, writing in 1972, pronounced Darwin to be quite correct when he contended a century before that, like man, "the lower animals manifestly feel pleasure and pain, happiness and misery." Some behaviorists, Simpson admitted, deplore such assertions, but it is no less than a "commonsense interpretation of thousands of observations and I, for one, do not think that commonsense needs to be banished from science." Like Darwin before him, Simpson recognized in the similarity of the human and animal mind a sign and a measure of their common embodiment in nature and their common evolution through natural selection.[5]

About the same time, Harvard biologist and evolutionary theorist Ernst Mayr defended the proposition that animals were sufficiently aware or conscious to exhibit color preferences, as Darwin himself had suggested. In *The Descent of Man* Darwin had advanced his theory of sexual selection in an effort to account for morphological differences between the sexes of certain species, differences that natural selection seemed unable to explain. If natural selection is to work, an organism must possess a difference that would provide it and its progeny with an advantage in the struggle for existence. Since Darwin could not discern any advantage in the different colors or plumages of the sexes in some species of birds, he needed a theory in addition to natural selection.

Among birds, according to the theory of sexual selection, the female generally makes a choice of colors or some other difference among her potential mates. Those choices, the theory argues, account for the differences in appearance of males and females.

Ernst Mayr offered three reasons why he saw a connection between female choice and the presence of the colors in males. The first was that sexual colors, that is, those that distinguish the two sexes, do not occur in species that court in the dark. Second, colors do not develop on body parts that are invisible, such as the wings of hummingbirds. And third, sexual colors are generally best seen from the direction which a female would assume during courtship.[6] In sum, some consciousness or awareness must be present.

To philosopher Mary Midgley, animal consciousness was at once more obvious and more profound. Forthrightly she sought to counter the frequent assertion that, in the absence of language, an animal's thoughts or mental activity could not be known by a human being. Language is not the only source of information, she pointed out. Patterns of behavior when closely observed provide insight into inner feelings. "The species barrier is, in itself, irrelevant," she contended. "Numbers of one species do in fact often succeed in understanding members of another well enough for both prediction and a personal bond. Nothing more is necessary" to prove the case, she insisted, undoubtedly thinking of the relation of many human beings with dogs and cats. It was an analogy, incidentially , that was a favorite with Darwin.[7]

Social scientists, unlike biologists or philosophers, were reluctant to acknowledge the continuity between the human and animal mind, which Darwin had been so intent upon establishing. As Rockefeller University animal physiologist Donald Griffin wryly remarked in 1978, even most students of animal behavior denied animal consciousness, either to avoid anthropomorphism or because they despaired of ever finding out if it existed. If there is "any scientific view that has hindered investigation of important subjects," he complained in 1984, "it is the behavioristic taboo against considering conscious experiences of animals and men."[8]

As early as 1978 Griffin had suggested that "a necessary first step is to open our eyes to these possibilities and begin to adopt in ethology some of the approaches customary in humanistic scholarship." After all, he quite rightly pointed out, "humanists are accustomed to dealing effectively with evidence that does not allow absolutely rigorous conclusions about causes and results." He thought that "cognitive ethologists could make good use of these kinds of critical scholarship, which have been successful in the humanistic disciplines." Indeed, two years before, in 1976, Griffin himself took up the task of reconsidering the continuity of mental evolution in his book *The Question of Animal*

Awareness. It meant abandoning the rigid anthropomorphism insisted upon by Lloyd Morgan, he admitted. But whatever need a young discipline may once have felt for such a "scientific" concept, the passage of time and the maturity of professional psychology had long since removed. Moreover, recent ethological studies now provided radically fresh arguments for loosening constraints once thought indispensable.[9]

Griffin opened his argument from a platform of organic theory. "Neurophysiologists have so far discovered no fundamental differences between the structure or function of neurons and synapses in men and animals," he noted. "Hence unless one denies the reality of human mental experience, it is actually parsimonious to assume that mental experiences are as similar from species to species as are the neurophysiological processes with which they are identical." The inescapable implication, he concluded, is "qualititative evolutionary continuity (though not identity) of mental experiences among multicellular animals." He rejected charges of anthropomorphism on the ground that they rested on the "questionable assumption that human mental experiences are the only kind that can conceivably exist." Such a belief in the uniquesness of a single species he stigmatized as "not only unparsimonious; it is conceited. It seems more likely than not that mental experiences, like many other characters, are widespread, at least among multicellar animals, but differ greatly in nature and complexity." After all, he continued, drawing on evolutionary theory, "the hypothesis that some animals are indeed aware of what they do, and of internal images that affect their behavior, simplifies our view of the universe by removing the need to maintain an unparsimonious assumption that our species is qualitatively unique in this important attribute."

Consciousness in animals reflected another evolutionary principle. It enabled "animals to react appropriately to physical, biological and social events and signals from the surrounding world with which this behavior interacts." Griffin's enthusiasm or commitment to the concept of animal consciousness is evident in his admonition that "we should not overlook the broader possibility that some animals may also have mental experiences about which they do not communicate." He failed to say, however, how those silent experiences would be revealed.[10]

As Griffin himself acknowledged, his theoretical arguments rested upon a mass of recent observations of wild animals' behavior that appeared to be conscious. As early as 1965, the prominent Berkeley physical anthropologist Sherwood Washburn remarked on African baboons' behavior patterns that varied according to character of the predator. When carnivores like lions threated the baboons they fled *up* trees; if man was the predator, however, they fled *down* from trees. "This difference persists even in groups protected in parks," Washburn noted, "even though, for at least some, it is probably years since the group was shot at." In short, the baboons had learned not only to fear man, but how to protect themselves against the means of predation peculiar to men, and that valuable information was being acted upon even by those who had no direct experience with predatory human beings.

Washburn described another incident in which baboons who had become accustomed to cars traveling through their range suddenly refused to come near a car after a visiting parasitologist shot two of them from a car. "Eight months later," Washburn reported, "it was still impossible to approach the group in a car." The significant point, Washburn stressed, was that it was "most unlikely that even a majority of the animals saw what happened." Yet they picked up the clue from those that had. "It is highly adaptive for animals to learn what to fear without having to experience events directly themselves," he explained.[11]

Primatologists Michael Chance and Clifford Jolly in 1970 ascribed to some savannah baboons what they called "displacement activities" such as yawning, shrugging shoulders, wiping the muzzle with the hand, scratching, fiddling with food objects, along with "frantic grooming of the partner while being harassed, and rapid copulations." They described those actions as "out of context behavior," betokening feelings of uncertainty and insecurity.

Even more redolent of human consciousness was their description of the reaction of wild chimpanzees coming upon a stuffed leopard that Chance and Jolly had "planted." After a moment of dead silence, the troop let out "a burst of yelling and barking," followed by much charging about by each member. Some fled, only to return a little later, joining the group in leaping around, charging the unmoving leopard with torn off branches and sticks, and sometimes throwing sticks and other "weapons" at the leopard. Throughout, "the bloodcurdling barking was loud enough to waken a human neighborhood" a quarter of a mile away. In between the charges, periods of stillness occurred only to be "followed by periods of seeking and giving reassurance by holding out of hands to be kissed, touching of their neighbors and homo and hetro pseudo-copulation. Voiding of diarrhoea and enormous amounts of intense body scratching took place," they reported. Ultimately, and significantly, the attacks on the leopard resulted in its head being detached from the body, the perpetrator of which rolled the head around. Another member of the troop then "seized the tail and they all rushed off into the bush with its body."[12]

An even more remarkable instance of interaction and cooperation among chimpanzees that seemed to mimic human consciousness was told by primatologist Jane Goodall and psychiatrist David Hamburg. Several male chimps watched for more than two minutes, without moving, as a single male pursued a young baboon from tree to tree. Only when the victim desperately leaped to

the ground and its direction of escape became clear did the group move to capture it. Like Chance and Jolly, Goodall and Hamburg in their dealings with
wild chimpanzees were impressed by both the excitement and the expressions
of reassurance displayed by the animals when they confronted a large pile of
food. "Three or four adults may pat each other, embrace, hold hands, press their
mouths against one another, and utter loud screams for several minutes before
calming down sufficiently to start feeding," they reported. The implications
seemed obvious. "This kind of behavior," they wrote, "is similar to that shown
by a human child, who, when told of a special treat, may fling his arms ecstat-
ically around the bearer of the good news and squeal with delight."[13] Nor were
apes the only animals in which cooperation and emotion seemed to reflect
consciousness.

Donald Griffin tells of watching a group of five lionesses carry out an elaborate scheme to bring down a wildebeest. Three of the lionesses positioned themselves in such a way as to distract the attention of the wildebeests while a fourth
hid in a small ravine over which the wildebeests fled when the fifth lioness star-
tled them. The lioness in the ravine then leaped up and seized one of the animals
as it ran over the trap. All five lionesses then shared the catch. Similar evidence
of planning ahead was reported by primatologist Frans de Waal in observing
chimpanzees on an island in a zoo. "An adult male may spend minutes searching for the heaviest stone on his side of the island, far away from the rest of the
group," de Waal wrote, "weighting each stone in his hand each time he finds
a potentially bigger one." The stone finally selected is then carried to the other
side of the island, where the chimpanzee begins an intimidation display toward
his rival, with the stone as a possible weapon. "We may assume that the male
knew all along that he was going to challenge the other," de Waal concludes.[14]

Both de Waal and Griffin recognized that in these and other instances they
were inferring intentionality in the animals' behavior. Can one believe that animals, like human beings, actually *intend* to perform a particular act? In his book
The Question of Animal Awareness, Griffin confronted the issue squarely.
"Since both conspecifics and human observers can predict the future behavior
of an animal from its intention movement," Griffin pronounced it remarkably
strange "to assume that the animal executing the intention movement cannot
anticipate the next steps in its own behavior." Behaviorist psychologist John
Watson and biologist Jacques Loeb, he remarked, "may have led us down a sort
of blind alley, at the end of which we find ourself defending to the last, at least
by implication, a denial of mental experience to animals, a denial which we
cannot justify on any explicit basis except the presumed absence of communication with conscious intent."[15]

It is perhaps revealing of the state of thinking on the subject that Griffin has received stronger support for his position on intentionality from philosophers than from his fellow natural scientists, presumably because of fear of the sin of anthropomorphism. Belief in intentionality, it is true, has not been a majority stance among philosophers, but Mary Midgley in her book *Beast and Man* clearly assumes that animals, like human beings, often know what they are about. So does Berkeley philosopher John Searle. In fact, in the course of a discussion on the merits of sociobiology in 1978, Searle belabored the sociobiologists for failing to recognize the place of intentionality in animal behavior. "To say that an animal is eating, or is stalking prey, or is engaged in aggressive behavior is already to ascribe to its intentional states," Searle pointed out. He wondered why psychologists and biologists had given the idea of intentionality in animals so little attention. The answer, he thought, lay in their "illusion that behavior is somehow observable in a way that intentional states are not observable." To which Searle replied, "There is no way to observe behavior without observing intentional states."[16]

Another philosopher, C. Wade Savage, joined Midgley and Searle in accepting the idea of animal awareness and its corollary, the rejection of ethology's cardinal sin, anthropomorphism. He had no doubt that "chimpanzees and many animals less like humans than chimpanzees ascribe purposes, knowledge, and feelings to others, and sometimes to themselves." Searle went Savage one better, encountering no difficulty in perceiving intentionality even lower on the scale of life than Savage did: in the behavior of honeybees. Searle pointed out that once a honeybee worker ascertains a source of food she returns to the hive where she performs the famous "waggle" dance that informs the other workers where they could load up with nectar. "Her fellow workers recognize the features of the waggle dance," Searle noted, and therefore a basic intention has been given. The workers "then know where the source of the food is. In this account," Searle continued, "we have no intellectual discomfort at all in construing the waggle dance of the honeybee as a case of communication. It is very much like a human speech act," he concluded.[17]

Communication among honeybees through the agency of a special dance, first documented by the eminent Austrian ethologist Karl von Frisch in 1965, has been undoubtedly the most spectacular example of animal awareness. For not only does a bee through the dance inform her sisters of the direction in which to fly to tap the source of nectar she located, but the distance is also communicated in the dance. As Donald Griffin observed "such complex signalling would have been surprising enough in mammals. . . . To find symbolic communication in an insect was truly revolutionary. If an insect brain weighing only a few milligrams can manage flexible two-way communication, the possibility clearly arises that language-like behavior may occur in other animals as well." Or to put the issue another way: If bees, which diverged at least half a billion years ago from the line of descent that led to *Homo sapiens,* seem to have communication skills similar to those of human beings, such behavior may not be "the exclusive prerogative of any one species," Griffin archly remarked in 1976.[18]

Communication through the waggle dance by bees is not the only sign of their intentional behavior. Princeton ethologist James L. Gould, who is now a leading authority on bee communication, offers two striking examples. The first was originally reported by von Frisch. When experimenters with bees moved food sources successively farther away from the hive, the bees soon "recognized" what was happening and began flying directly to the new location in anticipation of the move. As Gould commented, "it is not easy for me to imagine a natural analogue of this situation for which evolution could conceivably have programmed the bees." The bees, in short, were reacting to circumstances rather than responding "mindlessly" to their genes.

Gould's second example was equally dramatic and mysterious. The alfalfa plant has a blossom in which the anthers, which carry the pollen, are so placed that when an insect enters the flower to obtain nectar, the anthers are released, striking a hard blow at anything in their path. Bumblebees can absorb such punishment because of their large size, but the smaller honeybees are pushed out of the flower by the force of the anthers. After a single such experience a particular honeybee will systematically avoid alfalfa blossoms. Only pain of starvation will drive it to seek out such blossoms again, but even then it will enter only those in which the anthers have already been sprung. If flowers with spent anthers cannot be located, a bee will chew through the side of the flower to circumvent the blows of the anthers.[19]

The implications of animal awareness for understanding the nature of human beings are at least unsettling and perhaps even profound. On the one hand, as Griffin himself has remarked, if the complex, responsive behavior now known to exist among honeybees is genetically based, as most ethologists seem compelled to believe, how can we be sure our own complex behavior is not genetically determined or at least rooted in our genes? On the other hand, in his latest book *Animal Thinking,* Griffin placed a new emphasis upon "versatility and adaptedness to the problems animals face in the natural world where their species has evolved. When animals adjust their behavior effectively to solve problems," he surmised, "they are likely to think and feel consciously to some degree." In sum, by 1984 he was no longer worrying about the source of animal behavior; it might be from evolution or from learning. The whole question, he was convinced, deserved continued investigation, but he had not departed from his earlier position that it was highly unlikely that awareness and thinking were "unique to a single species." The assumption of "a human monopoly on conscious thinking" had become "more and more difficult to defend as we learn about the ingenuity of animals in coping with problems in their normal lives."[20] In that opinion he was right back with Darwin.

Remarkable, even astonishing as the apparent awareness exhibited by honeybees may be, no ethologist or even philosopher argued for an evolutionary link between Hymenoptera and *Homo sapiens*. As Griffin himself pointed out, the evolutionary distance was too great. (Griffin, however, could not refrain from suggesting a similarity between bees and human beings that might account for the advanced means of communication they had in common; their complex social organization. It would provide a basis for the evolution of language.) Yet the similarity in mental activity, once the specter of anthropomorphism had been exorcised, was not easy to explain or dismiss. It pushed forward tantalizing questions about what it meant to be human. If a mental activity like intentionality could be demonstrated in animal species, both those close and those remote phylogenetically, how could such mental behavior be seen as clearly differentiating human nature from animal nature? Was it one more trait that human beings shared, albeit in differing degree, with animals? Were there no longer any differences in kind?

Over the centuries, astute and cynical critics of humanity alike have directed poisonous barbs at the unwholesome traits of human beings that not even animals exhibited. Of all the aninmals, Thomas Jefferson on occasion remarked, only man murdered his own kind. Others pointed to the peculiarly human proclivity to war; still others to the forcible rape of females. Yet today, thanks to ethological research, each of these allegedly human sins has been found to have its analog in species other than *Homo sapiens*. That finding does not make the intention identical or even nearly so; it does reduce a little further, however, the specificity, the particularity of those differences between human and animal nature that our species for so long has been intent upon uncovering. By closing the gap still further, it makes the task of identifying the differences just a little more difficult. That gulf between beast and man, to be sure, is far from closed; yet as ethologists pursue their studies of our fellow animals, the edges continue to move closer to one another.

We may share awareness or consciousness with animals, but do animals have any sense of themselves; do they recognize themselves as distinct beings, as surely we do? A consciousness of self has been a central element in what it means to be human. Our imaginative literature, our myriad of works in the social sciences, even the products of our natural sciences, indeed, our everyday thoughts reflect a deep awareness of self. Yet here, too, some investigators of animal behavior have found an overlap in the awareness of human beings and animals, an overlap that also fits closely into the Darwinian paradigm of continuity.

Darwin, George Romanes, and other early Darwinians talked easily about the "mind" of animals, and as late as 1908 Cornell psychologist Margaret Washburn could publish a book entitled *The Animal Mind*. But animal minds disappeared from psychology about the same time that animal awareness did. Only in recent years have psychologists and other students of human nature been willing to revive such concepts. Of those social scientists who have been so bold, none has been more ingenious or more persistent in measuring self-awareness or what he calls the "animal mind" than psychologist Gordon Gallup, Jr., of the State University of New York at Albany. For some years he has been devising ways to test animals, in the absence of language, for a sense of self-awareness, the aim being to ascertain if animals have "minds" as well as brains.

The problem Gallup set for himself is more complicated than identifying a sense of self-interest. Animals, after all, are born with that; the concept stands at the center of the Darwinian paradigm, inherent in the phrase "survival of the fittest." The heart of Gallup's quest is: Does an animal have a sense of self to the degree that it can distinguish itself from other members of its species? Practically, the problem translates into the question: What does an animal see when it looks at its reflection? According to Gallup's findings, almost all animals see just another of their species, if they see that. (Obviously, those animals that identify others primarily by a sense other than sight see little or nothing at all.) No animal, when first placed before a mirror, seems to recognize itself. Indeed, as Gallup points out, even some human beings, those who are severely mentally retarded, never learn to identify themselves in a mirror. Therefore Gallup had to allow his chimpanzee subjects to become familiar with a mirror.*

The procedure was simple enough. He installed a mirror in the cage for a period of ten days. Within two days the chimpanzee subjects began to recognize that the image in the mirror was not that of another chimp. Soon they began to groom themselves while looking in the mirror; they then began to check out parts of themselves they could not see before. With that training completed, Gallup anesthetized several chimps and painted marks over their eyes and ears,

---

*Wolfgang Köhler back in the 1920s also introduced his chimpanzees to mirrors. Köhler's chimps, like Gallup's, recognized themselves in time, but it is a measure of the changed outlook among psychologists then and now that Köhler felt compelled to warn the reader that "nothing is said ... as to the 'consciousness' of the animal, but only as to his 'behavior.'" Wolfgang Köhler, *The Mentality of Apes*, trans. Ella Winter (2nd ed., London: Kegan Paul, Trench, Trubner, 1927), 101–2.

parts of their bodies that could not be seen without a reflection. When the chimpanzees recovered from the drug and looked in the mirror, they immediately touched the paint spots with their fingers, smelled their fingers, and spent three times as much time before the mirror as they had before being marked. It is significant that chimps that had not been trained to use a mirror, though also marked, only gestured and vocalized as if the image in the mirror were that of another chimpanzee.[21]

An important by-product of Gallup's research was his discovery that of all the other subjects he tested in that way, only organgutans exhibited the chimpanzees' awareness of self. Modern molecular biology has established the evolutionary distance between apes and human beings. Chimpanzees and organgutans, along with gorillas, in this evaluation, come closest to human beings. All other primates, including monkeys, are much farther removed. Gallup was surprised, therefore, to find gorillas lacking that same sense of self-awareness that he had found in organgutans, since gorillas are slightly closer to human beings in the molecular structure of their heredity than organgutans.*

On the assumption that the exposure to a mirror was too short for certain more evolutionary distant primates like monkeys and gibbons to exhibit self-awareness, Gallup left a mirror in a cage of macaques for five months, but there was no change in their threatening behavior when they observed themselves in the mirror. Another researcher kept a mirror in a macaque's cage for a year without any indication of self-recognition. Nevertheless, monkeys can learn to take cues from mirrors. They will turn away from a mirror to obtain something they see in it, suggesting that they can move from seeing only another monkey to seeing reflections. But they do not take that ultimate step of recognizing themselves. Frans de Waal offers another anecdotal indication of the difference in the mentality of monkeys and chimpanzees. "Every zookeeper who happens to leave his broom in a baboon cage knows there is no way he can get it back without entering the cage," de Waal writes. "With chimpanzees it is simple. Show them an apple, point or nod at the broom, and they understand the deal, handing the object back through the bars."[22]

Gordon Gallup describes the gap between the perceptual abilities of the organgutan and the chimpanzee on the one hand, and that of the macaques or monkeys on the other, as a "step-wise change . . . akin to a psychological void of sorts. What is missing" in the monkeys, he concluded, "is a sense of self." Without that sense, he suggested, there can be no "mind." To Gallup, "con-

---

* According to developmental psychologists Jerome Kagan, self-awareness comparable with that of the chimpanzees first emerges in human children at age two. Among them it appears in language as well as in a recognition of reflections of facial marks. See his *The Nature of the Child* (New York: Basic Books, 1984), 136–37.

sciousness is being aware of your own existence, and mind is the ability to monitor your own mental states." In effect it means that the organism knows it is acting and feeling, that is, that it has a self. Thus, by Gallup's standard, chimpanzees and orangutans can be said to have "minds."[23]

Primatologists have also been exploring the chimpanzee mind. David Premack and Guy Woodruff have been investigating the chimpanzee's ability to recognize intention in another individual, in this case, in a member of another species. Their subject was a fourteen-year-old chimpanzee named "Sarah" with whom they had worked for years. The test was intended to ascertain if Sarah could infer the intention of a human actor. They showed the chimpanzee video presentations of a human being trying to obtain a banana in four different problem situations. (The situations were actually those that Wolfgang Köhler had made famous earlier in his work on chimpanzee intelligence.) Sarah was then asked to choose between two photos of possible solutions to each of the problems. In 24 tests (six times for each problem) Sarah chose correctly 21 times. (The three failures were all in connection with one problem; in the other three, her score was perfect.)

Premack and Woodruff then escalated the difficulty by setting up four "human" instead of "chimpanzee" problems. No bananas this time. The first video showed a human actor struggling to escape from a cage, for which the right answer was a picture of a key. The second depicted an actor shivering and kicking a heater, for which the correct answer was a lighted paper cone. The third presented an actor seeking to play a phonograph with the electrical plug out of the wall outlet. The fourth displayed an actor unable to wash a dirty floor because a hose was not properly attached to the faucet. This time Sarah made no mistakes at all. The researchers then raised the level of difficulty by increasing the number of solutions. Thus, for the escape-from-the cage problem the choice of answers was now a broken key, a twisted key, or an intact key; in the case of the electrical cord, it was now plugged in, but cut; unplugged and intact; and plugged in and intact. On the three problems she failed only once out of eight trials: she chose a twisted key over an intact one. In these acts she had had no previous training, but she had frequently observed human beings in these situations. Premack and Woodruff interpreted Sarah's actions as imputing "at least two states of mind to the human actor, namely intention or purpose on the one hand, and knowledge or belief on the other." The animal had clearly recognized that the human beings wanted something.

In a final experiment they were able to show that Sarah could act upon her own feelings. The researchers displayed to her video scenes of two of her trainers confronting problems. One of the trainers was believed to be liked by her while the other was disliked. When given a choice of photos of possible solutions to their respective problems Sarah almost invariably selected incorrect solutions for the disliked trainer while providing correct ones for the trainer she liked.[24]

One does not have to go all the way with Gallup and Premack and Woodruff in concluding that certain great apes have minds to recognize that their work is yet another indication that a mental ability—an awareness of self—generally thought to be confined to human beings has an analogue in at least two other species. That the species in question are the closest to *Homo sapiens* would not surprise Darwin. Given his conception of mental evolution and man's place in it, he probably would have predicted it.

Intentionality, self-awareness, and inference are not the only human attributes that recent studies in animal behavior have found in non-human animals. Ever since social scientists, following the lead of anthropologists, adopted the concept of culture, they have increasingly identified it as a peculiarly human characteristic. The differences between animals and human beings, Franz Boas contended, were "so striking that little or no diversity of opinion exists." The principal reason for that unanimity, he added, was that only human beings communicated through language and used "utensils of varied application. . . . The higher apes employ now and then limbs of trees or stones for defence," he continued, "but the use of complex utensils is not found in any representative of the animal series."[25]

Alfred Kroeber phrased the differences more precisely, introducing the word "civilization" in place of culture, but the message was the same. "The beast has mentality, we have bodies; but in civilization man has something that no animal has." His fellow anthropologist Robert Lowie spelled out specifically what was uniquely human in civilization or culture. "Anything and everything a man . . . acquires from his social group is called a part of its 'culture,'" he explained in 1929. Learning the ways of one's fellows is a peculiarity of mankind's, he emphasized. "For even the highest apes have nothing of the sort." They may fashion crude tools to obtain food that is out of reach, he conceded, and that can be called "an invention—raw material for culture. If his neighbors imitated him, if he taught them this trick and they all passed it on to their offspring," he explained, "chimpanzees would be on the highroad to culture. But they do nothing of the sort." Contrary to popular opinion, he remarked, apes are not imitators, and the "inventor" of the tool "cares not a fig whether his brilliant idea becomes a part of chimpanzee behavior in the future. That is why apes hover on the outskirts of culture but never quite get there."[26]

Not all social scientists of the time drew such a sharp and rigid line between cultural man and biological animals. Bryn Mawr sociologists Hornell Hart and Adel Pantzen in 1925 asked "Have Subhuman Animals Culture?" and ended with a guarded affirmative answer.[27] Their expression of faith in the cultural as well as the biological continuity between humanity and animals, however, was rare. Only much later in the century, thanks to the explosion in ethological knowledge, would their affirmation of animal culture seem more than idiosyncratic.

From its theoretical origins, notably in Alfred Kroeber's seminal 1917 essay, "The Superorganic," culture's distinguishing element, as Lowie stressed, was its capacity to be passed from generation to generation. The cultural information might be no more than a new tool or be as broad as a ritual, but all that either required in order to be part of culture was for a parent or a tribe to teach it to a new generation. Biological information, in contrast, needed to go through the genes; its process for bringing about change was painfully slow, indeed so slow as to be unrecorded over the whole range of human history. Culture, on the other hand, as social scientists liked to remark, was Lamarckian in character. Habit was sufficient to pass the idea or practice along within a single generation. Animals and human beings, so the argument ran, changed through biology, as Darwin had taught; only human beings could change through culture as well. That was, in the end, the fundamental difference between animals and man, and which Boas, Kroeber, and Lowie had delineated so brilliantly. Now however, even in regard to culture, the findings of the new ethology seemed to be blurring the distinctions. *Homo sapiens,* it was becoming apparent, was not the only species to change itself through cultural Lamarckianism.

Perhaps the most commonly observed cultural transformations by animals is their use of tools or contrivances in gaining food or accomplishing some other action. Chimpanzees in the wild, for example, have been commonly observed using a blade of grass to extract termites from their underground nests. Boas and Lowie, it is true, recognized that certain animals used tools, but the termite-hunting chimpanzees did more than invent a useful tool. The adults teach the young how to manipulate the grass blades to extract the termite food. In short, they were passing on a practice and, in the process, *changing* a species' behavior by adding to its behavioral repertoire. (They were also, incidentally, displaying intentionality.) A less familiar pattern of some chimpanzees is "leaf-sponging," a means of slaking thirst. A chimp crushes a handful of leaves in its mouth and then uses the mass as a sponge to soak up water from a hole in a tree or another narrow container. An animal may do this a dozen times to satisfy its thirst. "Chimpanzees are the only non-human species known to use tools in the wild to facilitate water intake," commented one psychologist. Since not all chimpanzees exhibit this behavior, it is clearly learned.[28]

Although examples of behavior patterns that can be described as cultural are fairly numerous among chimpanzees, they are not the only animals to acquire behavior patterns from each other. Donald Griffin, for example, has reported on a certain kind of crow that picks up and drops whelks (a form of mollusk) from the air onto rocks below as a way of cracking the shells in order to feed on the meat inside. Since only some crows follow this practice it apparently is a learned one, and probably one that has been passed from individual to individual, just as human beings have picked up advantageous practices by observing others of their species. A similar transfer of a practice—one that Robert Lowie had specificlly denied animals could accomplish—was actually observed in modern Great Britain among certain domestic birds. Apparently a great tit discovered that it could penetrate with its bill the aluminum cap on milk bottles left on the steps of houses. Within a short time, as the practice spread among the birds, scores of milk bottles had been penetrated and the milk and cream consumed.

No one has observed the spread of sea otters' use of tools, but their usage also seems to have spread through learning, rather than through genes. Sea otters frequently place flat stones on their bellies, as they float on their backs. They then use the stones as anvils upon which they crack the shells of mussels and other mollusks for the food inside. The technique clearly opened up a fresh source of food for the animal and so offered an incentive for its spreading just as the opened milk bottles offered a similar enticement to the great tits.

Cynthia Moss, a young elephant ethologist, has documented a practice that comes close to the transfer of ritual among animals. After witnessing the killing of an elephant by human hunters, she watched the reaction of the elephants to the stricken beast. The mother broke off one of her tusks in seeking to raise the felled animal, while another member of the group tried to stuff grass in its mouth. Once the group seemed to recognize that the animal was dead, some members touched it with their trunks and feet, kicking loose dirt over the body. Others brought branches to lay on the carcass. By nightfall, the body had been nearly covered by branches and earth; the group of elephants did not leave the scene until dawn. The last to go was the mother, who several times touched her daughter with her hind foot, then brought up her trunk to touch her broken tusk before shuffling off to join the others.

It is difficult to know whether to interpret such behavior as an example of cultural transmission or of animal awareness. Like all instances of protocultural behavior, it is most likely both. Such proto-ritualistic behavior is probably passed on from generation to generation, though it obviously requires a remarkable degree of awareness to have been performed in the first place.

Undoubtedly the most striking and convincing instance of cultural transmission among animals is that begun by Imo, a Japanese macaque living on a primatological research island. At about two years of age, Imo was observed carrying a potato, the food provided to the animals, to a brook to wash off the sand. In subsequent years the practice of washing the potatoes slowly spread through the group of monkeys to become a custom. Soon the custom spread to using the surrounding sea. Potato-washing in *salt* water became an established part of the custom, taught by mothers to their offspring as part of the routine of eating potatoes. Significantly, in the beginning only the young monkeys and females learned it. Five years after Imo began her innovation, nearly 80 percent of the young animals, age two to seven years, washed their potatoes; only 18 percent of those over seven, that is those older than Imo when she initiated the practice, washed their potatoes. Of the older ones that picked up the practice, all were females.

Apparently, the transfer from generation to generation was deliberate. Mothers began by bringing their offspring into the water as they clung to the belly, even though earlier generations has stayed away from the water. At six months the babies were picking potatoes out of the water; by a year or more they had mastered the washing technique. Along the way they also learned to like the salt taste imparted by the sea water, for they could be observed dipping the potatoes back into the salt water between bites. As with the sea otters, the macaques had discovered a new source of food with less expenditure of energy, just as an early hominid might have found an improved way of maximizing resources through a cultural tranfser.

The innovative genius of Imo did not end with the potato trick, any more than the cultural innovations of early human beings did. When the monkeys' keepers decided to change the food from potatoes to wheat grains in order to evaluate the digital dexterity of the animals in picking out the grains from the sand, Imo came to the rescue once again. She built upon her early use of water by carrying a handful of sand and grain to the water and dropping it. The sand sank and the grain floated. She then scooped up the grains of wheat and ate them. (Primatologists marvel at Imo's solution, for this time around she departed from traditional macaque behavior. The innovation required her to "throw away" food, as it were, in order to improve it; hardly a common practice in her species.) This practice, too, soon spread; another instance of incipient cultural transfer.[29]

Even the most striking instances of transfer to behavior patterns among animals are, of course, far removed from what Boas wrote about or Kroeber called civilization. It is very likely that, as ethologists continue their researches and observations in the wild, other examples of cultural behavior will be uncovered among animals, but none will come close to the most primitive forms of cultural tranference among human beings. We are indeed the prime cultural animal. And that qualification once again carries us back to Darwin's vision: the recognition of continuity in mentality between us and animals. Even culture, that characteristic once thought to be unique to human nature, can now be seen as having its origins, however rudimentary, in our animal ancestors.

The possibility of including culture in evolution, as we have seen in previous chapters, intrigues several anthropologists, but none has linked his thinking so closely to Darwinian evolution and sociobiology as the Australian Derek Freeman. Strongly influenced by biologist John T. Bonner's *Evolution of Culture in Animals* (1980), Freeman traces "the origins of the human cultural capacity . . . back into early biological evolution." He detects the emergence of a "paradigm in which it becomes possible to view culture in an evolutionary setting and to take account of both the genetic and exogenetic in a way that gives due regard to the crucial importance of each of these fundamental aspects of human behavior and evolution."

Higher animals, Freeman remarks, exhibit a behavioral ability which biologist Bonner has called "primitive behavioral flexibility." Most animals have only one reflex action with which to respond to a situation, but among some higher animals that reflex is accompanied by a brain mediated ability to respond variously.* With that kind of open program of behavior, Freeman contends, "a choice is made by the brain or in other parts of the nervous system between two or more responses to produce what Bonner calls 'multiple choice behavior.'" Thus among some animals, Freeman argues, cultural behavior or habit can be seen as the attainment of a new niche of existence, one that developed out of the animals' experimentation with multiple choice behavior. Out of this evolutionary innovation, Freeman concludes, the cultural adaptations of the human species arose. From this kind of beginning the brain of the early hominids evolved to a point that made rudimentary traditions possible, such as have been shown to exist in populations of Japanese macaques and chimpanzees. Put another way, Freeman sees the emergence of a two-tracked inheritance "characterized by the interaction of its genetic and cultural components." He sees in evolution "a long existent and deep symbiosis between the genetic and the cultural," with the capacity to produce culture arising from natural selection because that capacity enhanced reproductive success. Thus emerges, in Freeman's own emphasized words, "a view of human evolution in which *the genetic and exogenetic are distinct but interacting parts of a single system.*"[30]

*Lewis Henry Morgan, the early American anthropologist, over a century ago provided an example of this response in behavior to changed circumstances in his study of the American beaver. He showed that in swampish territory beavers constructed canals to facilitate the movement of the branches and logs they had chewed down, but in areas where the banks of rivers were above their reach they constructed "slides" or chutes in the banks down which they could push their logs into the stream for easier conveyance to their lodges or dams. See Chapter 7 of his *The American Beaver and His Works* (Philadelphia: J. B. Lippincott, 1868).

If, in the light of modern ethology, modern Darwinian Derek Freeman has stressed the animal origins of human behavior, other observers of human nature have used that same light to invert the comparison by emphasizing the humanity in animals. Probably no better indication of that reversal, or of the persistence of the Darwinian Imperative into our own time, can be found than the emergence of the animal rights movement. For readers of this book it can come as no surprise that philosophical defenses of animal rights have drawn explicitly upon Darwin and his insistence upon continuity between man and beast. Philosopher Tom Regan, in perhaps the most sophisticated defense of animal rights, opens his case with Darwin's familiar remark that "there is no fundamental difference between man and the higher mammals in their mental faculties." From Darwin, Regan moves to evolution, pointing out that, "given evolutionary theory and given the demonstration of the survival values of consciousness the human case provides, we have every reason to suppose the members of other species are also conscious." Evolutionary theory may not *prove* consciousness in animals, he conceded, but it does provide "a theoretical basis for making this attribution to animals *independently of their ability to use language.*" In Regan's opinion, Darwin's ascription to mammalian animals the emotions of "terror, suspicion, affection and jealousy . . . cannot rationally be dismissed as the breezy anthropomorphism of a mind with its scientific guard down."[31] And it is true that the inclusion of animals within human morality was quite in line, though not identical with Darwin's original intention to bridge the gulf between human beings and the other animals with whom they shared a planet and an ancestry.

Among the ideas advanced by Harvard ethologist Edward Wilson in his various writings on sociobiology, none has been more controversial, not to say more denounced, than his assertion that human morality has been shaped by our biological origins. So vehement has the attack been that Wilson himself has gradually backed away. That Darwin himself, as we have seen, made a similar point has not deflected critics. But a spirited defense of Darwin's position in *Darwin and the Emergence of Evolutionary Theories of Mind and Behavior* by University of Chicago historian of science Robert Richards has recently provided yet another measure of the persistence of the Darwinian Imperative.

The book, impressively researched and argued, analyzes the impact on early American social scientists of Darwin's ideas on mind and behavior. Richards develops a case for a biological basis of human ethics, the beginnings of which, of course, he discerns in "Darwin's analysis of the instincts of social insects, which provided him the biological mechanism for human altruism." Although the book is a history of ideas, in a lengthy appendix Richards takes it upon himself to defend the original Darwinian view that man was "authentically moral" and that the altruism that has "seeped deeply into human hereditary stock" originated in the animal ancestry of human beings. Persistent opponents of sociobiology like natural scientists Stephen Jay Gould and Richard Lewontin have consistently denounced the foolishness, even the danger of seeking a biological or evolutionary basis for human actions. Explicitly opposing such views, Richards describes his book as deliberately seeking to restore that older Darwinian ethical image in order "to bring out its bright moral features, to show that if our morality has profound roots in our animal past and has evolved by natural selection, this conviction hardly demeans our humanity, rather it elevates our biology, our evolutionary human and moral biology."[32]

Only a few social scientists have been as forthrightly Darwinian in thinking about the continuity between animals and human beings as Robert Richards or Derek Freeman, and even among natural scientists Donald Griffin stands among only a small minority in boldly talking appreciatively about animal consciousness. Yet the movement that began three decades ago to follow out the implications of Darwinian evolutionary thought and to restore biology to the defintion of man seems likely to persist and, perhaps, to advance further in the direction Darwin had pointed, a direction which still delineates a conception of human nature more radical than many can accommodate.

Notes

1. Invoking the Darwinian Imperative

1. Charles Darwin, *The Origin of Species by Means of Natural Selection or the Preservation of Favored Races in the Struggle for Life and the Descent of Man and Selection in Relation to Sex* (New York: Modern Library, n.d.), 373.

2. Quoted in James R. Moore, "Darwin of Down: The Evolutionist as Squarson-Naturalist," David Kohn, ed., *The Darwinian Heritage*, (Princeton: Princeton Univ. Press, 1985), 453; Paul H. Barrett, trans., *Metaphysics, Materialism and the Evolution of Mind: Early Writings of Charles Darwin* (Chicago: Phoenix ed. Univ. of Chicago Press, 1980), 21, 32; quoted in Loren Eiseley, *Darwin's Century: Evolution and the Men Who Discovered It* (Garden City: Anchor Books, Doubleday, 1961), 352.

3. Darwin, *Origin of Species and Descent of Man*, 446, 494, 449, 460, 477.

4. Ibid., 460.

5. Ibid., 919–20.

6. Ibid., 482–83.

7. Ibid., 490, 400.

8. Ibid., 471–72, 919; Robert J. Richards, *Darwin and the Emergence of Evolutionary Theories of Mind and Behavior* (Chicago: Univ. of Chicago Press, 1987), 612.

9. Darwin, *Origin*, 919, 501–2.

10. Quoted in Cynthia Eagle Russett, *Darwin in America. The Intellectual Response, 1865–1912* (San Francisco: W. H. Freeman, 1976), 88–90.

11. On the new view of social Darwinism see especially Robert C. Bannister, *Social Darwinism: Science and Myth in Anglo-American Social Thought* (Philadelphia: Temple Univ. Press, 1979); Howard L. Kaye, *The Social Meaning of Modern Biology: From Social Darwinism to Sociobiology* (New Haven: Yale Univ. Press, 1984), and Peter J. Bowler, *The Eclipse of Darwinism: Anti-Darwinian Evolution Theories in the Decades around 1900* (Baltimore: Johns Hopkins Univ. Press, 1983), 18; Lester F. Ward, "Social and Biological Struggle," *American Journal of Sociology* 13 (Nov. 1907): 292.

12. Darwin, *Origin*, 501–2.

13. Ibid., 471–94.

14. Quoted in John S. Haller, Jr., "Race and the Concept of Progress in Nineteenth Century American Ethnology," *American Anthropologist* 73 (June 1971); 716.

15. Edward Alsworth Ross, *Social Control: A Survey of the Foundations of Order* (New York: Macmillian, 1901), 324; Charles A. Ellwood, *Sociology and Modern Social Problems* (New York: American Book Company, 1910), 42.

16. Darwin, *Origin*, 539.

17. Ibid., 541, 556.

18. Ibid., 556; Nancy Stepan, *The Idea of Race in Science: Great Britain 1800–1960* (London: Macmillan, 1982), 54–55, 77; Bannister, *Social Darwinism*, 199.

19. Gerrit L. Lansing, "Chinese Immigration: A Sociological Study," *The Popular Science Monthly* 20 (April 1882), 723; William Wilson Elwang, *The Negroes of Columbia, Missouri: A Concrete Study of the Race Problem* (Columbia, Mo.: Dept. of Sociology, Univ. of Missouri, 1904), 64.

20. W. E. Burghardt DuBois, "The Negro in Literature and Art," American Academy of Political and Social Science, *Annals* 49 (Sept. 1913), 233.

21. George Ward Stocking, Jr., "American Social Scientists and Race Theory—1890-1915," Unpublished Ph.D dissertation, Univ. of Pennsylvania, 1960, p. 590.

22. Franklin Henry Giddings, *The Principles of Sociology* (New York: Macmillan, 1896), 328–29.

23. Ibid., 238.

24. Edward Alsworth Ross, *Foundations of Sociology* (2nd ed., New York: Macmillan, 1905), 353–54, 309–10, 356n.

25. Comment in *American Journal of Sociology* 12 (March 1907), 715.

26. Charles Horton Cooley, *Social Organization* (New York: Scribners, 1924, orig. 1909), 28, 219–20.

27. D. K. Shute, "Racial Anatomical Peculiarities," *American Anthropologist* 9 (April 1896), 126–27; John Roach Straton, "Will Education Solve the Race Problem?," *North American Review* 170 (June 1900), 798; Joseph Alexander Tillinghast, "The Negro in Africa and America," *Publications of the American Economic Association*, 3rd Series 3 (May 1902), 632–33.

28. R. Meade Bache, "Reaction Time with Reference to Race," *Psychological Review* 2 (Sept. 1895), 484.

29. Darwin, *Origin*, 499.

30. Lester Frank Ward, "Eugenics, Euthenics, Eudemics," *American Journal of Sociology* 27 (May 1913), 741–50.

31. Lester Frank Ward, *Pure Sociology: A Treatist on the Origins and Spontaneous Development of Society* (New York: Macmillan, 1903), 499; Spencer quotation in Bowler, *Eclipse of Darwinism*, 71.

32. For a superb analysis of the issue see Bowler, *Eclipse of Darwinism*, 60–66; Vernon L. Kellogg, *Darwinism Today* (New York: Henry Holt, 1907), 5.

33. Quoted in Bowler, *Eclipse of Darwinism*, 204.

34. Joseph Le Conte, "The Factors of Evolution," *Monist* 1 (1890–91), 334.

35. Quoted in Haller, *Eugenics*, 60.

36. Fredrick L. Hoffman, "Race Traits and Tendencies of the American Negro," *Publications of the American Economic Association* (Aug. 1896), 311–12; George Oscar Ferguson, Jr., "The Mental Status of the American Negro," *Scientific Monthly* 12 (June 1921), 543.

37. Paul S. Reinsch, "The Negro Race and European Civilization," *American Journal of Sociology* 11 (Sept. 1905), 148.

38. Darwin, *Origin*, 398, 872.

39. Ibid., 873.

40. Ibid.

41. Ibid., 875.

42. M. A. Hardaker, "Science and the Woman Question," *Popular Science Monthly* 20 (March 1882), 578–83, emphasis in original.

43. R. Meade Bache, "Reaction Time with Reference to Race," *Psychological Review* 2 (Sept. 1895), 482; Dewey quoted in Rosalind Rosenberg, *Beyond Separate Spheres: Intellectual Roots of Modern Feminism* (New Haven: Yale Univ. Press, 1982), 22.

44. William I. Thomas, "On a Difference in the Metabolism of the Sexes," *American Journal of Sociology* 3 (July 1897), 31–32, 39–40.

45. Thomas, "Difference in the Metabolism," 61; Darwin, *Origin*, 873.

46. Quoted in Stephen Jay Gould, *Mismeasure of Man* (New York: W. W. Norton, 1981), 118.

47. Quoted in Willystine Goodsell, *The Education of Women: Its Social Background and Its Problems* (New York: Macmillan, 1924), 63, 65–66.

48. Wells, "Social Darwinism," 716; Cattell quoted in Rosenberg, *Beyond Separate Spheres*, 89.

49. Thorndike quoted in Stephanie A. Shields, "Functionalism, Darwinism, and the Psychology of Women: Study in Social Myth," *American Psychologist* 30 (July 1975), 750; David Snedden, "Probable Economic Future of American Women," *American Journal of Sociology* 24 (March 1919), 547.

2. Instinct, Eugenics, and Intelligence

1. Charles Darwin, *The Origin of Species by Means of Natural Selection or the Preservation of Favored Races in the Struggle for Life and The Descent of Man and Selection in Relation to Sex*, (New York: Modern Library, n.d.), 499.

2. William James, *The Principles of Psychology* (2 vols., Cambridge, Mass.: Harvard Univ. Press, 1981), 1004, 1026, 1028, 1030, 1034.

3. Ibid., 1039.

4. Ibid., 1040–41.

5. Charles A. Ellwood, "The Theory of Imitation in Social Psychology," *American Journal of Sociology* 6 (May 1901), 735; John E. Boodin, "Mind as Instinct," *Psychological Review* 13 (March 1906), 139.

6. James R. Angell, "William James," *Psychological Review* 18 (November 1911), 80.

7. William McDougall, "The Use and Abuse of Instinct in Social Psychology," *Journal of Abnormal Psychology and Social Psychology*, 16 (Dec. 1921–March 1922), 307–8; Hamilton Cravens, *The Triumph of Evolution: American Scientists and the Heredity-Environment Controversy, 1900–1941* (Philadelphia: Univ. of Pennsylvania Press, 1978), 77.

8. Alfred L. Kroeber, "The Morals of Uncivilized People," *American Anthropologist*, n.s. 12 (July–Sept., 1910), 439–40.

9. Robert Fletcher, "The New School of Criminal Anthropology," *American Anthropologist* 4 (July 1891), 207; Boies quoted in Arthur E. Fink, *Cause of Crime; Biological Theories in the United States, 1800–1915* (Philadephia: Univ. of Pennsylvania Press, 1938), 164.

10. Ellwood, "Theory of Imitation," 736.

11. Robert L. Dugdale, *"The Jukes: A Study in Crime, Pauperism, Disease, and Heredity* (4th ed., New York: G. P. Putnam's Sons, 1910), 11.

12. Ibid., 65, 55, 66.

13. Henry Herbert Goodard, *The Kallikak Family: A Study in the Heredity of Feeblemindedness* (New York: Macmillan, 1912), 18.

14. Ibid., 53–54.

15. Stephen Jay Gould, *Mismeasure of Man* (New York: W. W. Norton, 1981), 161–64.

16. Henry Herbert Goddard, *Feeblemindedness: Its Causes and Consequences* (1914; rep. New York: Arno Press, 1973), 556, emphasis in original; see also p. ix where Goddard confessed his continuing doubts, despite the latest evidence for a single character. "Even now we are far from believing the case settled. The problem is too deep to be thus easily disposed of." Though these sentences follow directly upon a quotation from Goddard that Gould uses, they do not appear in *Mismeasure of Man.*

17. Goddard, *Feeblemindedness,* 573.

18. Vernon Kellogg, *Human Life as the Biologist Sees It* (New York: Henry Holt, 1922), 81; East quoted in Mark H. Haller, *Eugenics: Hereditarian Attitudes in American Thought* (New Brunswick, N.J.: Rutgers Univ. Press, 1963), 167–68.

19. H. S. Jennings, *The Biological Basis of Human Nature* (New York: W. W. Norton, 1930), 17, 240, 156, 8.

20. *American Sociological Review* 5 (Aug. 1940), 658.

21. Darwin, *Descent of Man,* 918.

22. Quoted in Haller, *Eugenics,* 62.

23. G. Archibald Reid, "The Biological Foundations of Sociology," *American Journal of Sociology* 11 (Jan. 1906), 549.

24. Edward L. Thorndike, "Eugenics: With Special Reference to Intellect and Character," *Scientific Monthly* 83 (Aug. 1913), 130, 134; H. C. McComas, "The Heredity of Mental Traits," *Psychological Bulletin* 11 (Sept. 15, 1914), 380; Maynard M. Metcalf, "Eugenics and Euthenics," *Popular Science Monthly* 84 (April 1914), 388.

25. Ales Hrdlicka, *Physical Anthropology. Its Scope and Aims; Its History and Present Status in the United States.* (Philadelphia: Wistar Institution of Anatomy and Biology, 1919), 25.

26. William McDougall, *Is America Safe for Democracy?* (New York: Charles Scribner, 1921), 12. Italics in original.

27. Quoted in Haller, *Eugenics,* 49.

28. Quoted in Fink, *Causes of Crime,* 196.

29. Quoted in Rudolph Vecoli, Jr., "Sterilization: A Progressive Measure?," *Wisconsin Magazine of History* 43 (1960), 196.

30. Florence Mateer, "Mental Heredity and Eugenics," *Psychological Bulletin* 10 (June 15, 1913), 227.

31. *Buck v. Bell* 274 U.S. 200, 206, 208.

32. Mark DeWolfe Howe, ed., *Holmes-Laski Letters: The Correspondence of Mr. Justice Holmes and Harold J. Laski*, (2 vols., Cambridge Mass.: Harvard Univ. Press, 1953), 942; James Bishop Peabody, ed., *The Holmes-Einstein Letters: Correspondence of Mr. Justice Holmes and Lewis Einstein 1903–1935* (London: Macmillan, 1964), 267.

33. Edgard A. Doll, "The Average Mental Age of Adults," *Journal of Applied Psychology* 3 (Dec. 1919), 320; Albert Ernest Jenks, "The Relation of Anthropology to Americanization," *Scientific Monthly* 12 (March 1921), 241–42.

34. Lewis Madison Terman, "Were We Born That Way?," *World's Work*, 44 (Oct. 1922), 660.

35. Kimball Young, *Mental Differences in Certain Immigrant Groups. Psychological Tests of South Europeans in Typical California Schools with Bearing on the Educational Policy and on the Problems of Racial Contacts in This Country*, (Eugene Ore.: Univ. Press, 1922), 81.

36. Thomas R. Garth, "The Problem of Racial Psychology," *Journal of Abnormal Psychology and Social Psychology* 17 (July–Sept. 1922), 218.

37. Terman, "Were We Born That Way?," 660.

38. Quoted in Gould, *Mismeasure of Man*, 183.

39. McDougall, *Is American Safe?*, 66.

40. Quoted in Daniel J. Kevles, "Testing the Army's Intelligence: Psychologists and the Military in World War I," *Journal of American History* 55 (Dec. 1868), 571.

41. Robert Yerkes to Lewis Terman, May 30, 1921. Terman Papers, Standard Libraries.

42. Brigham, *Intelligence of Americans*, xxi.

43. Ibid., 210.

44. See, for example, Allan Chase, *The Legacy of Malthus: The Social Costs of the New Scientific Racism* (New York: Alfred A. Knopf, 1977), 289–301; Kenneth M. Ludmerer, *Genetics and American Society: A Historical Appraisal* (Baltimore: Johns Hopkins Univ. Press, 1972), 106–7; and Stephen Jay Gould, *Mismeasure of Man*, 231–32.

45. A valuable but too often overlooked study that expands upon the interpretation taken in the text is Robert A. Divine, *American Immigration Policy, 1924–1952* (New Haven: Yale Univ. Press, 1951).

46. Franz Samelson, "Putting Psychology on the Map: Ideology and Intelligence Testing," in Allan R. Buss, ed., *Psychology in Social Context* (New York: Irvington Publishers, 1979), 131–33.

47. Mark Snyderman and R. J. Herrnstein, "Intelligence Tests and the Immigration Act of 1924," *American Psychologist* 38 (Sept. 1983), 994.

48. Quoted in ibid., 136.

49. R. S. Woodworth, "Racial Differences and Mental Traits," *Science* 31 (Feb. 4, 1910), 185.

3. Laying the Foundation

1. Quoted in Malcolm J. Kottler, "Charles Darwin and Alfred Russel Wallace: Two Decades of Debate over Natural Selection," in David Kohn, ed., *The Darwinian Heritage* (Princeton: Princeton Univ. Press, 1985), 420–421.

2. Quoted in Loren Eiseley, *Darwin's Century: Evolution and the Men Who Discovered It* (Garden City: Anchor Books; Doubleday, 1961), 303, 312–13.

3. Ibid., 313.

3. Reports of the Immigration Commission, "Changes in Bodily Form of Descendants of Immigrants," 61st Congress, 2nd Session, Senate Document #208 (Washington, D.C.: Government Printing Office, 1911), 5.

4. Ibid.

5. Ibid., 76.

6. Boas to J. W. Jenks, June 20, 1910, Franz Boas Papers.

7. Boas to Dr. Cav Antonio Stella, Sept. 30, 1910, Franz Boas Papers; Report (1911), pp. 64, 70–74.

8. W. W. Husband to Boas, June 22, 1910, Franz Boas Papers; Reports (1911), pp. 41–44.

9. Clipping from *Leslie's Weekley*, March 10, 1910, p. 231, in Franz Boas Papers; "You Cannot Change Human Nature," *Outlook*, March 18, 1911, p. 579.

10. Paul R. Radosavljevich, "Professor Boas' New Theory of the Form of the Head—A Critical Contribution to School Anthropology," *American Anthropologist* 13 (July–Sept. 1911), 408–11; Maurice Fishberg, "Remarks on Radosavljevich's Critical Contribution to 'School Anthropology,'" *American Anthropology* 14 (Jan.–March 1912), 131–141; Robert H. Lowie, "Dr. Radoslavljevich's 'Critique' of Professor Boas," *Science* 35 (April 5, 1912), 537–40; John Swanton to Boas, Jan. 5, 1912, Franz Boas Papers.

11. Pitirim Sorokin, *Contemporary Sociological Theories* (New York: Harper and Bros., 1928), 132, 133n.; Bernice A. Kaplan, "Environment and Human Plasticity," *American Anthropologist* 56 (Octo. 1954), 784. For a more recent critical evaluation of Boas's findings see J. M. Tanner, "Boas' Contributions to Knowledge of Human Growth and Form," in Walter Goldschmidt, ed., *The Anthropology of Franz Boas* (San Francisco: Howard Chandler, 1959), 76–111. Tanner calls attention to the slight degree of actual convergence of head shapes, and Boas's skepticism toward his own findings. He notes, too, that when later Boas published his raw data he entitled the work *Contribution to the Study of Inheritance*.

12. Boas to Gustaf Retzius, May 3, 1910, Franz Boas Papers.

13. Quoted in George W. Stocking, Jr., "From Physics to Ethnology: Franz Boas' Arctic Expedition as a Problem in the Historiography of the Social Sciences," *Journal of the History of the Bahavioral Sciences* 1 (Jan. 1965), 64; Curtis M. Hinsley, Jr., *Savages and Scientists: The Smithsonian Institution and the Development of American Anthropology* (Washington, D.C.: Smithsonian Institution, 1981). 98–99; Franz Boas's letter in *Science* 9 (June 17, 1887), 589.

14. Boas letter in *Science*.

15. Franz Boas, "Human Faculty as Determined by Race," *Processings of the American Association for the Advancement of Science* 43 (1894), 224–25.

16. Geoge W. Stocking, Jr., ed., *The Shaping of American Anthropology 1883–1911. A Franz Boas Reader* (New York: Basic Books, 1974), 242, 239–40.

17. Franz Boas, "on Alternating Sounds," *American Anthropologist* 2 (Jan. 1889), 47–56. The quotation from Stocking appears in his *Race, Culture, and Evolution: Essays in the History of Anthropology* (New York: Free Press, 1986), 159.

18. Quoted in George Ward Stocking, Jr., "American Social Scientists and Race Theory: 1890–1915," Unpublished Ph.D. dissertation, Univ. of Pennsylvania, 1960, p. 549.

19. Stocking, ed., *Shaping of American Anthropology,* 236–37.

20. Quoted in Stocking, "American Social Scientists," 548.

21. "Poetry and Music of Some North American Tribes," *Science* 9 (April 22, 1887), 383.

22. Stocking, ed., *Shaping of American Anthropology,* 239–40.

23. Ibid., 217–18.

24. Stocking, "American Social Scientists," 569–571n; Stocking, *Race, Culture, and Evolution,* 203.

25. Theodor Waitz, *Introduction to Anthropology* (New York: AMS Press, 1975; 1863 English trans. of *Anthropologie der Naturvölker,* 1859), 352.

26. Ibid., 351.

27. Stocking, ed., *Shaping of American Anthropology,* 235; Franz Boas, "Psychological Problems in Anthropology," *American Journal of Psychology* 21 (July 1910), 373; Franz Boas, "Race," in *Encyclopedia of the Social Sciences,* XII, p. 34.

28. Quoted in Rosalind Rosenberg, *Beyond Separate Spheres: Intellectual Roots of Modern Feminism* (New Haven: Yale Univ. Press, 1982), 227.

29. Stocking, ed., *Shaping American Anthropology,* 41–43.

30. Franz Boas to Abraham Jacobi, Jan. 2, 1882, and undated c. April 10, 1882, Franz Boas Papers.

31. Franz Boas to Nicholas Murry Butler, June 18, 1902, Franz Boas Papers.

32. Stocking, ed., *Shaping of American Anthropology,* 207.

33. Ibid., 213, Franz Boas to B. Neuberger, Aug. 13, 1910, Franz Boas Papers.

34. Franz Boas to J. W. Jenks, Jan. 28, 1910, Franz Boas Papers.

35. Franz Boas, "The Negro and the Demands of Modern Life, Ethnic and Anatomical Considerations," *Charities* 15 (Oct. 7, 1905), 85; Quoted in Edward Beardsley, "The American Scientist as Social Activist: Franz Boas, Burt G. Wilder, and the Fight for Racial Justice, 1900–1915," *Isis* 64 (March 1973), 60–61; Franz Boas to Richard Watson Gilder, May 22, 1907, Franz Boas Papers.

36. Stocking, ed., *Shaping of American Anthropology,* 313–14; Quoted in Marshall Hyatt, "Franz Boas and the Struggle for Black Equality; The Dynamics of Ethnicity," *Perspectives in American History,* n.s. 2 (1985), 295.

37. Boas, "Negro and Demands of Modern Life," 85–86.

38. Ibid., 87; Stocking, ed., *Shaping of American Anthropology,* 328–29; Boas's Foreword to Mary White Ovington, *Half a Man: The Status of the Negro in New York* (New York, 1911), vii.

39. Franz Boas to Natalie Curtis, Aug. 20, 1903, Franz Boas Papers.

40. Paper on work and Organization of American Ethnology, no date, but near end of Dec. 1903, Franz Boas Papers.

41. Stocking, ed., *Shaping of American Anthropology,* 213.

42. Franz Boas, "The Problem of the American Negro," *Yale Review* 10 (Jan. 1921), 392–93.

43. Ibid., 393–95.

44. Stocking, *Race, Culture and Evolution,* 231–32; George W. Stocking, Jr., "Anthropology as Kulturkampf: Science and Politics in the Career of Franz Boas," in Walter Goldschmidt, ed., *Uses of Anthropology* (Washington: American Anthropological Association, No. 11, 1979), 47.

45. Ibid., 45–46.

46. Franz Boas, "Some Recent Criticisms of Physical Anthropology," *American Anthropologist*, n.s. 1 (Jan. 1899), 100–101; Boas, *Mind of Primitive Man*, 76.

47. Franz Boas, "Race and Progress," *Science* 74 (July 3, 1931), 4.

48. John Watson, *Behaviorism* (New York: W. W. Norton, 1924, 1925), 82.

49. End of May 1915, a single sheet, typed and in handwriting, Franz Boas Papers.

50. Quoted in Hamilton Cravens, *The Triumph of Evolution: American Scientists and The Heredity-Environment Controversy 1900–1944* (Philadelphia: Univ. of Pennsylvania Press, 1978), 107.

4. In the Wake of Boas

1. Josiah Morse, "A Comparison of White and Colored Children Measured by the Binet Scale of Intelligence," *Popular Science Monthly* 84 (Jan. 1914), 75; Carl Kelsey, *The Physical Basis of Society* (New York: D. Appleton, 1916), 287–89.

2. Charles A. Ellwood, "The Theory of Imitation in Social Psychology," *American Journal of Sociology* 6 (May 1901), 573.

3. Charles A. Ellwood, Review of William Benjamin Smith, *The Color Line*, in *American Journal of Sociology* 11 (Jan. 1906), 572.

4. Carl Kelsey, *The Negro Farmer* (Chicago: Jennings and Pye, 1903), 5–7.

5. Carl Kelsey, Comment on paper by D. Collin Wells, in *American Journal of Sociology* 12 (March 1907), 711.

6. Carl Kelsey, "Influence of Heredity and Environment Upon Race Improvement," *Annals of the American Academy of Political and Social Science* 34 (July 1909), 6–8.

7. Carl Kelsey, Review of Boas's *Mind of Primitive Man*, in *Annals of the American Academy of Political and Social Science* 46 (March 1913), 203–4.

8. Carl Kelsey, *The Physical Basis of Society* (New York: Appleton, 1916), 291–92, 301–2.

9. Howard W. Odum, *Social and Mental Traits of the Negro. Research into the Conditions of the Negro Race in Southern Towns. A Study in Race Traits, Tendencies and Prospects* (New York: Columbia Univ. Press, 1910), 294, 47, emphasis in original.

10. Howard W. Odum, "Negro Children in the Public Schools of Philadelphia," *Annals of the American Academy of Political and Social Science* 49 (Sept. 1913), 205–6, 200.

11. Thomas to Boas, May 14, 1907, Boas Papers; William I. Thomas, "The Psychology of Race-Prejudice," *American Journal of Sociology* 9 (March 1904), 608, 610–11.

12. William I. Thomas, *Sex and Society: Studies in the Social Psychology of Sex* (Chicago: Univ. of Chicago Press, 1907), 257–58, 262, 312; Thomas to Boas, May 14, 1907, Boas Papers.

13. Alfred L. Kroeber, "The Morals of Uncivilized People," *American Anthropologist*, n.s. 12 (July–Sept. 1910), 446, 443–45.

14. Kroeber, "Morals of Unicivilized," 446.

15. Kroeber, "Eighteen Professions," 285.

16. Alfred L. Kroeber, "Inheritance by Magic," *American Anthropologist*, n.s. 18 (Jan.–March 1916), 26–27.

17. Alfred L. Kroeber, "The Superorganic," in Kroeber, *The Nature of Culture* (Chicago: Univ. of Chicago, 1952), 47.

18. Ibid., 42.
19. Kroeber, "Eighteen Professions," 288; Kroeber, *Nature of Culture*, 37, 34.
20. Ibid., 34–35.
21. Ibid., Kroeber, "Eighteen Professions," 287.
22. Kroeber, "Eighteen Professions," 287.
23. Ibid., 285.
24. Ibid., 286; Kroeber, *Nature of Culture*, 49–50; George W. Stocking, Jr., *Race, Culture, and Evolution: Essays in the History of Anthropology* (New York: Free Press, 1968), 231–32.
25. Alfred L. Kroeber, "The Superorganic," *American Anthropologist* 19 (April–June 1917), 207; David Bidney, *Theoretical Anthropology* (New York: Columbia Univ. Press, 1953), 37.
26. Kroeber, *Nature of Culture*, 49, 41, 40.
27. Emile Durkheim, *The Rules of Sociological Method* (8th ed., Chicago: Univ. of Chicago Press, 1938; orig. pub. 1895), 104, 102.
29. Kroeber, "Eighteen Professions," 287; Durkheim, *The Rules*, 110.
30. Edward Sapir, "Do We Need a 'Superorganic'?," *American Anthropologist*, n.s. 19 (July–Sept. 1917), 444–46.
31. Alfred L. Kroeber to Alexander A. Goldenweiser, Oct. 23, 1917, Kroeber Papers.
32. Kroeber, *Nature of Culture*, 22.
33. Kroeber, "Heredity Without Magic," 296.
34. Ibid., 36; Kroeber, *Nature of Culture*, 41–42, 48.
35. Kroeber, *Nature of Culture*, 47.
36. Ibid., 47–48.
37. Kroeber, "Inheritance by Magic," 40.
38. Robert H. Lowie, *Culture and Ethnology* (New York: Boni and Liveright, 1917), 29, 46.
39. Robert H. Lowie, "The Universalist Fallacy," *New Republic* 12 (Nov. 17, 1917), 4–5.
40. Lowie, *Culture and Ethnology*, 66.
41. B. Laufer, Review of *Culture and Ethnology*, in *American Anthropologist*, n.s. 20 (Jan.–March, 1918), 90.
42. George Elliott Howard, "The Social Cost of Southern Race Prejudice," *American Journal of Sociology* 22 (March 1917), 579–80.
43. E. B. Reuter, "The Superiority of the Mulatto," *American Journal of Sociology* 23 (July 1917), 92–93; Ellsworth Faris, "The Mental Capacity of Savages," *American Journal of Sociology* 23 (March 1918), 618.
44. Faris, "Mental Capacity of Savages," 603, 618–19.
45. Charles A. Ellwood, "Theories of Cultural Evolution," *American Journal of Sociology* 33 (May 1918), 790, 800; Charles A. Ellwood, "Mental Patterns in Social Evolution," American Sociological Society, *Papers and Proceedings* 27 (1922–23), 99–100.

5. Does Sex Tell Us Anything?

1. Charles Darwin, *The Descent of Man* (New York: Modern Library, n.d.), 873. For further references to sexual differences among human beings see pp. 472n. and 867.

2. Quoted in Rosalind Rosenberg, *Beyond Separate Spheres: Intellectual Roots of Modern Feminism,* (New Haven:, Yale Univ. Press, 1982), 63.

3. Eliza Burt Gamble, *The Evolution of Woman: An Inquiry into the Dogma of Her Inferiority to Man* (New York: G. P. Putnam's Sons, 1894), v–vi, 28–31, 37,44.

4. Ibid., 61, 60.

5. Lester F. Ward, *Dynamic Sociology, or Applied Social Science, as Based Upon Statical Sociology and the Less Complex Sciences* (2 vols., New York: D. Appleton, 1883), 615–19.

6. Lester F. Ward, *Pure Sociology: A Treatise on the Origins and Spontaneous Development of Society* (New York: Macmillan, 1903), 296.

7. Ibid., 323, 327.

8. Ibid., 331.

9. Ibid., 336, 345.

10. Ibid., 376.

11. Charlotte Perkins Gilman, *The Man-Made World; Or Our Androcentric Culture* (New York: Charlton, 1911), 250.

12. Lydia Kingsmill Commander, "The Self-Supporting Woman and the Family," *American Journal of Sociology* 14 (May 1909), 752, 757.

13. Cordelia C. Nevers, "Dr. Jastrow on Community of Ideas of Men and Women," *Psychological Review* 2 (July 1895), 366.

14. May Whiton Calkins, "Community of Ideas of Men and Women," *Psychological Review* 3 (July 1896), 430.

15. Amy Tanner, "The Community of Ideas of Men and Women," *Psychological Review* 3 (July 1896), 549–50.

16. Frances A. Kellor, "Psychological and Environmental Study of Women Criminals, II," *American Journal of Sociology* 5 (March 1900), 679.

17. Kellor, "Psychological Study, II," 672.

18. Helen Bradford Thompson, *The Mental Traits of Sex: An Experimental Investigation of the Normal Mind in Men and Women* (Chicago: Univ. of Chicago Press 1903), 171–172.

19. Ibid., 172.

20. Ibid., 173.

21. Ibid., 173–74.

22. Ibid., 175–76.

23. Ibid., 176–77.

24. Ibid., 181–82.

25. Ibid., 182.

26. William I. Thomas, "The Relation of Sex to Primitive Social Control," *American Journal of Sociology* 3 (May 1898), 760–62, 776; William I. Thomas, "Sex in Primitive Industry," *American Journal of Sociology* 4 (Jan. 1899), 776.

27. Quoted in Rosenberg, *Beyond Separate Spheres,* 124.

28. William I. Thomas, *Sex and Society: Studies in the Social Psychology of Sex* (Chicago: Univ. of Chicago Press 1907), 310.

29. Ibid., 312–13.

30. Franz Boas, "Human Faculty as Determined by Race" (1894), in George W. Stocking, Jr., ed., *The Shaping of American Anthropology 1883–1911: A Franz Boas*

Reader (New York: Basic Books, 1874), 233; Edward Alsworth Ross, *Foundations of Sociology* (2nd ed., New York: Macmillan, 1905), 323–24.

31. Mary Roberts Coolidge, *Why Women Are So* (New York: Henry Holt, 1912), 301, 304, 310.

32. Ibid., 306.

33. Helen Thompson Woolley, "The Psychology of Sex," *Psychological Bulletin* 11 (Oct. 15, 1914), 365–66.

34. Robert H. Lowie and Leta Stetter Hollingworth, "Science and Feminism," *Scientific Monthly* 3 (Sept. 1916), 284.

35. Carl Kelsey, *The Physical Basis of Society* (New York: D. Appleton, 1916), 319–21, 325–27.

36. Ibid., 328–29.

37. Thomas, *Sex and Society*, 311; Kelsey, *Physical Basis of Society*, 30.

38. Elsie Clews Parsons, "Higher Education of Women and the Family," *American Journal of Sociology* 14 (May 1909), 763; Rosenberg, *Beyond Separate Spheres*, 177.

39. Leta S. Hollingworth, "Variability as Related to Sex Differences in Achievement," *American Journal of Sociology* 19 (Jan. 1914), 529; Leta S. Hollingworth, "Social Devices for Impelling Women to Bear and Rear Children," *American Journal of Sociology* 22 (July 1916), 21.

40. Hollingworth, "Variability," 528.

41. Hollingworth, "Social Devices," 20, 29.

42. Leta S. Hollingworth, "Sex Differences in Mental Traits," *Psychological Bulletin* 13 (Oct. 15, 1916), 383.

43. Leta S. Hollingworth, "Comparison of the Sexes in Mental Traits," *Psychological Bulletin* 15 (Dec. 1918), 431; Leta S. Hollingworth, "Comparison of the Sexes in Mental Traits," *Psychological Bulletin* 16 (Oct. 1919), 373.

44. Hollingworth to Terman, Jan. 3, 1921 (but actually 1922). Lewis Terman Papers, Stanford Univ. Library.

45. Leta S. Hollingworth, *Gifted Children: Their Nature and Nurture* (New York: Macmillan, 1926), 68.

46. Ibid., 354–56.

47. Ibid., 357, 360.

48. Ibid., 198–99.

49. "Introduction" in Box 14, Folder 9, of Lewis Terman Papers, Stanford Univ. Library.

50. Lewis Madison Terman, "Were We Born That Way?," *World's Work* 44 (Oct. 1922), 660; S. L. and L. W. Pressey, "Further Data with Regard to Sex Differences," *Journal of Applied Psychology* 5 (March 1921), 81–82.

51. Ibid., 82–83; Willystine Goodsell, *The Education of Women: Its Social Background and Its Problems* (New York: Macmillan, 1924) 69–76; Chauncey N. Allen, "Studies in Sex Differences," *Psychological Bulletin* 24 (May 1927), 296–99.

52. Ruth Reed, "Changing Conceptions of the Maternal Instinct," *Journal of Abnormal and Social Psychology* 18 (April–June 1923), 83, 86.

53. William Fielding Ogburn, *Social Change with Respect to Culture and Original Nature* (New York: B. W. Huebsch, 1922), 29–31.

54. Goodsell, *Education of Women*, 83–84.

55. Charles Going Woodhouse, "The Changing Status of Women," *American Journal of Sociology* 35 (May 1930), 1096.

56. Goodsell, *The Education of Women*, 58n., 57.

57. Allen, "Studies in Sex Differences," 301.

58. C. N. Allen, "Recent Studies in Sex Differences," *Psychological Bulletin* 27 (May 1930), 401.

59. Margaret Mead, *Coming of Age in Samoa: A Psychological Study of Primitive Youth for Western Civilization* (New York: William Morrow, 1928), 197–98.

60. Margaret Mead, *Sex and Temperament in Three Primitive Societies* (New York: Morrow, 1963, orig. 1935), ix, xiii, 280, xiv.

61. Mead, *Sex and Temperament*, 280; Margaret Mead, *Male and Female: A Study of the Sexes in a Changing World* (New York: William Morrow, 1949), 216, 320.

62. Mead, *Male and Female*, 310; Rosenberg, *Beyond Separate Spheres*, 236.

63. Mead, *Sex and Temperament*, preface to 1963 edition; Mary Catherine Bateson, *With a Daughter's Eye: A Memoir of Margaret Mead and Gregory Bateson* (New York: William Morrow, 1984), 139.

64. Rosenberg, *Beyond Separate Spheres*, 243–44; Mead, *Sex and Temperament*, 1950 preface reprinted in 1963 edition, n.p.

65. Margaret Mead, *The Changing Culture of an Indian Tribe* (New York: Capricorn Books, 1966, orig. 1932), 133–34.

66. Mead, *Sex and Temperament*, 314, 316.

67. Mead, *Male and Female*, 216–217.

68. Ibid., 160.

69. Margaret Mead, "Alternatives to War," in Morton Fried, Marvin Harris, and Robert Murphy, eds., *The Anthropology of Armed Conflict and Aggression* (Garden City, N.Y.: Natural History Press, 1968), 220.

6. Decoupling Behavior from Nature

1. Quoted in Hamilton Cravens, *The Triumph of Evolution: American Scientists and the Hereditary-Environment Controversy 1900–1941* (Philadelphia: Univ. of Pennsylvania Press, 1978), 244.

2. Carl Murchison, *Criminal Intelligence* (Worcester, Mass.: Clark Univ. Press, 1926), 28, 54, 58.

3. J. E. Wallace Wallin, "Who Is Feeble-Minded?," *Journal of Criminal Law and Criminology* 6 (Jan. 1916), 706n. J. E. Wallace Wallin, "Who Is Feeble-Minded? A Reply to Mr. Kohs," *Journal of Criminal Law and Criminality* 7 (May 1916), 72–73.

4. J. E. Wallace Wallin, *Problems of Subnormality* (Yonker-on-Hudson, N.Y.: World Book Company, 1917), 275; Quoted in Mark H. Haller, *Eugenics: Hereditarian Attitudes in American Thought* (New Brunswick, N.J.: 1963), 114.

5. Irene Case and Kate Lewis, "Environment as a Factor in Feeblemindedness," *American Journal of Sociology* 23 (March 1918), 668.

6. Loc. cit.

7. Ibid., 662, 668.

8. Quoted in Haller, *Eugenics*, 122; Henry Herbert Goddard, *Juvenile Delinquency* (New York: Dodd, Mead, 1921), 119; Henry H. Goddard, "Who is a Moron?," *Scientific Monthly* 24 (Jan. 1927), 45.

9. Henry H. Goddard, "Feeblemindedness: A Question of Definition," *Journal of Psycho-Asthenics* 33 (1928), 224; Henry H. Goddard, "In Defense of the Kallikak Study," *Science* 95 (June 5, 1942), 575. Scholars who have misread Goddard's "recantation" include Stephen Jay Gould, *Mismeasure of Man* (New York: W. W. Norton, 1981), 174, and Haller, *Eugenics*, 119.

10. See Daniel J. Kevles, *In the Name of Eugenics: Genetics and the Uses of Human Heredity.* (New York: Alfred A. Knopf, 1985), 141–42; R. S. Woodworth, *Heredity and Environment: A Critical Study of Recently Published Material on Twins and Foster Children* (New York: Social Science Research Council, n.d., c. 1941), 86.

11. Adolphus Miller, "The Psychological Limit of Eugenics," *Popular Science Monthly* 84 (April 1914), 392, 396; Erville B. Woods, "Heredity and Opportunity," *American Journal of Sociology* 26 (July 1920), 5.

12. Warren S. Thompson, "Eugenics, as Viewed by a Sociologist," American Sociological Society, *Papers and Proceedings* 28 (1924), 68–69.

13. Warren S. Thompson, "Eugenics and the Social Good," *Journal of Social Forces* 3 (March 1925), 419.

14. J. B. Eggen, "The Fallacy of Eugenics," *Journal of Social Forces* 5 (Sept. 1926), 107–8.

15. A. L. Kroeber, "Inheritance by Magic," *American Anthropologist*, n.s. 18 (Jan.-March 1916), 38, 35.

16. Woods, "Heredity and Opportunity," 20–21; Miller, "Psychological Limits," 391–92.

17. Franz Boas, "Eugenics," *Scientific Monthly* 3 (Nov. 1916), 473, 476.

18. Ibid., 475.

19. Ibid., 477.

20. Ibid., 478.

21. Edwin Grant Conklin, *Heredity and Environment in the Development of Man* (5th ed., Princeton: Princeton Univ. Press, 1923) 305–6.

22. Quoted in Kenneth M. Ludmerer, *Genetics and American Society: A Historical Appraisal* (Baltimore: Johns Hopkins Univ. Press, 1972), 84, 139.

23. Ibid., 84, 124–25; Haller, *Eugenics*, 168–69.

24. Lewis M. Terman to Charles B. Davenport, June 29, 1931, Terman Papers.

25. John B. Watson, "Psychology as the Behaviorist Views It," *Psychological Review* 20 (March 1913), 170, 158.

26. Quoted in John M. O'Donnell, *The Origins of Behaviorism: American Psychology, 1870–1920* (New York: New York Univ. Press, 1985), 201; Watson, "Psychology as the Behaviorist Views It," 175–76; John B. Watson, *Behavior: An Introduction to Comparative Psychology* (New York: Henry Holt, 1914), 27, 317.

27. Herbert S. Jennings, John B. Watson, Adolf Meyer, and William I. Thomas, *Suggestions of Modern Science Concerning Education* (New York: Macmillan, 1917), 72–73.

28. John B. Watson, *Psychology from the Standpoint of a Behaviorist* (Philadelphia: J. B. Lippincott, 1919), 257–58, 261.

29. John B. Watson, *Behaviorism* (New York: W. W. Norton, 1924, 1925), 103–4.

30. See Franz Samelson, "Struggle for Scientific Authority: The Reception of Watson's Behaviorism," *Journal of the History of the Behavioral Sciences* 17 (1981), 415, and Robert Boakes, *From Darwin to Behaviorism: Psychology and the Minds of Animals* (Cambridge, Eng.: Cambridge Univ. Press, 1984), 226.

31. James R. Angell, "Behavior as a Category of Psychology," *Psychological Review* 20 (July 1913), 255.

32. Quoted in O'Donnell, *Origins of Behaviorism*, 204.

33. Quoted in George Ward Stocking, Jr., "American Social Scientists and Race Theory—1890–1915," Unpublished Ph.D. dissertation, Univ. of Pennsylvania, 1960, p. 588n.

34. Charles H. Judd, "Evolution and Consciousness," *Psychological Review* 17 (March 1910), 88–89.

35. Knight Dunlap, "Are There Any Instincts," *Journal of Abnormal Psychology* 14 (Dec. 1914), 307, 311.

36. J. R. Kantor, "A Functional Interpretation of Human Instincts," *Psychological Review* 27 (Jan. 1920), 54.

37. J. R. Kantor, "The Problem of Instincts and Its Relation to Social Psychology," *Journal of Abnormal Psychology and Social Psychology* 18 (April–June 1923), 75, 77.

38. Quoted in Boakes, *From Darwinism to Behaviorism*, 225.

39. Ibid., 217–18. See also Gilbert Gottlieb, "Zing-yang Kuo: Radical Scientific Philospher and Innovative Experimentalist (1898–1970)," *Journal of Comparative and Physiological Psychology* 80 (July 1972), 1–10.

40. Zing yang Kuo, "How Are Our Instincts Acquired?," *Psychological Review* 29 (Sept. 1922), 344–45, 347n, 350; Gottlieb, "Zing-yang Kuo," 1; Zing Yang Kuo, "A Psychology Without Heredity," *Psychological Review* 31 (Nov. 1924), 445, 438.

41. Zing Yang Kuo, "The Net Result of the Anti-Heredity Movement in Psychology," *Psychological Review* 36 (May, 1919), 196–97.

42. L. L. Bernard, *Instinct: A Study in Social Psychology* (New York: Henry Holt, 1924), 529; L. L. Bernard, "The Misuse of Instinct in the Social Sciences," *Psychological Review* 28 (March 1921), 96, 99.

43. Robert H. Gault, "Psychology in Social Relations," *American Journal of Sociology* 22 (May 1917), 737–38.

44. Woods, "Heredity and Opportunity," 13, emphasis in original.

45. Robert H. Lowie, "The Universalist Fallacy," *New Republic* 13 (Nov. 17, 1917), Part II, p. 5; Bernard, *Instinct*, 221.

46. Ellsworth Faris, "Are Instincts Data or Hypotheses?," *American Journal of Sociology* 27 (Sept. 1921), 193–94, 195; Ellsworth Faris, "The Nature of Human Nature," American Sociological Society, *Papers and Proceedings* 20 (1926), 26.

47. Bernard, *Instinct*, 326; Faris, "Are Instincts Data or Hypotheses?," 193.

48. E. B. Reuter, "The Relation of Biology and Sociology," *American Journal of Sociology* 32 (March 1927), 706.

49. Cravens, *Triumph of Evolution*, 192–93, 191; William MacDougall, "The Use and Abuse of Instinct in Social Psychology," *Journal of Abnormal Psychology and Social Psychology* 16 (Dec. 1921), 298n.

50. Faris, "Are Instincts Data or Hypotheses?," 188–89; Kantor, "Functional Interpretation of Human Instincts," 57.

51. McDougall, "The Use and Abuse of Instinct," 333.

52. Lauren G. Wispé and James N. Thompson, Jr., "The War Between the Words: Biological Versus Social Evolution and Some Related Issues," *American Psychologist* 31 (May 1976), 346.

53. Edward Chase Tolman, "Can Instincts Be Given Up in Psychology?," *Journal of Abnormal Psychology and Social Psychology* 17 (July–Sept. 1922), 152, emphasis in original; H. G. Wyatt, "The Recent Anti-Instinctive Attitude in Social Psychology," *Psychological Review* 34 (March 1927), 126–31.

54. Max Schoen, "Instinct and Man. A Preliminary Note on Psychological Terminology," *Psychological Review* 34 (march 1927), 121, emphasis in original; John M. Fletcher, "An Old Solution of the New Problem of Instinct," *Psychological Review* 36 (Jan. 1929), 44–45.

55. D. G. Marquis, "The Criterion of Innate Behavior," *Psychological Review* 37 (July 1930), 88; Paul A. Witty and Harvey C. Lehman, "The Instinct Hypothesis Versus the Maturation Hypothesis," *Psychological Review* 40 (Jan. 1933), 33–59.

56. David L. Krantz and David Allen, "The Rise and Fall of McDougall's Instinct Doctrine," *Journal of the History of the Behavioral Sciences* 3 (Oct. 1967), 331–32; R. J. Herrnstein, "Nature as Nurture: Behaviorism and the Instinct Doctrine," *Behaviorism* 1 (1972), 24–25.

57. Jennings, *Biological Basis of Human Nature*, 284.

7. Decoupling Intelligence from Race

1. Franz Samelson, "World War I Intelligence Testing and the Development of Psychology," *Journal of the History of the Behavioral Sciences* 13 (1977), 277-78.

2. Quoted from Robert Yerkes in ibid., 279.

3. Robert M. Yerkes, "Testing the Human Mind," *Atlantic Monthly* 131 (March 1923), 363.

4. Walter Lippmann, "The Mental Age of Americans," *New Republic* 32 (Oct. 23, 1922), 213.

5. Walter Lippmann, "The Mystery of the 'A,'" *New Republic* 32 (Nov. 1, 1922), 247; "Reliability of Intelligence Tests," *New Republic* 32 (Nov. 8, 1922), 276–77.

6. Walter Lippman, "A Future for the Tests," *New Republic* 33 (Nov. 29, 1922), 10.

7. Lippmann, "Mystery of the 'A,'" 247.

8. James W. Bridges and Lillian E. Coler, "The Relation of Intelligence to Social Status," *Psychological Review* 24 (Jan. 1917), 31.

9. S. L. Pressey and Ruth Ralston, "The Relation of the General Intelligence of School Children to the Occupation of Their Fathers," *Journal of Applied Psychology* 3 (Dec. 1919), 370, 370n., 372.

10. Luella Winfred Pressey, "The Influence of (a) Inadequate Schooling and (b) Poor Environment upon Results with Tests of Intelligence," *Journal of Applied Psychology* 4 (March 1920), 93, 91–92, 95–96.

11. Thomas R. Garth, "A Comparison of Mental Abilities of Mixed and Full-Blooded Indians on the Basis of Education," *Psychological Review* 29 (May 1922), 223.

12. William McDougall, *Is American Safe for Democracy?* (New York: Charles Scribner, 1921), 66.

13. Ada H. Arlitt, "On the Need for Caution in Establishing Race Norms," *Journal of Applied Psychology* 5 (June 1921), 182–83.

14. Thomas R. Garth, "White, Indian and Negro Work Curves," *Journal of Applied Psychology* 5 (March 1921), 16.

15. Garth, "Comparison of Mental Abilities," 223.

16. Edwin G. Boring, "Facts and Fancies of Immigration," *New Republic*, April 25, 1923, p. 246; Boring quoted in Samelson, "Putting Psychology on the Map," 148–49; A. J. Snow, *American Journal of Psychology* 34 (April 1923), 305.

17. Kimball Young, Review, *Science* 57 (June 8, 1923), 669–70; Franz Boas, "The Problem of the American Negro," *Yale Review* 10 (Jan. 1921), 389.

18. William C. Bagley, *Determinism in Education* (Baltimore: Warwick and York, 1925), 125; Alexander Goldenweiser, "Race and Culture in the Modern World," *Journal of Social Forces* 3 (Nov. 1924), 128–29.

19. J. R. Kantor, "Anthropology, Race, Psychology, and Culture," *American Anthropologist*, n.s. 27 (April, 1925), 269fn.

20. Kimball Young, "The History of Mental Testing," *Pedogogical Seminary* 31 (March 1924), 47.

21. Carl C. Brigham, "Intelligence Tests of Immigrant Groups," *Psychological Review* 37 (March 1930), 165.

22. Thomas R. Garth, "A Review of Racial Psychology," *Psychological Review* 22 (June 1925), 350; Thomas R. Garth, " A Review of Race Psychology," *Psychological Bulletin* 27 (May 1930), 356.

23. G. H. Estabrooks, "The Enigma of Racial Intelligence," *Journal of Genetic Psychology* 35 (March 1928), 137–39.

24. Dale Yoder, "Present Status of the Question of Racial Differences," *Journal of Educational Psychology* 19 (Oct. 1928), 470.

25. Paul A. Witty and Harvey C. Lehman, "Racial Differences: The Dogma of Superiority," *Journal of Social Psychology* 1 (Aug. 1930), 394, 405, 398.

26. Garth, "A Review of Race Psychology," 348.

27. Thomas Russell Garth, *Race Psychology: A Study of Racial Mental Differences* (New York: Whittlesey House, 1931), 206–7.

28. H. S. Jennings, "Heredity and Environment," *Scientific Monthly* 19 (Sept. 1924), 237.

29. Otto Klineberg, *Race Differences* (New York: Harper and Bros., 1935), 194. The original work was "A Study of Psychological differences between 'Racial' and National Groups in Europe," *Archives of Psychology*, No. 132 (Sept. 1931).

30. Otto Klineberg, "An Experimental Study of Speed and Other Factors in 'Racial' Differences," *Archives of Psychology* 15 (Jan. 1928), 49–50.

31. Yerkes, "Testing the Human Mind," 363–364; Klineberg, *Race Differences*, 183–84.

32. Ibid., 165–66.

33. Klineberg, "An Experimental Study of Speed," 34–35.

34. Ibid., 104.

35. Ibid., 57.

36. Ibid., 62–63.

37. Klineberg, *Race Differences*, 189.

38. Ibid., 255.

39. Ibid., 226.
40. Ibid., 273-74, 262, 287-88.
41. Klineberg, *Race Differences*, 290.
42. Ibid., 299-300.
43. Ibid., 345.
44. Klineberg, "Study of Psychological Differences," 35; Klineberg, *Race Differences*, 154; "Study of Psychological Differences," 53, 30.

8. Why Did Culture Triumph?

1. Charles B. Davenport, "Euthenics and Eugenics," *Popular Science Monthly* 78 (Jan. 1911), 20; Otto Klineberg, "An Experimental Study of Speed and Other Factors in 'Racial' Differences," *Archives of Psychology* 15 (Jan. 1928), 5; letter from Otto Klineberg to Carl N. Degler, Feb. 9, 1985, in Degler's possession.
2. William C. Bagley, *Determinism in Education* (Baltimore: Warwick and York, 1925), 45, 129-31.
3. Dale Yoder, "Present Status of the Question of Racial Differences" *Journal of Education Psychology* 19 (Oct. 1928), 463; Alexander Goldenweiser, "Cultural Anthropology," in Harry Elmer Barnes, ed., *The History and Prospects of the Social Sciences* (New York: Alfred A. Knopf, 1925), 462-63; Joseph Peterson, "Methods of Investing Comparative Abilities in Races," *Annals of the American Academy of Political and Social Science* 140 (Nov. 1928), 184.
4. Thomas Russell Garth, *Race Psychology: A Study of Racial Mental Differences* (New York: Whittlesey House, 1931), 84-85; Thomas Russell Garth, "The Results of Some Tests on Mixed and Full Blood Indians," *Journal of Applied Psychology* 5 (Dec. 1921), 359.
5. Garth, "The Results of Some Tests," 359.
6. Paul A. Witty and Harvey C. Lehman, "Racial Differences: The Dogma of Superiority," *Journal of Social Psychology* 1 (Aug. 1930), 405. Emphasis in original.
7. E. George Payne, "Negroes in the Public Elementary Schools of the North," *Annals of the American Academy of Political and Social Science,* 140 (Nov. 1928), 233.
8. Paul A. Witty and Martin D. Jenkins, "The Case of 'B'—A Gifted Negro Girl," *Journal of Social Psychology* 6 (Feb. 1935), 118, 123, 118, 124.
9. Quoted in Clarence J. Karier, *Scientists of the Mind: Intellectual Founders of Modern Psychology* (Urbana, Ill.: Univ. of Illinois Press, 1986), 99.
10. Pitirim Sorokin, *Comtemporary Sociological Theories* (New York: Harper and Bros., 1928), 298n, 304n, 303.
11. Robert Lowie, *American Anthropologist,* n.s. 30 (April-June 1928), 317; Frank H. Hankins, *The Racial Basis of Civilization: A Critique of the Nordic Doctrine* (New York: Alfred A. Knopf, 1926), ix; Frank H. Hankins, "Individual Differences and Their Significance for Social Theory," American Sociological Society, *Papers and Proceedings* 17 (1922), 35.
12. Hankins, "Individual differences," 38-39; Hankins, *Racial Basis,* 247, 365.
14. Quoted in Robert C. Bannister, *Sociology and Scientism: The American Quest for Objectivity, 1880-1940* (Chapel Hill: Univ. of North Carolina Press, 1987), 137; Howard W. Odum, *American Sociology: The Story of Sociology in the United States Through*

1950 (New York: Longman, Green, 1951), 189; Frank H. Hankins, Review of book, *American Sociological Review* 2 (Feb. 1937), 129–30. See also his "German Policies for Increasing Births," *American Journal of Sociology* 42 (March 1937), 630–52; Brewton Berry, "The Concept of Race in Sociology Textbooks," *Journal of Social Forces* 18 (March 1940), 411.

15. Yoder, "Present Status of the Question," 463; Witty and Lehman, "Racial Differences," 394, 398.

16. H. S. Jennings, *The Biological Basis of Human Nature* (New York: W. W. Norton, 1930), 130.

17. Gladys C. Schwesinger, *Heredity and Environment: Studies in the Genesis of Psychological Characteristics* (New York: Macmillan, 1933), 457–58, 459.

18. Donald Young, *Annals of the American Academy of Political and Social Science* 140 (Nov. 1928), vii.

19. William B. Thomas, "Black Intellectuals, Intelligence Testing in the 1930s, and the Sociology of Knowledge," *Teachers College Record* 85 (Spring 1984), 481; William B. Thomas, "Black Intellectuals' Critique of Early Mental Testing: A Little-Known Saga of the 1920's" *American Journal of Education* 90 (May 1982), 282–83; Charles H. Thompson, "The Educational Achievements of Negro Culture," *Annals of the American Academy of Political and Social Science,* 140, 204–5.

20. Ellsworth Faris, "The Subjective Aspect of Culture," American Sociological Society, *Papers and Proceedings* 19 (1925), 43.

21. Joseph Peterson and Lyle H. Lanier, *Studies in the Comparative Abilities of Whites and Negroes* (Baltimore: Williams and Wilkins, 1929; Mental Measurement Monographs No. 5), 99.

22. Guy B. Johnson, "The Negro Migration and Its Consequences," *Journal of Social Forces* 2 (March 1924), 408.

23. Martha McLear, "Sectional Differences as Shown by Academic Ratings and Army Tests," *School and Society* 15 (June 1922), 676–78.

24. S. L. Wang, "A Demonstration of the Language Difficulty Involved in Comparing Racial Groups by Means of Verbal Intelligence Tests," *Journal of Applied Psychology* 10 (No. 1, 1926), 105.

25. Hsiao Hung Hsiao, "The Mentality of the Chinese and Japanese," *Journal of Applied Psychology* 13 (No. 1, 1929), 30–31.

26. Hamilton Cravens, *The Triumph of Evolution: American Scientists and the Heredity-Environmental Controversy 1900–1940* (Philadelphia: Univ. of Pennsylvania Press, 1978), 231.

27. Franz Samelson, "From 'Race Psychology' to 'Studies in Prejudice,'" *Journal of the History of the Behavioral Sciences,* 14 (July 1978), 270–73. See also his article, "World War I Intelligence Testing and the Development of Psychology," *Journal of the History of the Behavioral Sciences* 13 (1977), 274–82. See also Hamilton Cravens, *The Triumph of Evolution.*

28. Ralph H. Gundlach, "The Psychologists' Understanding of Social Issues," *Psychological Bulletin* 37 (Oct. 1940), 616; Donald K. Pickens, *Eugenics and the Progressives* (Nashville: Vanderbilt Univ. Press, 1968), 205–7; Richard Weiss, "Ethnicity and Reform: Minorities and the Ambience of the Depression Years," *Journal of American History* 66 (Dec. 1979), 575.

29. Popenoe quoted in Kenneth M. Ludmerer, *Genetics and American Society: A Historical Appraisal* (Baltimore: Johns Hopkins Univ. Press, 1972), 117–18, 128; Marie E. Kopp, "Legal and Medical Aspects of Eugenic Sterilization in Germany," *American Sociological Review* I (Oct. 1936), 766, 761, 770.

30. Popenoe quoted in Pickens, *Eugenics and Progressives,* 99n.; Weiss, "Ethnicity and Reform," 571–72; Leta S. Hollingworth, "The Problem of Comparing Races," National Society for the Study of Education, *Thirty-Ninth Yearbook. Intelligence: Its Nature and Nurture. Part I. Comparative and Critical Exposition,* Guy Montrose Whipple, ed. (Bloomington, Ill.: Public School Publishing, 1940), 257–58. It is revealing (and typical) that Hollingworth should follow the remark quoted in the text with the observation that "the anthropologists failed to add that neither does science provide a basis for not discriminating against any people on the ground of rational inferiority."

31. William B. Provine, "Geneticists and the Biology of Race Crossing," *Science* 182 (Nov. 23, 1973), 795. See also his article "Geneticists and Race," *American Zoologist* 26 (1986), 857–87.

32. Provine, "Geneticists and the Biology of Race Crossing," 796.

33. Brewton Berry, "The Concept of Race in Sociology Textbooks," *Journal of Social Forces* 18 (March 1940), 414–15.

34. George A. Lundberg, "The Biology of Population Cycles," *Journal of Social Forces* 9 (March 1931), 407.

35. Ruth Benedict, *Patterns of Culture* (Boston: Houghton Mifflin, 1934), 10–14, 25, 19, 15.

36. Ibid., 254–55.

37. Leslie A. White, *The Science of Culture. A Study of Man and Civilization* (New York: Farrar, Strauss, 1949), 163, 164–65, 168.

38. Ibid., 181.

39. Leslie White, "Individuality and Individualism: A Cultural Interpretation," *Texas Quarterly* 6 (1963), 120–21.

40. White, *Science of Culture,* 149; M. F. Ashley Montagu, ed., *Man and Aggression* (New York: Oxford Univ. Press, 1968), 9.

41. George Ward Stocking, Jr., "American Social Scientists and Race Theory, 1890–1915," Unpublished Ph.D. dissertation, Univ. of Pennsylvania, 1960, p. 3.

42. Robert E. L. Faris, "The Ability Dimension in Human Society," *American Sociological Review* 26 (Dec. 1961), 837–38, 842.

9. Biology Redivivus

1. Gunnar Myrdal, *An American Dilemma: The Negro Problem and Modern Democracy* (2 vols., New York: Harper Bros., 1944), 92, 1024.

2. Howard W. Odum, *American Sociology: The Story of Sociology in the United States Through 1950* (New York: Longman, Green, 1951), 326.

3. R. S. Woodworth, *Heredity and Environment: A Critical Study of Recently Published Materials on Twins and Foster Children* (New York: Social Science Research Council, n.d.), 1–2, 86–87.

4. W. S. Hunter, "Summary Comments on the Heredity-Environment Symposium," *Psychological Review* 54 (Sept. 1947), 348.

5. Leonard Carmichael, "The Growth of the Sensory Control of Behavior Before Birth," *Psychological Review* 54 (Sept. 1947), 323, 376, 323.

6. Clyde Kluckhohn, "An Anthropologist Looks at Psychology," *American Psychologist* 3 (Oct. 1948), 442.

7. Alfred L. Kroeber, *An Anthropologist Looks at History*, T. Kroeber, ed., (Berkeley: Univ. of California Press, 1963), 212-13, 204.

8. David Bidney, *Theoretical Anthropology* (New York: Columbia Univ. Press, 1953), 150, 154, 148.

9. Earl W. Count, "The Biological Basis of Human Sociality," *American Anthropologist* 60 (1958), 1049, 1066.

10. *Psychological Bulletin* 49 (Nov. 1952), 598-627.

11. *American Psychologist* 15 (March 1960), 219.

12. Keller and Marian Breland, "The Misbehavior of Organisms," *American Psychologist* 16 (Nov. 1961), 681-84.

13. H. C. McCurdy, "William McDougall," in B. Wolman, ed., *Historical Roots of Contemporary Psychology* (New York: Harper and Row, 1968), 123.

14. Jerry Hirsch, "Behavior-Genetics, or 'Experimental' Analsysis: The Challenge of Science versus the Lure of Technology," *American Psychologist* 22 (Feb. 1967), 118; R. L. Eaton, "An Historical Look at Ethology: A Shot in the Arm for Comparative Psychology," *Journal of the History of the Behavioral Sciences* 6 (April 1970), 187.

15. Bruce E. Eckland, "Genetics and Sociology: A Reconsideration," *American Sociological Review* 32 (April 1967), 173.

16. Allan Mazur and Leon S. Robertson, *Biology and Social Behavior* (New York: Free Press, 1972), 157, 7; John T. Doby, "Man the Species and the Individual: A Sociological Perspective," *Social Forces* 49 (Sept. 1970), 4-5.

17. Albert Somit, "Toward a More Biologically-Oriented Political Science: Ethology and Psychopharmacology," *Midwest Journal of Political Science* 12 (Nov. 1968), 550-67; Thomas Landon Thorson, *Biopolitics* (New York: Holt, Rinehart and Winston, 1970), 96, 210; Roger D. Masters, "The Impact of Ethology on Political Science," in Albert Somit, ed., *Biology and Politics: Recent Explorations* (The Hague: Mouton, 1976), 199.

18. Marian Lowe, "Social Bodies: The Interaction of Culture and Woman's Biology," in Ruth Hubbard, Mary Sue Henifin, and Barbara Fried, eds., *Biological Woman—The Convenient Myth: A Collection of Feminist Essays and a Comprehensive Bibliography* (Cambridge, Mass.: Shenkman, 1982), 109.

19. *Psychological Bulletin* 49 (Nov. 1952), 642-44.

20. T. C. Schneirla, "A Consideration of Some Conceptual Trends in Comparative Psychology," *Psychological Bulletin* 49 (Sept. 1952), 568; Robert B. Lockard, "Reflections on the Fall of Comparative Psychology. Is There a Message for Us All?," *American Psychologist* 26 (Feb. 1971), 172-73.

21. F. A. Beach, "The Descent of Instinct," *Psychological Review* 62 (Nov. 1955), 403; "The Perpetuation and Evolution of Biological Science," *American Psychologist* 21 (Oct. 1966), 948. See Beach's article, "The Shark Was a Boojum," *American Psychologist* 5 (April 1950) 115-24 for an early criticism of excessive reliance on experiments with rats.

22. Donald W. Ball, "Biological Bases of Human Society," Jack D. Douglas et al., eds., *Introduction to Sociology: Situations and Structures* (New York: Free Press, 1973), 124; Doby, "Man the Species," passim.

23. Albert Somit, "Review Article: Biopolitics," in Somit, *Biology and Politics*, 4–5; Roger D. Masters, "The Impact of Ethology on Political Science," in Somit, *Biology and Politics*, 198–99; Fred W. Willhoite, Jr., "Ethology and the Tradition of Political Thought," *Journal of Politics* 33 (Aug. 1971), 628–29.

24. Earl W. Count, "Beyond Anthropology: Toward a Man-Science," *American Anthropologist* 74 (Dec. 1972), 1358–59.

25. Erik Trinkaus, "A History of *Homo Erectus* and *Homo Sapiens* in America," in Frank Spencer, *A History of American Physical Anthropology, 1930–1980* (New York: Academic, 1982), 266–67; Laura Thompson, "Steps Toward a Unified Anthropology," *Current Anthropology* 8 (Feb.–April 1967), 68, 73fn.

26. Victor Turner, "Body, Brain, and Culture," *Zygon* 18 (Sept. 1983), 222.

27. William C. Boyd, "Four Achievements of the Genetical Method in Physical Anthropology," *American Anthropologist* 65 (April 1963), 243–52.

28. Gardner Lindzey, "Genetics and the Social Sciences," *Social Science Research Council Items* 18 (Sept. 1964), 29–30, 34: Hirsch, "Behavior-Genetic," 118–19.

29. Somit, "Toward a More Biologically-Oriented Political Science ," 560.

30. Leonard Carmichael, "The Growth of the Sensory Control of Behavior Before Birth," *Psychological Review* 54 (Sept. 1947), 322–23; Hirsch, "Behavior-Genetic," 119–21; Eaton, "An Historical Look at Ethology," 176–77.

31. Elliott White, "Genetic Diversity and Political Life: Toward a Populational-Interaction Paradigm," *Journal of Politics* 34 (Nov. 1972), 1204–5; John C. Wahlke, "Pre-Behavioralism in Political Science," *American Political Science Review* 73 (March 1979), 24–25.

32. Roger D. Masters, "Jean-Jacques Is Alive and Well: Rousseau and Contemporary Sociobiology," *Daedalus* 107 (Summer 1978), 94–95; Roger D. Masters, "The Impact of Ethology on Political Science," in Somit, *Biology and Politics*, 207.

33. Glendon Schubert, "Politics as a Life Science: How and Why the Impact of Modern Biology Will Revolutionize the Study of Politcal Behavior," in Somit, *Biology and Politics*, 174–75, 164; Somit, "Introduction" to *Biology and Politics*, 9.

34. Steven A. Peterson, "Biopolitics: Lessons from History," *Journal of the History of the Behavioral Sciences* 12 (Oct. 1978), 19; Gerhard Lenski, *Human Societies: A Macrolevel Introduction to Sociology* (New York: McGraw-Hill, 1970), 190–91.

35. Turner, "Body, Brain and Culture," 221–22.

36. Mazur and Robertson, *Biology and Social Behavior*, 4; Eckland, "Genetics and Sociology," 174.

37. Jerome H. Barkow, "Darwinian Psychological Anthropology: A Biosocial Approach," *Current Anthropology* 14 (Oct. 1973), 375.

38. G. P. Murdock, "Anthropology's Mythology," *Proceedings of the Royal Anthropological Institute of Great Britain and Ireland for 1971* (1972), 20–23.

39. Eliot D. Chapple, *Culture and Biological Man: Explorations in Behavioral Anthropology* (New York: Holt, Rinehart and Winston, 1970), 55; George D. Spindler, ed., *The Making of Psychological Anthropology* (Berkeley: Univ. of California Press, 1979), 27–29.

40. Eckland, "Genetics and Sociology," 167; White, "Genetic Diversity," 1208–9; Somit, *Biology and Politics*, 4.

41. C. Loring Brace, "The Roots of the Race Concept in American Phsyical Anthropology," in Spencer, *History of American Physical Anthropology*, 24; Allan Mazur,

"Biosociology," in James E. Short, Jr., ed., *The State of Sociology: Problems and Prospects* (Beverly Hills, Calif.: Sage Publications, 1981), 141.

42. Lockard, "Reflections on the Fall of Comparative Psychology," 177; Willhoite, "Ethology and the Tradition of Political Thought," 518; Barkow, "Darwinian Psychological Anthropology," 374.

43. Joseph Lopreato, *Human Nature and Biocultural Evolution* (Boston: Allen and Unwin, 1984), 40; Donald W. Ball, "Biological Bases of Human Society," in Douglas, *Introduction to Sociology*, 118.

44. Thomas C. Wiegele, *Biopolitics: Search for a More Human Political Science* (Boulder, Colo.: Westview Press, 1979), 151; Lockard, "Reflections," 177; Frank B. Livingston, "Cultural Causes of Genetic Change," in George W. Barlow and James Silverberg, eds., *Sociobiology: Beyond Nature/Nurture. Reports, Definitions and Debate* (AAAS Selected Symposium 35, Boulder Colo.: Westview Press, 1980), 307; J. Hartung, "On Natural Selection and the Inheritance of Wealth," *Current Anthropology* 17 (Dec. 1976), 620.

45. Lenski, *Human Societies*, 48–49; Letter to the Editor, *Contemporary Sociology* 1 (July 1972), 306; Lee Ehrman and Peter A. Parsons, *Behavior Genetics and Evolution* (New York: McGraw-Hill, 1981), 389.

46. Wahlke, "Prebehavioralism in Political Science," 26.

47. Somit, *Biology and Politics*, 315–16.

48. Fred H. Willhoite, Jr., "Primates and Political Authority: A Biobehavioral Perspective," *American Political Science Review* 70 (Dec. 1976), 1110; Wiegele, *Biopolitics*, 145.

49. Ibid., 4–5; Roger D. Masters, "Sociobiology: Science or Myth," *Journal of Social and Biological Structures* 2 (July 1979), 245; Roger D. Masters, "The Value and Limits of Sociobiology: Toward a Revival of Natural Right," Elliott White, ed., *Sociobiology and Human Politics* (Lexington, Mass.: Lexington Books, 1981), 145; Mazur and Robertson, *Biology and Social Behavior*, 4–5.

50. Lindzey, "Genetics and the Social Sciences," 32.

51. Wahlke and Corning quoted in Wiegele, *Biopolitics*, 27, 14–15; John Langton, "Darwinism and the Behavioral Theory of Sociocultural Evolution," *American Journal of Sociology* 85 (Sept. 1979), 288.

52. Walter L. Wallace, *Principles of Scientific Sociology* (New York: Aldine Publishing, 1983), 6, 4; Lopreato, *Human Nature*, 18; Murdock, "Anthropology's Mythology," 19.

53. *Ethology and Sociobiology* 1 (1979), 1.

10. The Case of the Origin of the Incest Taboo

1. Quoted in Joseph Shepher, *Incest: A Biosocial View* (New York: Academic, 1983), 43, 46, 45.

2. Quoted in ibid., 46.

3. Ibid.

4. Quoted in William Arens, *The Original Sin: Incest and Its Meaning* (New York: Oxford Univ. Press, 1986), 61.

5. Ibid., 73.

6. Quoted in Garden Lindzey, "Some Remarks Concerning Incest, the Incest Taboo, and Psychoanalytical Theory," *American Psychologist* 22 (Dec. 1967), 1057.

7. In *Basic Writings of Sigmund Freud,* A. A. Brill, ed. (Modern Library ed., n.d.), 617n. 1.

8. Bronislaw Malinowski, *Sex and Repression in Savage Society* (New York: Meridien, 1955, orig. 1927), 216.

9. Malinowski, *Sex and Repression,* 217, 164, 174, 217.

10. Quoted in Shepher, *Incest,* 139; Malinowski, *Sex and Repression,* 210.

11. Leslie A. White, "The Definition and Prohibition of Incest," *American Anthropologist* 50 (July–Sept. 1948), 418, 423–24, 433–34.

12. Ibid., 428.

13. Ibid., 433.

14. Ibid., 425. Emphasis added.

15. Ibid., 425.

16. Claude Levi-Strauss, *Les Structures Elementaires de la Parente* (Paris: Presses Universitaires de France, 1949), 485, 16, 18.

17. The 1940 quotation from Lowie appears in Shepher, *Incest,* 49. See also George B. Vetter, "The Incest Taboo," *Journal of Abnormal Psychology and Social Psychology* 23 (July–Sept. 1928), for another example of the persistence of biological explanations for the incest taboo in the twenties. Vetter thought that the taboo originated as a way of avoiding intra-family conflict, as many culturally oriented anthropologists of the time also did. But unlike them, he referred to primate behavior as accomplishing a similar end. He thought animal behavior patterns were a better source of explanations than Freud's primal crime, which he ridiculed as silly enough to be rejected by any sophomore (p. 237). At the same time, significantly enough, he rejected any suggestion that the taboo's origin had any connection with genetic disadvantages.

18. Ibid., 18–25. I have used the translation from Claude Levi-Strauss, *The Elementary Structures of Kinship* (rev. ed., Boston: Beacon Press, 1969), since the text is the same in both editions.

19. Levi-Strauss, *Elementary Structures,* 30–31.

20. Claude Levi-Strauss, *Structural Anthropology,* Claire Jacobson, trans. (New York: Basic Books, 1963), 50.

21. Mariam Kreeselman Slater, "Ecological Factors in the Origin of Incest," *American Anthropologist* 61 (Dec. 1959), 1057.

22. Charles Darwin, *The Descent of Man* (New York: Modern Library, n.d.), 895–96.

23. David F. Aberle, Urie Bronfenbrenner, Eckhard H. Hess, Daniel R. Miller, David M. Schneider, and James N. Spuhler, "The Incest Taboo and the Mating Patterns of Animals," *American Anthropologist* 65 (April 1963), 256–57.

24. Gardner Lindzey, "Some Remarks Concerning Incest, the Incest Taboo, and Psychoanalytical Theory," *American Psychologist* 22 (Dec. 1967), 1056, 1051–55.

25. Donald Stone Sade, "Inhibition of Son-Mother Mating Among Free-Ranging Rhesus Monkeys," *Science and Psychoanalysis* 12 (1968), 18–38; Jane van Lawick-Goodall, *In the Shadow of Man* (Boston: Houghton Mifflin, 1971), 182.

26. Norbert Bischof, "Comparative Ethology of Incest Avoidance," in Robin Fox, ed., *Biosocial Anthropology* (London: Malaby, 1975), 42.

27. See, for example, the doubts of primatologist Jane B. Lancaster, "Sex and Gender in Evolutionary Perspective," in Herant A. Katchadourian, *Human Sexuality: A Comparative and Developmental Perspective* (Berkeley: Univ. of California Press, 1979), 70–71. By 1986, however, one student of animal behavior concluded that "incest avoidance is the rule in most mammalian species. . . ." She went on to note that the mechanisms for recognizing close relatives are only "slowly becoming apparent. This is a hotly disputed topic just beginning to be understood as evidence in species from insects to mammals reveals that animals recognize degrees of kinship, and act toward related individuals thus recognized in ways peculiar to each species." Bettyann Kevles, *Females of the Species: Sex and Survival in the Animal Kingdom* (Cambridge, Mass.: Harvard Univ. Press, 1986), 115, 112.

28. Vernon Reynolds, *The Biology of Human Action* (Reading and San Francisco: W. H. Freeman, 1976), 63–65; Jane Goodall, *The Chimpanzees of Gombe: Patterns of Behavior* (Cambridge, Mass.: Belknap Press of Harvard Univ. Press, 1986), 470.

29. K. Kortmulder, "An Ethological Theory of the Incest Taboo and Exogamy," *Current Anthropology* 9 (Dec. 1968), 438; the comment occurs on page 444. An interesting sidelight on the role of familiarity in mating is displayed in regard to hybrids among mammals. Annie Gray in her study of matings among hybrids notes that stallions are reluctant to mate with she-asses to breed hinnies. Therefore, she points out, "male horse foals intended for hinny breeding are usually kept among asses almost from birth." She cites a similar solution in regard to vicuna and alpaca matings: "Male vicunas intended for mating with alpacas are usually suckled and reared by a female alpaca." In these "unnatural" cases, apparently, familiarity does stimulate sexual interest. Annie P. Gray, *Mammalian Hybrids* (2nd rev. ed., Slough: Commonwealth Agricultural Bureaux, 1972), 95,163.

30. Arthur P. Wolf, "Childhood Association, Sexual Attraction, and the Incest Taboo: A Chinese Case," *American Anthropologist* 68 (Aug. 1966), 883–98; quotation on 897.

31. Arthur P. Wolf, "Childhood Association and Sexual Attraction: A Further Test of the Westermarck Hypothesis," *American Anthropologist* 72 (June 1970), 503–15, esp, 503–4; see also his intermediate study: "Adopt a Daughter-in-Law, Marry a Sister: A Chinese Solution to the Problem of the Incest Taboo," *American Anthropologist* 70 (Oct. 1968), 864–74 and his book on the subject, *Marriage and Adoption in China, 1854–1945* (Stanford, Calif.: Stanford Univ. Press, 1980), especially p. 143 where he explicitly adopts Westemarck's explanation for the origin of the incest taboo and rejects specifically the arguments of Frazer, Freud, Levi-Strauss, and Leslie White.

32. Yonina Talmon, "Mate Selection in Collective Settlements," *American Sociological Review* 29 (Aug. 1964), 491–508.

33. Seymour Parker, "The Precultural Basis of the Incest Taboo: Toward a Biosocial Theory," *American Anthropologist* 78 (June 1976), 285–305; Shepher, *Incest*, 55,52.

34. Roy H. Bixler, "Incest Avoidance as a Function of Environment and Heredity," *Current Anthropology* 22 (Dec. 1981), 275.

35. William J. Demarest, "Incest Avoidance among Human and non Human Primates," in Suzanne Chevalier-Skolnikoff and Frank E. Poirier, eds., *Primate Bio-Social Development: Biological, Social and Ecological Determinants* (New York: Garland, 1977), 339–40.

36. Justine McCabe, "FBD Marriage: Further Support for the Westermarck Hypothesis of the Incest Taboo?," *American Anthropologist* 85 (March 1983), 50–69.

37. R. C. Solomon, reporting for "Group Three," in Gunther S. Stent, ed., *Morality as a Biological Phenomenon: The Presuppositions of Sociobiological Research* (rev.ed., Berkeley: Univ. of California Press, 1980), 272, emphasis in original.

38. Bixler, "Incest Avoidance," 639–54.

39. Frank B. Livingstone, "Genetics, Ecology and the Origins of Incest and Exogamy," *Current Anthropology* 10 (Feb. 1969), 45–61; the quotation appears on 59.

40. Solomon, Reporting for "Group Three," 273.

41. Melvin Ember, "On the Origins and Extension of the Incest Taboo," *Behavior Science Research* 10 (1975), 250.

42. John R. Searle, "Sociobiology and the Explanation of Behavior," in Michael S. Gregory, Anita Silvers, and Diane Sutch, eds., *Sociobiology and Human Nature* (San Francisco: Jossey-Boss, 1978), 164–82.

43. Philip Kitcher, *Vaulting Ambition: Sociobiology and the Quest for Human Nature* (Cambridge, Mass.: MIT Press, 1985), 279–80.

44. Shepher, *Incest,* 133. Joseph Lopreato, *Human Nature and Biocultural Evolution* (Boston: Allen and Unwin, 1984), 314–15.

45. Robin Fox, *Kinship and Marriage: An Anthropological Perspective* (Harmondsworth, Eng.; Penguin Books, 1976), 70–71.

46. Parker, "The Precultural Basis," 299.

47. Ember, "On the Origins," 276–77; Lindzey, "Some Remarks," 1052.

48. Aberle et al., "The Incest Taboo," 263.

49. Frank A. Beach, "Human Sexuality and Evolution," in William Montagna and William A. Sadler, eds., *Reproductive Behavior* (New York and London: Plenum, 1974), 352.

50. Parker, "Precultural Basis," 300.

51. Fox, *Kinship and Marriage,* 730.

52. Pierre van den Berghe and Gene M. Mesher, "Royal Incest and Inclusive Fitness," *American Ethnologist* 7 (May 1980), 313. Stephen Jay Gould, *Mismeasure of Man* (New York: W. W. Norton, 1981), 156.

53. Arens, Original Sin, 9.

54. Keith Hopkins, "Brother-Sister Marriage in Roman Egypt," *Comparative Studies in Society and History* 22 (July 1980), 303–54.

55. White, "The Definition . . . of Incest," 429.

56. Arens, *Original Sin,* 99; Kinsey quotation in David P. Barash, *The Hare and the Tortoise* (New York: Penguin Books, 1986), 86.

57. Ibid., 14, 137.

11. The Uses of Biology

1. Allan Mazur, "A Cross-Species Comparison of Status in Small Established Groups," *American Sociological Review* 38 (Oct. 1973), 516.

2. Michael R. A. Chance and Clifford J. Jolly, *Social Groups of Monkeys, Apes and Men* (New York: Dutton, 1970), 208–9.

3. Ibid., 208.

4. Allan Mazur, "A Cross-Species Comparison of Status in Small Established Groups," *American Sociological Review* 38 (Oct. 1973), 516, 526–27.

5. Donald W. Ball, "Biological Bases of Human Society," in Jack D. Dougals et al., eds., *Introduction to Sociology: Situations and Structures* (New York: Free Press, 1973), 131.

6. Fred H. Willhoite, Jr., "Ethology and the Tradition of Political Thought," *Journal of Politics* 33 (Aug. 1971), 638.

7. Fred H. Willhoite, Jr., "Primates and Political Authority: A Biobehavioral Perspective," *American Political Science Review* 70 (Dec. 1976), 1110, 1112, 1118–9, 1122–25.

8. Roger D. Masters, "Politics as a Biological Phenomenon," *Social Science Information* (1975), 35.

9. Stanley Milgrim, *Obedience to Authority: An Experimental View* (New York: Harper and Row, 1974), 33–35. chap. 13, pp. 123–25.

10. Jerome H. Barkow, "Prestige and Culture: A Biosocial Interpretation," *Current Anthropology* 16 (Dec. 1975), 553–55.

11. Ibid., 569.

12. Joseph Lopreato, *Human Nature and Biocultural Evolution* (Boston: Allen and Unwin, 1984), 111–12, 164.

13. William H. Durham, "Toward a Coevolutionary Theory of Human Biology and Culture," in Arthur L. Caplan, ed., *The Sociobiology Debate: Readings on Ethical and Scientific Issues* (New York: Harper and Row, 1978), 433, 435.

14. N. G. Blurton Jones, "Two Investigations of Human Behavior Guided by Evolutionary Theory," in Graham C. L. Davey, ed., *Animal Models of Human Behavior: Conceptual, Evolutionary, and Neurological Perspectives* (Chichester: John Wiley and Sons, 1983), 195–96, 198.

15. Joseph Shepher, *Incest: A Biosocial View* (New York: Academic, 1983), 94.

16. Kenneth Bock, *Human Nature and History: A Response to Sociobiology* (New York: Columbia Univ. Press, 1980), 84–85.

17. Robert J. Richards, *Darwin and the Emergence of Evolutionary Theories of Mind and Behavior* (Chicago: Chicago Univ. Press, 1987), 152–54.

18. Bettyann Kevles, *Female of the Species: Sex and Survival in the Animal Kingdom* (Cambridge, Mass.: Harvard Univ. Press, 1986), 189.

19. Davydd J. Greenwood, *The Taming of Evolution: The Persistence of Nonevolutionary Views in the Study of Humans* (Ithaca, N.Y.: Cornell Univ. Press, 1984), 146–47.

20. Peter J. Richerson and Robert Boyd, "A Dual Inheritance Model of Human Evolutionary Process I: Basic Postulates and a Simple Model," *Journal of Social and Biological Structures*, I (April 1978), 148–53; Richard C. Lewontin, "Sociobiology—A Caricature of Darwinism," in Frederick Suppe and Peter D. Asquith, eds., *Proceedings of the Philosophy of Science Association. Volume Two. Symposia.* (East Lansing, Mich.: Philosophy of Science Association, 1977), 29.

21. Martin Etter, "Sahlins and Sociobiology," *American Ethnologist* 5 (Feb. 1978), 162, 166; Davydd J. Greenwood and William A. Stini, *Nature, Culture, and Human History. A Bio-Cultural Introduction to Anthropology* (New York: Harper and Row, 1977), 378.

Notes to Pages 270–292 377

22. Pierre van den Berghe and D. P. Barash, "Inclusive Fitness and Human Family Structure," *American Anthropologist* 79 (1977), 817; Pierre van den Berghe, *Man in Society: A Biosocial View* (New York: Elsevier, 1978), 88–90.

23. Lopreato, *Human Nature and Biocultural Evolution*, 25–27.

24. Donald Symons, "Darwinism and Contemporary Marriage," in Kingsley Davis, ed., *Contemporary Marriage: Comparative Perspectives on a Changing Institution* (New York: Russell Sage Foundation, 1985), 135.

25. Jane Goodall, *The Chimpanzees of Gombe. Patterns of Behavior* (Cambridge, Mass.: Belknap Press of Harvard Univ. Press, 1986), 380–81.

26. Robert Axelrod, *The Evolution of Cooperation* (New York: Basic Books, 1984). See especially chap. 5, which is written in collaboration with the zoologist W. D. Hamilton. Evolutionary theorist Richard Dawkins in the second edition of his influential book *The Selfish Gene* (1989) expresses his warm appreciation of Axelrod's work.

27. Susan Bolin and Robert Bolin, "Comments," *American Sociological Review* 45 (Feb. 1980), 157.

28. Donald T. Campbell, "On the Conflict Between Biological and Social Evolution and Between Psychology and Moral Tradition," *American Psychologist* 30 (Dec. 1975), 1115–22; Donald T. Campbell, "On the Genetics of Altruism and the Counterhedonic Components in Human Culture," in Lauren Wispe, ed., *Altruism, Sympathy, and Helping: Psychological and Sociological Principles* (New York: Academic, 1978), 51–52.

29. Pierre L. van den Berghe, *Man in Society: A Biosocial View*, 2nd. ed. (New York: Elsevier, 1978), 58; Mary Midgley, *Beast and Man: The Roots of Human Nature* (Ithaca, N.Y.: Cornell Univ. Press, 1978), xviii; Robin Fox, *The Search for Society: Quest for a Biosocial Science and Morality* (New Brunswick, N.J.: Rutgers Univ. Press, 1989), 43.

30. John H. Kagel et al. "Experimental Studies of Consumer Demand Behavior Using Laboratory Animals," *Economic Inquiry* 13 (March 1975), 22–38, quotation on 22.

31. J. Hirschliefer, "Economics from a Biological Point of View," *Journal of Law and Economics* 20 (April 1977), 17–19, 26.

32. Jack Hirschliefer, *Economic Behaviour in Adversity* (Chicago: Univ. of Chicago Press, 1987), 191.

33. Roger D. Masters, "Of Marmots and Men: Animal Behavior and Human Altruism," in Wispe, *Altruism, Sympathy and Helping*, 60, 62–63, 70–71; "Evolutionary Biology, Political Theory and the State," *Journal of Social and Biological Structure* 5 (Oct. 1982), 282–83.

34. Roger D. Masters, "The Value—and Limits—of Sociobiology: Toward a Revival of Natural Right," in Elliott White, ed., *Sociobiology and Human Politics* (Lexington, Mass.: Lexington Books, 1981), 156–57.

35. Pierre van den Berghe and D. P. Barash, "Inclusive Fitness and Human Family Structure," *American Anthropologist* 77 (1977), 815–16.

36. Stephen H. Balch, "The Neutered Civil Servant: Eunuchs, Celibates, Abductees and the Maintenance of Organizational Loyalty," *Journal of Social and Biological Structures* 8 (Oct. 1985), 315–16, 318, 319–20.

37. John H. Beckstrom, *Sociobiology and the Law: The Biology of Altruism in the Courtroom of the Future* (Urbana: Univ. of Illinois Press, 1985), 128.

38. Ibid., chap. 1.

39. Ibid., pp. 119–22.

Notes to Pages 293–309

40. Joseph Losco, "Ultimate and Proximate Explanation: Explanatory Modes in Sociobiology and the Social Sciences," *Journal of Social and Biological Structures* 4 (Oct. 1981), 341; Peter K. Smith, "Biological, Psychological, and Historical Aspects of Reproduction and Child-Care," in Davey, *Animal Models of Human Behavior*, 172.

41. Linda Pollock, *Forgotten Children: Parent-Child Relations from 1500 to 1900* (Cambridge, Eng.: Cambridge Univ. Press, 1983), 36–38.

12. Biology and the Nature of Females

1. Alice Rossi, "A Biosocial Perspective on Parenting," *Daedalus* 106 (1977), 2–4. I have omitted the italics in the original. See Alice S. Rossi, "The Biosocial Side of Parenthood," *Human Nature* 1 (June 1978), 72, on separating herself from Wilson's sociobiology.

2. Alice S. Rossi, "Gender and Parenthood," *American Sociological Review* 49 (Feb. 1984), 1, 10.

3. Rossi, "A Biosocial Perspective," 4–5; Rossi, "Gender and Parenthood," 9; Rossi, "The Biosocial Side of Parenthood," 75.

4. Rossi, "A Biosocial Perspective," 18, 24, 2; Rossi, "Gender and Parenthood," 8; Janet Radcliffe Richards, *The Skeptical Feminist: A Philosophical Enquiry* (Boston: Routledge and Kegan Paul, 1980), 225; Allan Mazur, "Biosociology," in James E. Short, Jr., ed., *The State of Sociology: Problems and Prospects* (Beverly Hills, Calif.: Sage Publications, 1981), 143–44.

5. Melvin Konner, *The Tangled Wing: Biological Constraints on the Human Spirit* (New York: Holt, Rinehart and Winston, 1982), 319–20; Daniel G. Freedman, *Human Sociobiology: A Holistic Approach* (New York: Free Press, 1979), 21.

6. Judith K. Brown, "A Note on the Division of Labor by Sex," *American Anthropologist* 72 (Oct. 1970), 1073–78; "A Reconsideration of Ida Hahn's 'Dauernahrung und Frauarbeit,'" *Current Anthropology* 16 (Sept. 1975), 447–48.

7. Eleanor Emmons Maccoby and Carol Nagy Jacklin, *Psychology of Sex Differences* (Stanford, Calif.: Stanford Univ. Press, 1974), vii, 351–52.

8. Ibid., 368.

9. Ibid., 361.

10. Steven Goldberg, *The Inevitability of Patriarchy* (New York: William Morrow, 1973/74), 93.

11. Ibid., 74–75.

12. "Contrary to some popular assumptions," wrote anthropologist Michelle Zimbalist Rosaldo in 1974, "there is little reason to believe that there are, or once were, societies of primitive matriarchs, societies in which women predominated in the same way that men predominate in the societies we actually know. An asymmetry in the cultural evolutions of male and female, in the importance assigned to women and men, appears to be universal." Michelle Zimbalist Rosaldo, "Woman, Culture, and Society: A Theoretical Overview," in Michelle Zimbalist Rosaldo and Louise Lamphere, eds., *Woman, Culture, and Society* (Stanford, Calif.: Stanford Univ. Press, 1964), 19.

13. Goldberg, *Inevitability of Patriarchy*, 137–38.

14. Ibid., 146, 295.

15. James Q. Wilson and Richard J. Herrnstein, *Crime and Human Nature* (New York: Simon and Schuster, 1985), 124–25.

16. Mary Knudsen, "Comment," *Current Anthropology* 16 (Dec. 1975), 566; Glendon Schubert, "Politics as a Life Science: How and Why the Impact of Modern Biology Will Revolutionize the Study of Political Behavior," in Albert Somit, ed., *Biology and Politics: Recent Explorations* (The Hague: Mouton, 1976), 187.

17. Konner, *The Tangled Wing*, 126.

18. Melford E. Spiro, *Gender and Culture: Kibbutz Women Revisited* (New York: Schocken, 1980), 97–100.

19. Rossi, "Gender and Parenthood," 8–9.

20. Maccoby and Jacklin, *Psychology of Sex Differences*, 374.

21. Goldberg, *Inevitability of Patriarchy*, 146.

22. Doris F. Jonas and A. David Jonas, "Gender Difference in Mental Function: A Clue to the Origin of Language," *Current Anthropology* 16 (Dec. 1975), 626–30; See comments in *Current Anthropology* 17 (Sept. 1976), 521–26; (Dec. 1976), 744–49.

22. Donald Symons, *The Evolution of Human Sexuality* (New York: Oxford Univ. Press, 1979), 86–95, 111–12; Symons, "Darwinism and Contemporary Marriage," in Kingsley Davis, ed., *Contemporary Marriage: Comparative Perspectives on a Changing Institution* (New York: Russell Sage Foundation, 1985), 145–46.

23. David M. Buss, "Conflict Between the Sexes: Strategic Interference and the Evocation of Anger and Upset," *Journal of Personality and Social Psychology* 56 (May 1989), 743.

24. On orgasms see Shere Hite, *The Hite Report: A Nationwide Study on Female Sexuality* (New York: Macmillan, 1976), especially 131–251, and Seymour Fisher, *The Female Orgasm Psychology Physiology Fantasy* (New York: Basic Books, 1973); Symons, *Evolution of Human Sexuality*, 78–79, and p. 90 for the quotation from Mead.

25. John L. Fuller, "Genes, Brains, and Behavior," in Michael S. Gregory, Anita Silvers, and Diane Sutch, eds., *Sociobiology and Human Nature* (San Francisco: Jossey-Bass, 1978), 112.

26. Elizabeth H. Wolgast, *Equality and Rights of Women* (Ithaca, N.Y.: Cornell Univ. Press, 1980), 136–37, 129, 123.

27. Lynda Birke, *Women, Feminism and Biology: The Feminist Challenge* (Brighton, Eng.: Wheatshaft Books, 1986), 105–6.

13. Evolutionary Theory in Social Science

1. Charles J. Lumsden and Edward O. Wilson, *Promethean Fire: Reflections on the Origins of Mind* (Cambridge, Mass.: Harvard Univ. Press, 1983), 118.

2. James Silverberg, "Sociobiology, the New Synthesis? An Anthropologist's Perspective," in George W. Barlow and James Silverberg, eds., *Sociobiology: Beyond Nature/ Nurture. Reports, Definitions and Debate* (Boulder, Colo.: Westview Press, 1980), 39; Alexander Alland, Jr., *Evolution and Human Behavior* (Garden City, N.Y.: Natural History Press, 1967), 10; Roger D. Masters, "Functional Approaches to Analogical Comparison Between Species," in Mario von Cranach, ed., *Methods of Inference from Animal to Human Behavior* (n.p., 1973), 73; Martin Etter, "Sahlins and Sociobiology," *American Ethnologist* 5 (Feb. 1971), 168; Davydd J. Greenwood, *The Taming of Evolution: The*

Persistence of Nonevolutionary Views in the Study of Humans (Ithaca, N.Y.: Cornell Univ. Press, 1984), 12; Alice Rossi, "A Biosocial Perspective on Parenting," *Daedalus* 106 (1977), 24; Alice S. Rossi, "Gender and Parenthood," *American Sociological Review,* 49 (Feb. 1984), 10; Pierre van den Berghe and D. P. Barash, "Inclusive Fitness and Human Structure," *American Anthropologist* 79 (1977), 818–19.

3. Victor Turner, "Body, Brain and Culture," *Zygon* 18 (Sept. 1983), 224, 243.

4. William H. Durham, "The Coevolution of Human Biology and Culture," in N. Blurton Jones and V. Reynolds, eds., *Human Behavior and Adaptation* (Symposia of the Society for the Study of Human Biology, vol. 18, London: Taylor and Francis, 1978), 12, 17; William H. Durham, "Toward a Coevolutionary Theory of Human Biology and Culture," in Arthur L. Caplan, ed., *The Sociobiology Debate: Readings On Ethical and Scientific Issues* (New York: Harper and Row, 1978), 430; William H. Durham, "The Adaptive Significance of Cultural Behavior," *Human Ecology* 4 (Nov. 2, 1976), 108; Durham, "The Coevolution of Human Biology," 14–17.

5. John Langton, "Darwinism and the Behavioral Theory of Sociocultural Evolution: An Analysis," *American Journal of Sociology* 85 (Sept. 1979), 291–96.

6. Marion Blute, "Sociocultural Evolutionism: An Untried Theory," *Behavioral Sciences* 24 (Jan. 1979), 56–58.

7. Richard R. Nelson and Sidney G. Winter, *An Evolutionary Theory of Economic Change* (Cambridge, Mass.: Belknap Press of Harvard Univ. 1982), 9, 4, 43, 407.

8. Peter J. Richerson and Robert Boyd, "A Dual Inheritance Model of Human Evolutionary Process I: Basic Postulates and a Simple Model," *Journal of Social and Biological Structures* 1 (April 1978), 129; Robert Boyd and Peter J. Richerson, *Culture and the Evolutionary Process* (Chicago: Univ. of Chicago Press, 1985), 20; Joseph Lopreato, *Human Nature and Biocultural Evolution* (Boston: Allen and Unwin, 1984), xii, 337.

9. Alland, *Evolution and Human Behavior,* 196–98.

10. Thomas Landon Thorson, *Biopolitics* (New York: Holt, Rinehart and Winston, 1970), 143–44, 178.

11. Roger D. Masters, "Politics as a Biological Phenomenon," *Social Science Information,* 14(a) (1975), 9–10; Roger D. Masters, "The Value—and Limits—of Sociobiology: Toward a Revival of Natural Right," in Elliott White, ed., *Sociobiology and Human Politics* (Lexington, Mass.: Lexington Books, 1981), 139–40.

12. Thomas C. Wiegele, "Is a Revolution Brewing in the Social Sciences?," in Wiegele, ed., *Biology and the Social Sciences: An Emerging Revolution* (Boulder, Colo.: Westview Press, 1982), 3; J. Hirschliefer, "Economics from a Biological Point of View," *Journal of Law and Economics* 20 (April 1977), 52; Ray H. Bixler, "Incest Avoidance as a Function of Environment and Heredity," *Current Anthropology* 22 (Dec. 1981), 639; Rossi, "Gender and Parenthood," 15.

13. Lopreato, *Human Nature and Biocultural Evolution,* 154, 337; Vernon Reynolds, *Biology of Human Action* (San Francisco: W. H. Freeman, 1976), 323; Julian Jaynes and Marvin Bressler, "Evolutionary Universals, Continuities, and Alternatives," in J. F. Eisenberg and Wilton S. Dillon, eds., *Man and Beast: Comparative Social Behavior* (Washington, D.C.: Smithsonian Institution Press, 1971), 344.

14. S. L. Chorover, *From Genesis to Genocide* (Cambridge, Mass.: MIT Press, 1979), 108–9; Randall Collins, "Upheavals in Biological Theory Undermine Sociobiology," in Randall Collins, ed., *Sociological Theory 1983* (San Francisco: Jossey-Boss, 1983), 307; S. L. Washburn, "Human Behavior and the Behavior of Other Animals," *American Psy-*

chologist 33 (May 1978), 416, 410, 414; Nancy Stephan, *The Idea of Race in Science: Great Britain 1800–1960* (London: Macmillan, 1982), 189.

15. Gerald Holton, "The New Synthesis?," in Michael S. Gregory, Anita Silvers, and Diane Sutch, eds., *Sociobiology and Human Nature* (San Francisco: Jossey-Boss, 1978), 82n.; George W. Barlow, "The Development of Sociobiology: A Biologist's Perspective," in Barlow and Silverberg, *Sociobiology: Beyond Nature/Nurture,* 13.

16. Claude Levi-Strauss, *Le Regard Eloigne* (Paris: Plon, 1983), 57–58; Arthur L. Caplan, "A Critical Examination of Sociobiological Theory," in Barlow and Silverberg, *Sociobiology: Beyond Nature/Nurture,* 111.

17. Paul Heyer, *Nature, Human Nature, and Society: Marx, Darwin, Biology, and the Human Sciences* (Westport, Conn.: Greenwood Press, 1982), 220, 225; Roger D. Masters, "Is Sociobiology Reactionary? The Political Implications of Inclusive-Fitness Theory," *Quarterly Review of Biology* 57 (Sept. 1982), 288.

18. Rossi, "Gender and Parenthood," 11. In the original, the last sentence is emphasized. Melvin Konner, *The Tangled Wing. Biological Constraints on the Human Spirit* (New York: Holt, Rinehart and Winston, 1982), 104–5.

19. S. L. Washburn and P. C. Dolhinow, "Comparison of Human Behaviors," in D. W. Rajecki, ed., *Comparing Behavior: Studying Man Studying Animals* (Hillsdale, N.J.: Lawrence Erlbaum Associates, 1983), 38–40. Edward O. Wilson, *On Human Nature* (Cambridge, Mass.: Harvard Univ. Press, 1978), 48, 50.

20. Elliott White, "Genetic Diversity and Political Life: Toward a Populational-Interaction Paradigm," *Journal of Politics* 34 (Nov. 1972), 1211.

21. Stephen Jay Gould, *Mismeasure of Man* (New York: W. W. Norton, 1981), 330–31; Stephen Jay Gould, "Between You and Your Genes," *New York Review of Books* 31 (Aug. 16, 1984), 32; Rossi, "Gender and Parenthood," 10–11. At one point in his article in the *New York Review of Books,* Gould seems to come much closer to Rossi's view. For example, on p. 32 he writes, "No serious student of human behavior denies the potent influence of evolved biology upon our cultural lives. Our struggle is *to figure out how biology affects us,* not whether it does." (Emphasis added.)

22. Caplan, "A Critical Examination," 111; James Silverberg, "Sociobiology, the New Synthesis? An Anthropologist's Perspective," in Barlow and Silverberg, *Sociobiology: Beyond Nature/Nurture,* 55–58.

23. William H. Durham, "Toward a Coevolutionary Theory of Human Biology and Culture," in Arthur L. Caplan, ed., *The Sociobiology Debate: Readings on Ethical and Scientific Issues* (New York: Harper and Row, 1978), 111; Roger Trigg, *Shaping of Man: Philosophical Aspects of Sociobiology* (New York: Schocken, 1983), 138, 142.

24. Davydd J. Greenwood, *The Taming of Evolution: The Persistence of Nonevolutionary Views in the Study of Humans* (Ithaca, N.Y.: Cornell Univ. Press, 1984), 163, 23; Charles Darwin, *The Origin of Species by Means of Natural Selection or the Preservation of Favored Races in the Struggle for Life and The Descent of Man and Selection in Relation to Sex* (New York: Modern Library, n.d.), 500.

25. Jerome Kagan, *The Nature of the Child* (New York: Basic Books, 1984), 131, 152.

26. Kaye, *Social Meaning of Biology,* 151; Stuart A. Kauffman, "Constraints on the Sociobiologists' Program," in Suppe and Asquith, eds. *Proceedings 1976 Meeting of the Philosophy of Science Association,* 45–46; Roger Masters, *The Nature of Politics* (New Haven: Yale Univ. Press, 1989), 1.

14. Epilogue: Beyond Social Science

1. Loren Eiseley, *Darwin's Century. Evolution and the Men Who Discovered It* (Garden City, N.Y.: Anchor Books, Doubleday, 1961), 350, 306–7, 347.

2. C. Lloyd Morgan, *An Introduction to Comparative Psychology* (London and New York, 1898), 377.

3. James R. Angell, "Behavior as a Category of Psychology," *Psychological Review* 20 (July 1913), 256–57.

4. R. H. Waters, "Morgan's Canon and Anthropomorphism," *Psychological Review* 46 (Sept. 1939), 539–40; Joseph Wood Krutch, *The Great Chain of Life* (London: Eyre and Spottiswoode, 1957), x–xi.

5. George Gaylord Simpson, "The Evolutionary Concept of Man," in Bernard Campbell, ed., *Sexual Selection and the Descent of Man, 1871–1971* (Chicago: Aldine, 1972), 31.

6. Ernst Mayr, "Sexual Selection and Natural Selection," in Campbell, *Sexual Selection*, 93.

7. Mary Midgley, *Beast and Man: The Roots of Human Nature* (Ithaca, N.Y.: Cornell Univ. Press, 1978), 348.

8. Donald R. Griffin, "Humanistic Aspects of Ethology," in Michael S. Gregory, Anita Silvers, and Diane Sutch, eds., *Sociobiology and Human Nature* (San Francisco: Jossey-Boss, 1978), 250; Donald R. Griffin, *Animal Thinking* (Cambridge, Mass., Harvard Univ. Press, 1984), 20–22.

9. Griffin, "Humanistic Aspects," 251; Donald R. Griffin, *The Question of Animal Awareness: Evolutionary Continuity of Mental Experience* (New York: Rockefeller Univ. Press, 1976), 55.

10. Griffin, *Question of Animal Awareness*, 104, 101.

11. Sherwood L. Washburn and David A. Hamburg, "The Implications of Primate Research," in Irven DeVore, ed., *Primate Behavior: Field Studies of Monkeys and Apes* (New York: Holt, Rinehart and Winston, 1965), 619, 620.

12. Michael R. A. Chance and Clifford J. Jolly, *Social Groups of Monkeys, Apes and Men* (New York: E. P. Dutton, 1970), 75–76, 110–11.

13. Jane Goodall and David A. Hamburg, "Chimpanzee Behavior as a Model for the Behavior of Early Man," in Silvano Arieti, ed., *American Handbook of Psychiatry* (2nd ed., New York: Basic Books, 1975), 20, 27.

14. Griffin, *Animal Thinking*, 86–87; Frans de Waal, *Peacemaking Among Primates* (Cambridge, Mass.: Harvard Univ. Press, 1989), 38–39.

15. Griffin, *Question of Animal Awareness*, 44, 60.

16. Mary Midgely, *Beast and Man*, 246–51, 276–81; John R. Searle, "Sociobiology and the Explanation of Behavior," in Gregory, et al., *Sociobiology and Human Nature*, 176, 175.

17. Savage in "Commentary," *Behavioral and Brain Sciences* 1 (1978), 596; Searle, "Sociobiology and Explanation," 178.

18. Griffin, *Question of Animal Awareness*, 55.

19. Gould in "Commentary," *Behavioral and Brain Sciences* 1 (1978), 572–73; James L. Gould, *Ethology: The Mechanisms and Evolution of Behavior* (New York: W. W. Norton, 1982), 281.

20. Donald R. Griffin, "Humanistic Aspects of Ethology," in Gregory et al., *Sociobiology and Human Nature*, 253–55; Griffin, *Animal Thinking*, 47.

21. Gordon G. Gallup, Jr., "Toward a Comparative Psychology of Mind," in Roger L. Mellgren, ed., *Animal Cognition and Behavior* (Amsterdam, N.Y.: North-Holland Publishing), 476–78.

22. De Waal, *Peacemaking Among Primates,* 82.

23. Gallup, "Toward a Comparative Psychology," 490.

24. David Premack and Guy Woodruff, "Does the Chimpanzee Have a Theory of Mind?," *Behavioral and Brain Sciences* 1 (1978), 516–21, quotation on p. 518.

25. Franz Boas, *The Mind of Primitive Man* (New York: Macmillan, 1911), 96.

26. Alfred L. Kroeber, *The Nature of Culture* (Chicago: Univ. of Chicago Press, 1952), 27; Robert H. Lowie, *Are We Civilized? Human Culture in Perspective* (London: George Routledge and Sons, 1929), 5.

27. Hornell Hart and Adele Pantzer, "Have Subhuman Animals Culture?," *American Journal of Sociology* 30 (May 1925), 703, 706–7, 709.

28. W. C. McGrew, "Socialization and Object Manipulation of Wild Chimpanzees," in Suzanne Chevalier-Skolnikoff and Frank E. Poirier, eds., *Primates Bio-Social Development: Biological, Social, and Ecological Determinants* (New York: Garland, 1977), 251.

29. Hans Kummer, *Primate Societies: Group Techniques of Ecological Adaptation* (Chicago: Aldine-Atherton, 1971), 117–24.

30. Derek Freeman, *Margaret Mead and Samoa: The Making and Unmaking of an Anthropological Myth* (Cambridge, Mass.: Harvard Univ. Press, 1983), 298–300.

31. Tom Regan, *The Case for Animal Rights* (Berkeley: Univ. of California Press, 1983), 18–19.

32. Robert J. Richards, *Darwin and the Emergence of Evolutionary Theories of Mind and Behavior* (Chicago: Univ. of Chicago, 1987), 548.

Index

Animals: and altruism, 284; and communication, 337-38; consciousness in, 330-49; and the emergence of evolutionary thought, 6; and the family, 265; and intellect, 190n, 228; and kin selection, 290-91; in natural circumstances, 332; and reproductive success, 291; and sex differences, 306. *See also* Animal-human relationship

Anthropology: American school of, 72; and the animal-human relationship, 220; and the biology-culture relationship, 219-20, 230-31, 235-36, 237, 238, 242; biology's influence on, 250; Boas's influence on, 66-67; and cultural evolution theory, 312-13, 315; and culture, 82, 100-102, 103-4; emergence of, 62, 82, 103-4; and ethology, 227; and eugenics, 145; and incest, 245, 247, 250, 255, 258, 260-61, 266; and instinct, 35; and intellect, 176; and the Modern Synthesis theory, 230-31; mythology in, 235-36; and the nature/nurture conflict, 145, 187-88, 204-5; and Nazism, 203, 204-5; and political science, 274; and psychology, 219, 227; and race, 77, 176; as a science, 242; and social behavior, 286; and testing, 176; and universals, 286; and white supremacy, 77. *See also* Ethology; *name of specific person*

Anthropomorphism, 208, 330-31, 333. *See also* Animal-human relationship

Anti-social behavior, 35-41, 207-8. *See also* Criminality

Ardrey, Robert, 229

Arens, William, 267-68, 269

Aristotle, 3, 4, 5, 10, 289

Arlitt, Ada H., 172-73, 177

Army tests: black scholars' criticisms of the, 199; and intellect and race, 50-51, 167, 168, 169-70, 174, 176, 180, 199; and the nature/nurture conflict, 140, 144, 165; purpose of the, 50

Art, and the human sciences, 96

Assimilation, 67, 74-75, 78, 80, 201

Associationism, 33-34

Authority, 270-77

Awareness. *See* Consciousness

Axelrod, Robert, 284-85

"Baby tests," 181-82

Bache, R. Meade, 20, 28

Bagley, William, 175, 188

Balch, Stephen, 289

Baldwin, J. Mark, 34

Ball, Donald W., 228-29, 238, 272

Bannister, Robert, 15

Barash, David, 282-83, 289, 311

Barkow, Jerome, 235, 237, 275-76, 278, 300

Barlow, George, 318

Bateson, Gregory, 135

Beach, Frank, 218n, 222, 228, 265-66

Becker, Gary, 287n

Beckstrom, John, 290

Behaviorism: aim of, 152-53, 154; and the animal-human relationship, 216, 236, 332, 333; and the biology-culture relationship, 222-24, 232-33, 234, 236; and consciousness, 153-57, 331, 332; and emotion, 154; and experimentation, 159, 162; and genetics, 159, 160; and the group-individual relationship, 155, 160; and heredity, 155; and individuals, 163-64, 185-86; and instinct, 152-64, 216-17; and intellect, 185-86; and the maternal instinct, 161-62; and the nature/nurture conflict, 152-64; and social reform, 163. *See also* *name of specific discipline*

Benedict, Ruth, 133, 185, 206-7

Bentham, Jeremy, 247

Berghe, Pierre van den, 260, 266-67, 282, 286, 289, 311

Bernard, Jessie, 136

Bernard, Luther L., 156, 159, 160, 161-62, 163, 164, 195, 205, 218, 331

Bidney, David, 221

Binet, Alfred, 36, 37, 40

Binet tests, 139, 141-42, 145

Biological determinism, 309n

Biology: and altruism, 279-85, 288, 289, 291; and cooperation/competition, 285-86; and cultural evolution theory, 313; and dominance/submission behavior, 270-77; and incest, 246, 248, 255, 256-57; and the individual, 208; information breakthroughs in, 227-32; and intellect, 216-17; and kin selection, 279-85, 288-90; laws of, 95; limits of, 321; and the nature/nurture conflict, 178-79, 196; and psychology, 100-101; and reproductive success, 277-79, 288, 291-92; and sex differences, 106-7, 296, 298-99, 308-9; and the social sciences, 253, 261, 268; as static/dynamic, 316; and values, 301. *See also* Biology-culture relationship; Genetics; Heredity; Reproduction; Sociobiology

Biology-culture relationship: and adaptation, 236; and the animal-human relationship, 220, 221-24, 227, 230, 233-34, 233n, 236, 237-38; and behaviorism, 222-24, 232-33, 234, 236; and competition, 285-86; and decline in discussion about biology, 205-6; and equality, 236-37; and ethology, 238-39, 242-43; and experimentation, 220-21; and genetics, 242; and incest, 246-57, 258-59, 261, 262, 263, 264, 266-67, 268, 269; and

388

Equality (*continued*)
188–90, 191–92; and sex differences, 110,
111–12, 120, 121, 293–96, 302–3
Eriksen, Erik, 230
Estabrook, Arthur, 37
Ethnicity: and the hierarchy of humans, 48;
and intellect, 48–55, 165–66; and the
nature/nurture conflict, 201–2; politics of,
201–2; as the root of the concept of human
nature, 3; and testing, 151–152, 165–66. *See
also* Ethnology; Immigration; Intellect and
race; Race
Ethnology, 66–67, 101, 184–85. *See also*
Ethnicity; Race; *name of specific person*
Ethology: and anthropology, 227; and the
biology-culture relationship, 238–39, 242–
43; and cultural evolution theory, 317;
emergence of, 241, 315; and
experimentation, 243; founders of, 243; and
incest, 257; methodology of, 228, 333–34;
and political science, 227, 274; and
psychology, 224, 227; and the social
sciences, 225, 242, 270; and sociology, 227.
See also Animal-human relationship;
Anthropology; Biology-culture relationship
Etter, Martin, 282, 311
Eugenics: and acquired characters, 24, 42; and
the animal-human relationship, 147; benefits
of, 149–50; and class, 146, 149–50, 195; and
criminality, 44–48; criticisms of, 147; and
cultural evolution theory, 315, 321; and
culture, 99, 146, 188; death knell for, 150;
and education, 145; and elitism, 42–43; and
emotion, 149; and the environment, 42, 43;
and the feebleminded, 44–48; and genetics,
42, 145, 150; and government role, 42; and
heredity, 41–48, 146; and intellect, 44–48,
146; leaders in the movement for, 43; legal
cases about, 47–48; and the nature/nurture
conflict, 144–52, 192–93; and Nazism, 202–
5; and progress, 21; purpose of, 151; and
reform, 42, 43; regulation of, 151; and sex
differences, 129; as a social imperative, 21;
and social reform, 43, 145, 163; and the
social sciences, 43, 44, 46–47, 145, 147–48,
150–51; and testing, 145–52. *See also name
of specific discipline*
Evolution: as accidental, 20; definition of, 315;
length of time required by, 20
Evolutionary thought: as a child of the
Enlightenment, 5; emergence of, 5–6; as
materialistic, 23; and religion, 23
Exogamy (marrying out), 245–47, 248, 249,
250, 251, 255, 258–59, 265
Experience, 4–5, 67, 131, 132
Experimentation, 159, 162, 220–21, 243. *See
also name of specific person or study*

*The Expression of the Emotions in Man and
Animals* (Darwin), 10, 20–21

Family, 265–66
Faris, Ellsworth, 102–3, 161, 162–63, 198, 218
Faris, Robert, 209–10
Fatherhood, 110, 137, 219–20, 295
Feebleminded, 37, 38–41, 44–48, 127, 139–45,
165, 205, 217. *See also* Eugenics
Feingold, Gustave, 201
Females, nature of. *See* Sex differences
Feminism, 123–24, 137–38, 307–9
Ferguson, George, 25
Fighting instinct. *See* Aggression
Fisher, R. A., 230
Fletcher, Robert, 36
Fortes, Meyer, 230, 242
Fortune, Reo, 136
Fossey, Diane, 243
Fox, Robin, 229, 263, 266, 267, 286
Frazer, James, 247, 262
Frazier, E. Franklin, 198, 199
Freedman, Daniel, 296, 320n
Freeman, Derek, 347–48, 349
Freud, Sigmund, 248, 249, 251–59, 262, 264,
266, 268, 269
Frisch, Karl von, 227, 243, 337
Fuller, John, 307

Gallup, Gordon, Jr., 340–42
Galton, Francis, 41, 99, 145, 193
Gamble, Eliza Burt, 109–10
Garth, Thomas, 171–72, 173–74, 177–78, 180,
189–90
Gault, Robert, 160
General good, 9
Genetics: and acquired characters, 20–21, 22,
23–24; and the animal-human relationship,
347; and behaviorism, 159, 160; and the
biology-culture relationship, 206, 242; and
consciousness, 338, 347; and cultural
evolution theory, 318; and culture, 92–93,
135–36; and eugenics, 42, 145, 150; and the
feebleminded, 142–43n, 143; and the
framework of Darwinian thought, 23–24;
importance of, 22; and incest, 255, 261, 267;
information breakthroughs in, 231–32, 241;
and instinct, 160; and intellect, 165–66, 217;
as the key to understanding human
behavior, 38; and kin selection, 280; and
natural selection, 23–24; and the nature/
nurture conflict, 142–43n, 143, 145, 150,
160; and Nazism, 203, 204–5; and
psychology, 218; and race, 24–25, 93, 231–
32; and sex differences, 294, 296, 299; and
the social sciences, 217–18, 242; and

Socialization (*continued*)
and kin selection, 282; and sex differences, 118, 120–21, 122–23, 127–28, 132, 134, 303, 307; and social reform, 123; and testing, 128. *See also* Culture

Social reform: and acquired characters, 24; and behaviorism, 163; and cultural evolution theory, 319; and culture, 86, 192; and eugenics, 42, 43, 145, 163; and instinct, 163; and Myrdal's work, 215–16; and the nature/nurture conflict, 123, 145; and sex differences, 112–13, 123; and social Darwinism, 13, 112–13; and socialization, 123; and the social sciences, 13; and sociology, 145

Social sciences: and acquired characters, 21, 22–23; and adaptation, 14; and behaviorism, 156; and biology, 83, 217, 241–42, 253, 261, 268; and cooperation/cohesion, 14; and cultural evolution theory, 312–17; and culture, 62, 83, 84–86, 241–42; definition of the, 318; and ethology, 225, 242, 270; and eugenics, 43, 44, 46–47, 145, 147–48, 150–51; and genetics, 217–18, 242; and the human sciences, 97–98; and incest, 255–56, 261, 268–69; and independence of the disciplines, 240–41; and natural selection, 22–23; and the nature/nurture conflict, 145, 147–48, 150–51, 156, 310–11; and Nazism, 203; and race, 16–20, 21, 84–86; and reform, 13; as sciences, 241; and sex differences, 28, 30, 297; and social Darwinism, 13–14; and the social origins of social scientists, 201; and sociobiology, 315; unification of the, 314–17; women in the, 108–9. *See also name of specific person or discipline*

Sociobiology: and adaptation, 323, 326; and altruism, 323–24; and the animal-human relationship, 243, 326, 327; and competition, 285–86; and consciousness, 336–37; and cultural evolution theory, 317, 319; and culture, 310–12, 324, 326–27; and differences among societies, 320; emergence of, 218, 317; in France, 319; as imperialistic, 310, 323–24; and incest, 262, 266–67; and intellect, 320; journals about, 242; and kin selection, 315; and limits, 321; and morality, 324–26; and natural selection, 324; objections to/reservations about, 317–26; and reproductive success, 278, 326; and sex differences, 294, 296–97, 304–5, 306–7; and social Darwinism, 323; and the social sciences, 315; and universals, 320–21, 323. *See also* Biology-culture relationship; *name of specific discipline or person*

Sociology: and the animal-human relationship, 225, 237–38, 272; and behaviorism, 160, 162, 163; and the biology-culture relationship, 217, 224–25, 227, 228–29, 231, 236, 237–38, 239, 241; and cultural evolution theory, 313, 314, 316–17; and culture, 102, 103–4, 187–88, 209–10, 253; and dominance/submission behavior, 272; emergence of, 103–4; and the environment, 145–46; and ethology, 227; and eugenics, 43, 44, 145, 147; and genetics, 224, 231; and incest, 253; and instinct, 36, 160, 163; and intellect, 231; and the nature/nurture conflict, 145–46, 147, 160, 162, 193–94, 195–96; and race, 17, 195–96; as a science, 163n, 241–42; and sex differences, 113; and social Darwinism, 236; and social reform, 145; and sociobiology, 241–42; women in, 113. *See also name of specific person*

Somit, Albert, 225, 229, 232, 234, 236–37, 239
Sorokin, Pitirim, 65, 193, 196
Sound-blindness, 68–70
Spencer, Herbert, 11, 15–16, 22 , 24, 61, 69, 90, 98, 235
Spengler, J. J., 54
Spindler, George, 236, 325n
Spiro, Melford F., 302
Stepan, Nancy, 15, 318
Sterilization, 45–48, 151, 202, 203
Stini, William A., 282
Stocking, George, Jr., 16–17, 68, 70, 71, 77, 80, 95, 209
Stoddard, Lothrop, 168
Stone, Calvin, 218n
Strum, Shirley, 243
Suicide, 253, 262–63
Supernatural, 6–7, 330
Superorganic, 96, 98, 147, 208, 221, 235, 344
Surrogate mothers, 222–23
Survival of the fittest. *See* Natural selection
Symons, Donald, 283, 304–5, 306
Sympathy, 12, 109–10

Tabula rasa theory, 7, 229–30, 233, 234, 236
Talmon, Yonina, 259
Tanner, Amy, 114
Tarde, Gabriel, 34
Taussig, Frank, 35
Taylor, Harriet, 26
Temperament, 133–34
Terman, Lewis: and the Army tests, 51; and ethnicity, 49; and eugenics, 44, 151; and immigration restrictions, 54; liberalism of, 169n; as a proponent of testing, 167; and race, 50, 191; revision of Binet test by, 37; and sex differences, 127, 129, 130, 136; and the sovereignty of culture, 205
Testing: achievement vs. intelligence, 175; and anthropology, 176; assumptions made for, 168–69; and the "baby tests," 181–82;

Printed in the United Kingdom
by Lightning Source UK Ltd.
121968UK00001B/122/A